Nanostructured Magnetic Materials

Functionalized magnetic nanomaterials are used in data storage, biomedical, environmental, and heterogeneous catalysis applications, but there remain developmental challenges to overcome. *Nanostructured Magnetic Materials: Functionalization and Diverse Applications* covers different synthesis methods for magnetic nanomaterials and their functionalization strategies, and highlights recent progress, opportunities, and challenges to utilizing these materials in real-time applications.

- Reviews recent progress made in the surface functionalization of magnetic nanoparticles.
- Discusses physico-chemical characterization and synthesis techniques.
- Presents the effect of the external magnetic field.
- Details biological, energy, and environmental applications as well as future directions.

This reference will appeal to researchers, professionals, and advanced students in materials science and engineering and related fields.

Sathish-Kumar Kamaraj is a Research Professor Titular C in Instituto Politécnico Nacional (IPN)-Centro de Investigación en Ciencia Aplicada y Tecnología Avanzada, Unidad Altamira (CICATA-Altamira), Carretera Tampico-Puerto Industrial Altamira Km 14.5, C. Manzano, Industrial Altamira, 89600 Altamira, Tamps., Mexico.

Arun Thirumurugan is an Assistant Professor at Sede Vallenar, Universidad de Atacama, Vallenar Chile, working on the development of magnetic nanocomposites for energy storage and biological applications.

Sebastián Díaz de la Torre is a Research Professor at the Centro de Investigación e Innovación Tecnológica CIITEC, which belongs to Instituto Politécnico Nacional, Mexico, working on designing and developing advanced ceramic and composite materials through the mechanical alloying MA and spark plasma sintering SPS techniques.

Suresh Kannan Balasingam is a Research Professor in the Graduate School of Energy and Environment (KU-KIST Green School), Korea University, Seoul, Republic of Korea.

Shanmuga Sundar Dhanabalan works as a Researcher in the Functional Materials and Microsystems Group at the School of Engineering, RMIT University, Melbourne, Australia.

Emerging Materials and Technologies

Series Editor: Boris I. Kharissov

The *Emerging Materials and Technologies* series is devoted to highlighting publications centered on emerging advanced materials and novel technologies. Attention is paid to those newly discovered or applied materials with potential to solve pressing societal problems and improve quality of life, corresponding to environmental protection, medicine, communications, energy, transportation, advanced manufacturing, and related areas.

The series takes into account that, under present strong demands for energy, material, and cost savings, as well as heavy contamination problems and worldwide pandemic conditions, the area of emerging materials and related scalable technologies is a highly interdisciplinary field, with the need for researchers, professionals, and academics across the spectrum of engineering and technological disciplines. The main objective of this book series is to attract more attention to these materials and technologies and invite conversation among the international R&D community.

For more information about this series, please visit: www.routledge.com/Emerging-Materials-and-Technologies/book-series/CRCEMT

Nanostructured Magnetic Materials

Functionalization and Diverse Applications

Edited by
Sathish-Kumar Kamaraj
Arun Thirumurugan
Sebastián Díaz de la Torre
Suresh Kannan Balasingam
Shanmuga Sundar Dhanabalan

CRC Press
Taylor & Francis Group
Boca Raton London New York

CRC Press is an imprint of the
Taylor & Francis Group, an **informa** business

Designed cover image: Dan Schrodinger, Shutterstock

First edition published 2024
by CRC Press
2385 Executive Center Drive, Suite 320, Boca Raton, FL 33431

and by CRC Press
4 Park Square, Milton Park, Abingdon, Oxon, OX14 4RN

CRC Press is an imprint of Taylor & Francis Group, LLC

ISBN: 978-1-032-36982-2 (hbk)
ISBN: 978-1-032-37159-7 (pbk)
ISBN: 978-1-003-33558-0 (ebk)

DOI: 10.1201/9781003335580

Typeset in Times
by codeMantra

Contents

Preface

In the recent development of modern society, magnetic materials play a vital role in the advancement of every cutting-edge technology. These specific materials are applied to various devices, including motors, devices for the generation and transmission of power, electronics, medical devices, sensors, intelligent drug delivery systems, scientific equipment, etc. The scientific world is much attracted toward nanotechnology due to the remarkable advancements that have been made in this field over the past few decades. The nanoscale dimension of nanostructured materials results in various beneficial consequences, including confinement effects, large surface-to-volume ratios, promising transport capabilities, a wide range of physical characteristics, and fast access to a wide range of these qualities. The use of magnetic nanoparticles for photocatalytic, energy conversion and storage, water purification, detoxification, and environmental and biomedical applications has been investigated in depth. Due to the several value-added properties of the finished product, surface modification or functionalization of magnetic nanoparticles is quite desirable for any particular application. This modification is necessary due to the multifunctional nature of the final result. The surface properties of the nanomaterials are significant because they interact with the world around them. This interaction is especially significant for nanomaterials in which surface properties are notably distinct from their bulk properties due to changes in their physico-chemical properties. Hence, the investigation of the surface properties is one of the critical elements.

In recent years, the application of magnetic nanoparticles in the active layers of solar cells has increased power conversion efficiency. Due to the magnetic effect of nanoparticles, more charge carriers dissolved, increasing open circuit voltage. Technological advancements and morphological improvements are needed to enhance the energy conversion capacity of magnetic nanomaterials. The utilization of magnetic nanomaterials/nanocomposites for solar energy harvesting is one of the new directions of the renewable energy conversion process. Magnetic nanoparticles are excellent electroactive materials for electrochemical energy storage, especially in supercapacitors. The external magnetic field or the magnetic effect of the electrode/-device components can influence its electrochemical performance. Especially, the influence of an external magnetic field on the electrochemical performance of the material plays a very crucial role. Magnetic semiconducting nanoparticles have lately gained attention in photocatalysis due to their UV and visible light efficiency and magnetic recoverability. The material's photocatalytic performance is limited by charge recombination, surface area, and stability. This problem can be met through covalent or noncovalent functionalization, surface modification, nanocomposites with metal oxides, 2D materials, and other techniques. Traditional synthesis methods use hazardous reducing agents, capping agents, and surfactants to control process parameters. This causes an ecosystem imbalance. Green synthesis is a nontoxic, cost-effective, biocompatible, and ecologically acceptable means to create nanoparticles from plant extracts. In green synthesis, plant extracts are used to reduce, cap, and stabilize nanomaterials. Several methods exist for this. Magnetic nanoparticles (PMNs) made from plants are nontoxic, affordable, and environmentally beneficial. Failure to purify effluents and hazardous pollutants before releasing them into water bodies causes various diseases. Current technologies can't remove poisons from water bodies. PMNPs, which may remove toxins by more than 95%, are gaining popularity as a way to restore our environment to its natural state.

Every year, data storage device technique, mechanism, and materials improve. The quest for cheap data storage drove the swift decline in size and cost per gigabyte of data and the rapid development in areal density. Magnetic tape was replaced by hard disk drives, a kind of magnetic random-access memory (MRAM). Despite the rise of other data capturing and storage systems, magnetic data storage continues to expand. This allows additional reading and writing heads and disks per drive. Modern hard disk drives are based on perpendicular magnetic recording (PMR), heat-assisted magnetic recording (HAMR), two-dimensional magnetic recording (TDMR), and heated-dot magnetic recording (HDMR). The current standard for areal density is 10 TB/in2 (PMR) (HDMR). Future magnetic recording and storage technologies will use many magnetic materials and methods. The strategies for producing magnetic materials, surface functionalization to improve magnetic-based storage performance/areal density, and their advancement need an attention. The magnetic nanoparticles, among other types, have intrinsic properties successfully used for biomedical applications such as bioimaging, drug delivery, chemotherapy, magnetic hyperthermia, biosensors, and regenerative medicine.

The purpose of this book is to discuss the numerous bio- and physio-chemical approaches of long-term functionalization of magnetic nanoparticles and their corresponding applications. In addition, we go over the many physiochemical techniques for characterizing materials and their functional groups on the surface. The chapters in our book are all filled with information that we believe will be interesting to our readers.

Acknowledgments

First of all, we thank the almighty for providing us with good health and a valuable opportunity to complete this book successfully. In our journey toward this book, our heartfelt thanks to the series editor and advisory board for accepting our book and for their support and encouragement. We extend our sincere thanks to all authors and reviewers for their valuable contribution and genuine support to complete this book. We have great pleasure in acknowledging various publishers and authors for permitting us the copyright to use their figures and tables.

Sathish-Kumar Kamaraj would like to express his gratitude to the Director General of Instituto Politécnico Nacional (IPN) and Director of Centro de Investigación en Ciencia Aplicada y Tecnología Avanzada, Unidad Altamira (CICATA Altamira) for their constant support and facilities, able to promote the research activities. Thanks for the project SIP:20231443, Secretaría de Investigación y Posgrado (SIP) – IPN. Further extensions to the funding agency of the National Council of Humanities, Sciences and Technologies (CONAHCyT - México) and Secretary of Public Education (SEP- México). He extended his gratitude to Mrs Mounika Kamaraj and Bbg Aarudhra for the family support.

Arun Thirumurugan would like to express his gratitude to Dr. Justin Joseyphus (NIT-T, India), Prof. P.V Satyam (IOP, India), Dr. Ali Akbari-Fakhrabadi (FCFM, University of Chile, Chile), and Prof. RV Mangala Raja (University of Adolfo Ibanez, Santiago, Chile) for their kindness and guidance. Arun Thirumurugan would like to thank Dr. R. Udaya Bhaskar and Mauricio J. Morel (University of Atacama, Chile), Carolina Venegas, Yerko Reyes, and Juan Campos, Sede Vallenar, University of Atacama, Chile for their support. Arun Thirumurugan acknowledges ANID for the financial support through SA 77210070.

Sebastián Díaz de la Torre expresses his gratitude to the National Polytechnic Institute IPN of Mexico through the Technological Center for Research and Innovation CIITEC, from which most of his recent investigation's outcomes have been derived. SDT appreciates direct support from the General Direction of IPN, SIP, COFAA, Fundación Politécnico, and CONAHCyT. My gratitude, admiration, and love to Gisela González Corral, my wife, and working colleague in CIITEC-IPN. The general support and never-ending encouragement received from Dr. David Jaramillo Vigueras are very much appreciated. To Dr. Hiroki & Key Miyamoto of TRI-Osaka, Japan. To Dr. Hideo Shingu and Kei Ishihara of Kyoto University, Japan. To the Central European Institute of Technology CEITEC of the Brno University of Technology: Prof. RNDr. Ing. Petr Stepanek, Prof. Ing. Radimir Vrba, Prof. Dr. Ladislav Celko, Dr. Jozef Kaiser, Dr. Edgar Montufar, and Dr. Mariano Casas Luna. To the Autonomous University of Zacatecas, Mexico, colleagues from the School of Chemistry Science CQ-UAZ; Dr. Juan Manuel García, Javier Aguayo, and Manuel Macías are greatly recognized.

Shanmuga Sundar Dhanabalan would like to express his sincere thanks to Prof. Sivanantha Raja Avaninathan (Alagappa Chettiar Government College of Engineering & Technology, Karaikudi, Tamilnadu, India), Prof. Marcos Flores

Carrasco (FCFM, University of Chile, Chile), Prof. Sharath Sriram, and Prof. Madhu Bhaskaran (Functional Materials and Microsystem, RMIT University, Australia) for their continuous support, guidance, and encouragement.

Suresh Kannan Balasingam is grateful to Prof. Yongseok Jun (Korea University-Seoul Campus, Republic of Korea), Prof. S. Vasudevan (Central Electrochemical Research Institute, India), Prof. P. Manisankar (Ex. Vice Chancellor, Bharathidasan University, India), Prof. Jae Sung Lee (UNIST, Republic of Korea), Prof. S. N. Karthick (Bharathiar University, India), Dr. A. Mohammed Hussain (Research Scientist, Nissan Technical Center-North America) for their constant support and encouragement throughout the journey. He would also like to acknowledge Dr. Ananthakumar Ramadoss, School for Advanced Research in Polymers (SARP), Central Institute of Petrochemicals Engineering and Technology, India, and Dr. Ramesh Raju, Aalto University, Finland, for their extended support. Last but not least, heartfelt gratitude is extended to Mrs. Revathy Suresh Kannan for her patience, love, and moral support.

Sathish-Kumar Kamaraj
Arun Thirumurugan
Sebastián Díaz de la Torre
Suresh Kannan Balasingam
Shanmuga Sundar Dhanabalan

Contributors

Muqarrab Ahmed
State Key Laboratory of Chemical
  Engineering, Department of
  Chemical Engineering
Tsinghua University
Beijing, China

Saravanan Alamelu
Department of Biochemistry and
  Biotechnology, Faculty of Science
Annamalai University
Chidambaram, India

Thamer Alomayri
Department of Physics, Faculty of
  Applied Science
Umm Al-Qura University
Makkah, Saudi Arabia

Nadia Anwar
School of Materials Science and Engineering
Tsinghua University
Beijing, China

G. Sahaya Dennish Babu
Department of Physics
Chettinad College of Engineering and
  Technology
Karur, India

Suresh Kannan Balasingam
Department of Energy Environment
  Policy and Technology
Graduate School of Energy and
  Environment (KU-KIST Green
  School) Korea University,
Seoul, Republic of Korea.

S. Benazir Begum
Department of Zoology
V.O. Chidambaram College
Thoothukudi, India

F. Caballero-Briones
Instituto Politécnico Nacional, Materiales
  y Tecnologías para Energía, Salud y
  Medio Ambiente (GESMAT)
Altamira, México

Krishna Chattopadhyay
Department of Chemistry
University of Calcutta
Kolkata, India

N. Chidhambaram
Department of Physics
Rajah Serfoji Government College
  (Autonomous)
Thanjavur, India

J. Jenifer Annis Christy
Department of Zoology and
  Microbiology
Thiagarajar College
Madurai, India

Shanmuga Sundar Dhanabalan
Functional Materials and Microsystems
  research group
School of Engineering
RMIT University
Melbourne, Australia

R. Dhivya
Department of Chemistry
Sri Sarada Niketan College of Science
  for Women
Karur, India

S. Gobalakrishnan
Department of Nanotechnology
Noorul Islam Centre for Higher
  Education (Deemed to be University)
Kanyakumari, India

H. Gómez-Pozos
Computing and Electronics Academic
   Area, ICBI
Autonomous Hidalgo State
   University
Hidalgo, Mexico

A. G. Hernández
Industrial Engineering, Undergraduate
   School Campus Tepeji
Autonomous Hidalgo State University,
   ESTe-UAEH
Hidalgo, Mexico

O. Icten
Department of Chemistry, Faculty of
   Science
Hacettepe University
Ankara, Turkey

Ali Raza Ishaq
State Key Laboratory of Biocatalysis
   and Enzyme Engineering,
   Environmental Microbial
   Technology Center of Hubei
   Province
College of Life Sciences
Hubei University
Wuhan, China

A. Judith Jayarani
PG & Research Department of Physics
Bishop Heber College (Autonomous)
Tiruchirappalli, India

Nivedha Jayaseelan
Department of Biochemistry and
   Biotechnology, Faculty of Science
Annamalai University
Chidambaram, India

Sathish-Kumar Kamaraj
Sección de Posgrado
Instituto Politécnico
   Nacional-Centro de
   Investigación en Ciencia
   Aplicada y Tecnología Avanzada
Unidad Altamira (CICATA Altamira)
Altamira, Mexico

T. V. K. Karthik
Industrial Engineering,
   Undergraduate School Campus
   Tepeji,
Autonomous Hidalgo State University,
   ESTe-UAEH
Hidalgo, Mexico

M. Karthikeyan
Department of Zoology and
   Microbiology
Thiagarajar College
Madurai, India

N. Karthikeyan
Department of Physics
Ramco Institute of Technology
Virudhunagar, India

S. Jasmine Jecintha Kay
Department of Physics
Rajah Serfoji Government College
   (Autonomous)
Thanjavur, India

S.M. López-Estrada
Unidad de Investigación y Desarrollo
   Tecnológico (UNINDETEC)
Secretaría de Marina-Armada de
   México
Veracruz, México

Manas Mandal
Department of Chemistry
Sree Chaitanya College
Habra, India

G. Murali Manoj
Department of Physics
KPR Institute of Engineering and
  Technology
Coimbatore, India

M. Morales-Luna
Departamento de Ciencias
Tecnológico de Monterrey
Monterrey, México

R.M. Murugappan
Department of Zoology and
  Microbiology
Thiagarajar College
Madurai, India

Sivanantham Nallusamy
Department of Physics
K. Ramakrishnan College of
  Engineering
Tiruchirappalli, India

Durga Prasad Pabba
Departamento de Ingeniería
  Mecánica
Universidad Tecnologica Metropolitana
Santiago, Chile

M. Pérez-González
Área Académica de Matemáticas y
  Física, Instituto de Ciencias Básicas
  e Ingeniería
Universidad Autónoma del Estado de
  Hidalgo
Pachuca, Mexico

R. Subramaniyan Raja
Department of Physics
KPR Institute of Engineering and
  Technology
Coimbatore, India

F. Ruiz Perez
Instituto Politécnico Nacional,
  Materiales y Tecnologías para
  Energía, Salud y Medio Ambiente
  (GESMAT)
Altamira, México

R. Saidur
Research Center for Nano-Materials
  and Energy Technology (RCNMET),
  School of Engineering Technology
Sunway University
Bandar Sunway, Malaysia

S. Sham Sait
Department of zoology and
  Microbiology
Thiagarajar College
Madurai, India

Muhammad Munir Sajid
Henan Key Laboratory of Photovoltaic
  Materials, School of Physics
Henan Normal University
Xinxiang, China

Kalist Shagirtha
Department of Biochemistry
St. Josephs College of Arts and Science
Cuddalore, India

M. Shalini
Department of Physics
KPR Institute of Engineering and
  Technology
Coimbatore, India

Shajahan Shanavas
Department of Chemistry
Khalifa University of Science and
  Technology
Abu Dhabi, United Arab Emirates

H. Shankar
Department of Physics
KPR Institute of Engineering and
    Technology
Coimbatore, India

Manoj Kumar Srinivasan
Department of Biochemistry and
    Biotechnology, Faculty
    of Science
Annamalai University
Chidambaram, India

M. Swetha
Department of Chemistry
MVJ College of
    Engineering
Bangalore, India

S. Kulandai Tererse
Department of Chemistry
Nirmala College for
    Women
Coimbatore,
    India

R. Thenmozhi
Department of Chemistry
Sakthi College of Arts and Science for
    Women
Oddanchatram,
    India

Arun Thirumurugan
Sede Vallenar
Universidad de Atacama
Vallenar, Chile

Sebastián Díaz de la Torre
Departamento de Ciencia de
    Materiales e Ingeniería
Instituto Politécnico Nacional IPN –
    Centro de Investigación e Innovación
    Tecnológica CIITEC
Unidad Altamira (CICATA Altamira)
Altamira, Mexico

Kamalesh Balakumar Venkatesan
Department of Biochemistry and
    Biotechnology, Faculty of Science
Annamalai University
Chidambaram, India

Vasanthi Venkidusamy
Department of Physics
NIT-Tiruchirappalli
Tiruchirappalli, India

Haifa Zhai
Henan Key Laboratory of Photovoltaic
    Materials, School of Physics
School of Materials Science
    and Engineering
Henan Normal University
Xinxiang, China

1 Functionalized Magnetic Nanomaterials and Their Applications

Sathish-Kumar Kamaraj
Instituto Politécnico Nacional-Centro de
Investigación en Ciencia Aplicada y Tecnología
Avanzada, Unidad Altamira (CICATA Altamira)

Arun Thirumurugan
Sede Vallenar, Universidad de Atacama

Sebastián Díaz de la Torre
Instituto Politécnico Nacional IPN–Centro de Investigación
e Innovación Tecnológica (CIITEC-Azcapotzalco)

Suresh Kannan Balasingam
Korea University (KU)

Shanmuga Sundar Dhanabalan
RMIT University

CONTENTS

DOI: 10.1201/9781003335580-1

1

1.1 MAGNETIC NANOMATERIALS

Among the nanomaterials, magnetic nanomaterials (MNMs) are found to be interesting due to their specific magnetic characteristics depending on the crystal structure and magnetic nature of the material [1–5]. Furthermore, MNMs with optimal
magnetic characteristics will be required for some specific applications such as
magnetic separation, magnetic hyperthermia, magnetic data storage, and magnetic
field-assisted applications [6–8]. The phase formation of MNMs during the synthesis
or postsynthesis process with thermal annealing is a key parameter to achieve the
required magnetic phase of MNMs [9–14]. The size and stoichiometry of MNMs
could be tuned during the synthesis by adjusting the experimental parameters and
also by postthermal treatments [15–19]. Further surface modification MNMs could
be utilized for various applications with multifunctional characteristics [19–22]. The
interaction of the interparticles as well as the foreign material used for surface modification is critical in determining the synergistic impact in the final form of functionalized magnetic nanomaterials (f-MNMs) [23,24] (Figure 1.1).

As a result of the analysis done by the VOS viewer, a variety of magnetic nanomaterials are explored, each of which has its own unique set of functional, physical,
and chemical properties. As a consequence, these materials investigate a variety of
different applications in a number of different fields. In addition to this, an increase

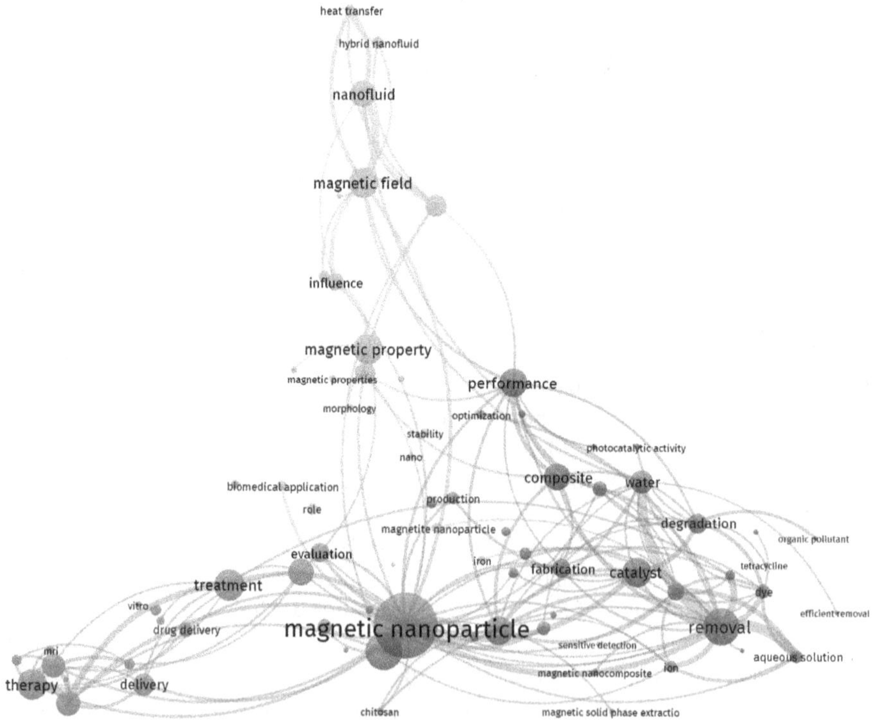

FIGURE 1.1 A bibliometric analysis using VOS viewer of publications retrieved by
SCOPUS.

in several disciplines has been noticed as a result of the scale par. In this context, we would organize and focus our book on the fundamental concepts related to the synthesis and functionalization of magnetic nanomaterials, and then we would diverge into the applications of these concepts in the fields of biomedical applications, energy conversion, energy storage, water treatment, and data storage. In this chapter, we discuss various synthesis methods, different surface modifier which could be used for the functionalization of MNPs, and the possible applications of functionalized MNPs.

1.2 SYNTHESIS OF MAGNETIC NANOMATERIALS

Numerous synthesis methods were available for the synthesis of MNM such as coprecipitation, chemical oxidation process, polyol process, sol-gel process, and ball milling, and they could be selected based on the nature of the MNMs [25–29]. The selection of the synthesis process for the synthesis of MNMs is important as the final particles' size and shape will be depending on the synthesis methods and the experimental parameters adopted.

1.2.1 COPRECIPITATION METHOD

To get a smaller-sized MNM, the coprecipitation method is suitable for any MNM. The advantage of this coprecipitation method is that it could produce mono-sized particles. By adjusting the starting precursors, we could adjust the stoichiometry of the final particles and also stoichiometry could be altered by doping of foreign metal ions. The particle size of the MNPs could be tuned by reaction temperature and post-thermal treatment. The coprecipitation method cannot be used where the larger-sized MNPs were required. However, the doping of metal ions into specific ferrite could be achieved through the coprecipitation method. A recent report discussed the successful stoichiometry modification by selective metal ion doping into Fe_3O_4. Studies [30] suggest that the stoichiometry variation could alter the magnetic characteristics, which could enhance the efficiency in a specific application of the samples, even though the particles are similar in size. The Mn- and Co-doped Fe_3O_4 are shown in Figure 1.2. The variation in the Curie transition temperature was estimated from the thermos gravimetric analysis, and the Hopkinson peak shifts between 529°C and 590°C were observed corresponding to the doping of the prepared ferrites [30].

1.2.2 CHEMICAL OXIDATION METHOD

To achieve larger-sized MNPs, a chemical oxidation process could be adopted as this process will have an opportunity to tune the particle's size by varying the oxidation or precipitant agent through a modified chemical oxidation process. Further size reduction can also be achieved in the modified chemical oxidation process by introducing the nucleating agent during the reaction. The major advantage of this modified chemical oxidation process is that it does not require a high-cost nucleating agent. Even with the metal ions, we could reduce the particle size.

Mn–Zn ferrite was attempted to synthesize large-sized particles through a modified chemical oxidation process, and the particle sizes were tuned by adjusting the KNO_3

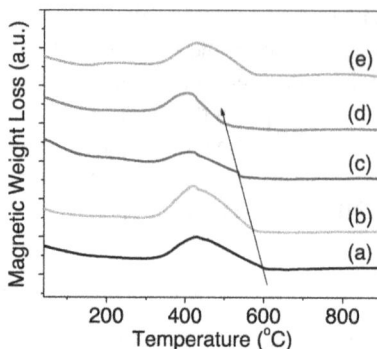

FIGURE 1.2   Thermomagnetic curve of (a) Fe_3O_4, (b) $Mn_{0.1}Fe_{2.9}O_4$, (c) $Mn_{0.3}Fe_{2.7}O_4$, (d) $Mn_{0.5}Fe_{2.5}O_4$, and (e) $Co_{0.3}Fe_{2.7}O_4$. Reprinted with permission from [30] Copyright (2020) Elsevier.

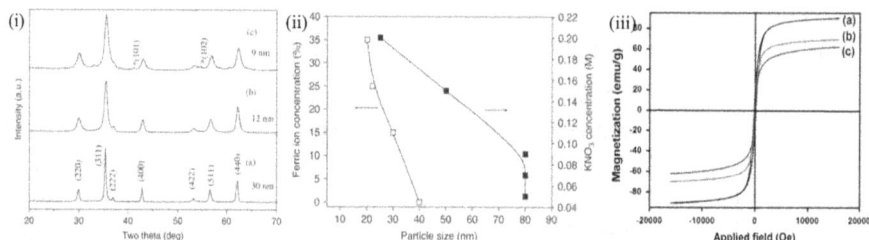

FIGURE 1.3   (i) XRD pattern of $MnZnFe_2O_4$ prepared with (a) 0.09, (b) 0.2 M, and (c) 35% of ferric ions, and (ii) particles size variation depending on KNO_3 and ferric ion concentrations. Reprinted with permission from [31] Copyright (2006) Elsevier and (iii) VSM hysteresis loop of Fe_3O_4 prepared with (a) 0% (b) 30%, and (c) 200% of ferric ions. Reprinted with permission from [32] Copyright (2006) Elsevier.

concentration [31]. Ferric ions were used to further reduce the particle size. The XRD pattern of the ferrites prepared with ferric ions (35%) and KNO_3 concentrations of 0.09 and 0.2 M is shown in Figure 1.3(i). The overall size variation with KNO_3 concentration and ferric ions is shown in Figure 1.3(ii). Similarly, Fe_3O_4 MNPs were developed through a modified chemical oxidation process with ferric ions for size reduction [32]. When the ferric ions employed in this approach were increased from 10% to 200%, the process converted into a coprecipitation process due to the 1:2 ratio of ferrous to ferric ions. The average particle sizes were tuned from 10 to 45 nm, and the variation in the magnetic characteristics due to the size reduction can be observed in Figure 1.3(iii). The saturation magnetization of the ferrites was tuned between 62 and 90 emu/g.

1.2.3   Polyol Process

The above-discussed synthesis process cannot be used to synthesize the metal or metal alloy MNS as both processes are involved in the water-based synthesis

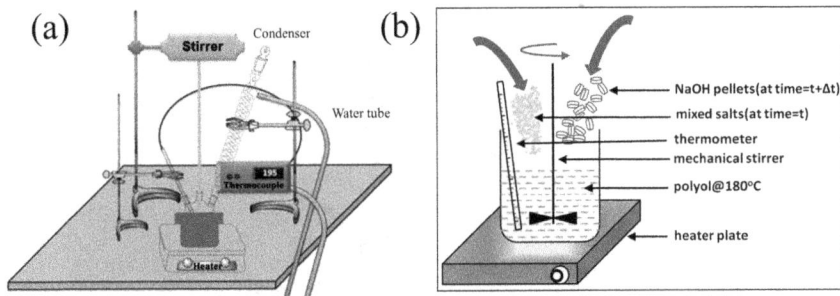

FIGURE 1.4 Polyol experimental setup for the synthesis of FePt and FeCo. Reproduced with permission from [33] Copyright (2015) Springer Nature and [34] Copyright (2013) Elsevier.

process. The use of water during the metal or metal alloy synthesis could easily form the metal hydroxide through the chemical reaction, and further thermal energy may produce the magnetic metal oxides or any other impurity phase. The polyol process could be utilized for the synthesis of metal or alloy MNMs without any impurity phase. One of the issues in the polyol process is it will be required some costly chemicals such as polyols and it is required a tricky washing procedure to remove the polyols and unreacted products. The selection of polyols for the synthesis of MNMs is important as it is reported that the particle size of Fe MNMs could be varied with different polyols such as propylene glycol (PG), ethylene glycol (EG), and trimethylene glycol (TMEG) [26]. The starting precursors, solvent, and reaction temperature could result in the purity of the metal NPs. Along with the previously listed characteristics, metal to hydroxyl ions play an important role in the reduction of metal NPs.

The synthesis process and time duration will be different for various MNPs. Fe MNPs can be reduced instantly, whereas the reduction of FePt could take more time to complete the reaction to form the alloy phase. Figure 1.4 shows the experimental setup, which could be used for the synthesis of alloy and metal MNPs. The particle size of metal NPs could be reduced by introducing nucleating agents such as Pt, Pd, and Ag before or during the chemical reaction. Fe MNPs' sizes were tuned between 10 and 90 nm through the polyol process with Pt precursor as a nucleating agent.

Other synthesis methods such as hydrothermal, microemulsion, solvothermal, reflex, and electrochemical processes can also be utilized for the development of various MNPs.

1.3 FUNCTIONALIZATION OF MAGNETIC NANOMATERIALS

Functionalization is important to improve the efficient characteristics of MNPs for any specific applications. Functionalization is also considered as a surface modification process but with the conjugation or any other biomolecules for better surface characteristics. The interaction between the surface modifier and MNPs is important

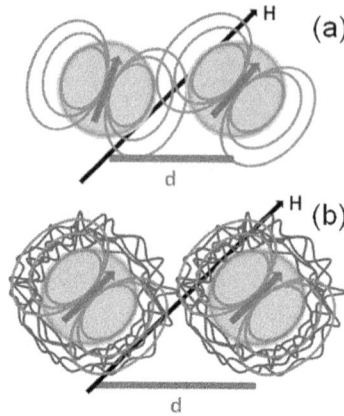

FIGURE 1.5   Different between the (a) nonfunctionalized and (b) functionalized particle's dipolar interactions. Reproduced with permission from [35] Copyright (2020) Elsevier.

to have stable f-MNPs. F-MNPs have several advantages such as nonagglomeration and enhance physicochemical characteristics over the bare MNPs. For the applications of f-MNPs in detoxification or any other heavy metal ion adsorption, MNPs could be functionalized with a specific surface modifier based on the selection of metal ions to be removed.

The functionalization requires a little optimization for better surface modification. Because some surface modifiers may not be able to effectively connect to the surface of MNPs. To avoid improper functionalization, MNPs could be pretreated with mild acids to achieve complete functionalization with a surface modifier. The improved surface attachment of the surface modifier with the acid treatment of Fe_3O_4 with hydrochloric acid was evident from several recent reports [20]. The functionalization could improve the interparticle interactions and reduce the dipolar interactions because of the increase in the distance between the two particles [35]. The removal of dipolar interaction due to the surface functionalization is explained schematically in Figure 1.5.

For any specific applications, agglomeration should be avoided to utilize a maximum contribution from particles. Numerous polymeric compounds could be utilized to avoid agglomeration through the surface modification of MNPs.

1.4   APPLICATIONS OF FUNCTIONALIZED MAGNETIC NANOMATERIALS

Functionalized MNPs could be used for various applications including energy storage, energy conversion, water purification, degradation of pollutants, magnetoptical, detoxification, hyperthermia, and other bio-related applications. As we discussed in earlier sections, the synthesis methods and surface modifier could be selected for any particular applications to obtain the synergic effect from MNPs and also from the surface modifier. Various synthesis methods, materials used for the functionalization, and their applications are shown in Figure 1.6.

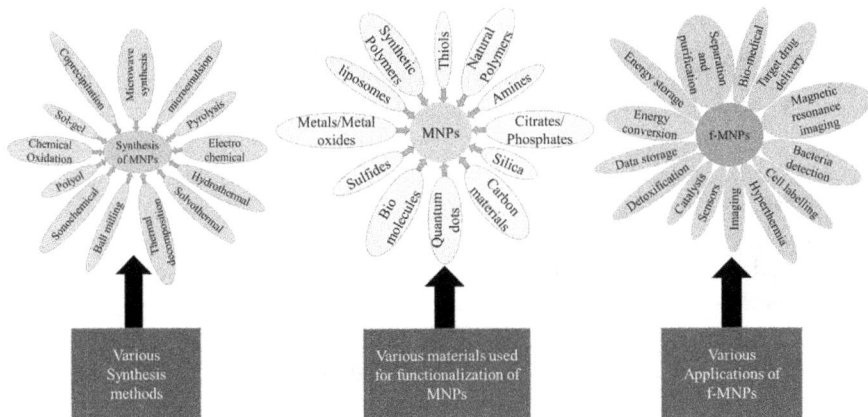

FIGURE 1.6 Various synthesis methods, materials used for functionalization, and the applications of functionalized MNPs.

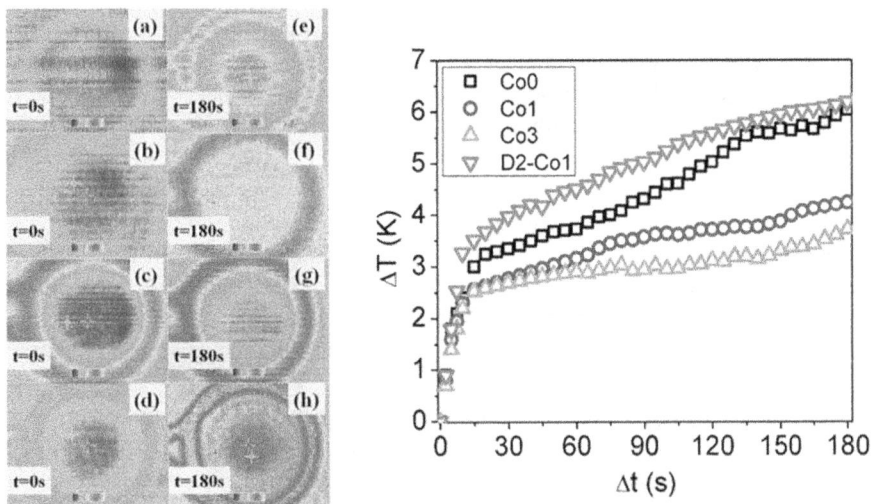

FIGURE 1.7 Thermograms of (a and e) Fe_3O_4, (b and f) $Co_{0.1}Fe_{2.9}O_4$, (c and g) $Co_{0.3}Fe_{2.7}O_4$, and (d and h) dextran-modified $Co_{0.1}Fe_{2.9}O_4$ and heating characteristics with time. Reproduced with permission from [35] Copyright (2020) Elsevier.

Co-doped Fe_3O_4 and dextran-modified Co-doped Fe_3O_4 MNPs were evaluated for their heating characteristics for their potential utilization in biomedical applications. The observed thermograms of Fe_3O_4 (Co0), $Co_{0.1}Fe_{2.9}O_4$ (Co1), $Co_{0.3}Fe_{2.7}O_4$ (Co3), and dextran-modified $Co_{0.1}Fe_{2.9}O_4$ (D2-Co1) MNPs are shown in Figure 1.7. The time-dependent heating behaviors of the unmodified and surface-modified MNPs are shown in Figure 1.7. The enhancement of heating characteristics with surface functionalization is evident through the enhancement of effective specific absorption rate which is estimated from the infrared thermography.

1.5   FROM FUNDAMENTALS OF SURFACE CHARACTERIZATION TECHNIQUES INTO DIVERSE APPLICATIONS

Nanotechnology has been advancing rapidly in the last decade, which has attracted the interest of the scientific community. In point of fact, the advances in nanotechnology have made it possible to obtain new synthetic materials for use in a diverse array of applications, including medicine, energy conversion, energy storage, technology, industry, water treatment, and data storage, to name a few. Because of the growing need to understand the properties of emerging nanomaterials, surface characterisation approaches have grown in significance.

Surfaces are important because of their interaction with the surrounding environment, particularly because the surface properties of nanomaterials differ significantly from their bulk properties due to differences in their physical structure and chemistry. This makes studying surfaces a relevant topic. On the other hand, magnetic nanoparticles have attracted a significant amount of attention over the past 20 years due to their application in the field of cancer treatment as a method of controlled medication administration. The capping of nanoparticles has proven to be a significant obstacle in the synthesis and characterization of magnetic nanomaterials such as iron oxide. This is because some of the nanoparticles are highly reactive and susceptible to oxidation, both of which can alter the magnetic behavior of the material. The complexity of nanoparticles' surface properties necessitates the use of methodologies typically used for characterizing surfaces to the investigation of the properties of nanomaterials.

Other methods, such as doping or the synthesis of composites, have also been established in order to achieve modified properties, in addition to the various types of synthesis that have been created for magnetic nanomaterials in order to obtain these modified properties. As a consequence of this, surface characterizations should be reliable approaches for obtaining information regarding the chemical, structural, morphological, magnetic, and electrical properties of the innovative nanomaterials that are continually being developed.

In the chapter titled "Surface Characterization Techniques," we make an effort to provide surface characterization techniques that are both unique and useful to the study of magnetic nanoparticles. In addition, procedures are divided into component parts, and their underlying physical concepts are discussed for enhanced comprehension. This chapter details the prerequisites for sample preparation, and the benefits and drawbacks of the approaches are discussed.

Nanostructured materials instantly provide a number of benefits, including large surface-to-volume ratios, promising transport capabilities, a variety of physical properties, and confinement effects as a result of the nanoscale size. These nanomaterials have been explored in depth for use in applications linked to energy, including solar cells, catalysts, thermoelectric, lithium-ion batteries, supercapacitors, and hydrogen storage devices. In more recent times, the presence of magnetic nanoparticles in the active layer of solar cell devices has greatly contributed to the overall power conversion efficiency of the devices. Because of the magnetic field that is inherent to the nanoparticles, the devices had a greater number of charge carriers that had become dissociated, which resulted in an increased open circuit voltage. It is emphasized that innovative methods and structures will likely be necessary in order to greatly improve the capacity of magnetic nanostructured materials to convert energy. The uses of magnetic functional nanomaterials and nanocomposites including magnetic nanoparticles for converting solar energy will be the primary focus of this chapter.

For electrochemical energy storage, particularly in the context of supercapacitor application, magnetic nanoparticles are utilized as efficient electrodes. The electrochemical performance of an electrode that is based on magnetic nanomaterials can be influenced either by an external magnetic field or by the magnetism that is possessed by the materials that make up the electrode. This chapter discusses recent developments in research on magnetic nanomaterials, their composite materials for the use of electrochemical energy storage, and the influence of an external magnetic field on the electrochemical performance of the material. In conclusion, we discuss the opportunities and difficulties presented by these magnetic nanomaterials as a potential contender for the storage of energy.

In the field of photocatalysis, magnetic semiconducting nanoparticles have recently garnered increased attention due to their superior efficiency in UV and visible light, as well as the fact that they are magnetically recoverable and reusable. The high rate of charge recombination, the magnetic nanomaterial's surface area, and its stability all act as limitations on the photocatalytic performance of the material. This challenge can be met through the functionalization of magnetic nanoparticles, which can be accomplished through either covalent or noncovalent functionalization, surface modification, production of nano composites with metal oxides, 2D materials, and other approaches. On the other hand, traditional methods of synthesis involve the use of potentially toxic reducing agents, capping agents, and surfactants in order to control the various parameters of the synthesis process. This can also generate an environmental imbalance in the ecosystem. When it comes to creating nanoparticles with plant extracts, one of the alternative ways that is also nontoxic, cost-effective, biocompatible, and environmentally friendly is known as the green synthesis method. In the green synthesis process, the extract of a plant (seed, fruit, leaf, peel, etc.) is utilized as a reducing, capping, and stabilizing agent to regulate the size and shape of the nanomaterials. This can be done in a number of different ways.

The production of magnetic nanoparticles from plant-based materials (plant-based magnetic nanoparticles, PMNPs) is not only nontoxic but also economical and friendly to the environment. The failure to properly cleanse effluents and hazardous wastes before releasing them into neighboring water bodies is the underlying cause of many different diseases. The elimination of toxins from water bodies using technologies that are currently prevalent is ineffective. The use of PMNPs, which have been reported to reach an efficiency of removing contaminants of more than 95%, has gained traction as a means of returning our environment to the natural and unspoiled state it once occupied. As a result, the purpose of this study is to discuss the feasible applications of phytogenic MNPs in the treatment of wastewater.

Every year, the process, mechanism, and materials that are utilized in the development of data storage devices see improvements. The desire for data storage at cheap prices was the driving force behind the quick fall in size and cost per gigabyte of data, as well as the rapid growth in areal density that accompanied it. The technology used for magnetic recording and storing progressed from magnetic tape to hard disk drives (HDD), which are a type of magnetic random-access memory (MRAM). In spite of the emergence of various methods for the recording and storing of data, magnetic-based data storage is continuing to grow in area density. This allows more reading and writing heads as well as disks to be packed onto a single drive. Researchers have made consistent efforts to increase the capacity of magnetic recording media, which has resulted in the development of modern hard disk drives that are based on perpendicular magnetic recording (PMR), heat-assisted magnetic recording (HAMR), two-dimensional magnetic recording (TDMR), and heated-dot magnetic recording (HDMR). The areal density of contemporary technology has increased to 10 TB/in^2 from the previous standard of 1 TB/in^2 (PMR) (HDMR). The development of future magnetic recording and storage technology will involve utilization of a wide variety of magnetic materials and approaches. This chapter discusses the various methods for developing magnetic materials, surface functionalization which is required to improve the performance/ areal density of magnetic-based storage tools, and their respective progresses. In addition, the various methods for developing magnetic materials are broken down into their respective subtopics.

1.6 SUMMARY, SCOPE, AND FUTURE DIRECTIONS

The magnetic characteristics of the MNMs could be tuned by adjusting the synthesis experimental parameters, utilizing the surface modification/functionalization and postthermal treatment. By selecting the suitable surface modifier, we can make MNMs as multifunctional NM. The nonmagnetic fraction in hybrid MNMs after the surface functionalization can reduce the saturation magnetization of MNM. However, other magnetic characteristics could be improved by a suitable selection of surface modifiers or by making effective interparticle interactions. The functionalization strategies are important key parameters to achieve the required magnetic or any other multifunctional characteristics for any specific application. Functionalized MNMs could be used in various applications such as photocatalysts, energy production/conversion/storage, magnetic separation, sensors, and biomedical applications. Recently, a few studies have shown a remarkable improvement in the efficiency of f-MNM in electrochemical applications with an external magnetic field. The magnetic characteristics of f-MNMs and the amount of external magnetic field can play an important role in the final efficiency. The magnetic characteristics of f-MNM can be advantageous over the separation and removal after the completion of specific applications. The collected f-MNMs can be evaluated for the next cyclic performance to evaluate the reusability of f-MNM. The magnetic field-assisted applications need more attention, as they have to be further explored in various applications. There is a lot of scope on the functionalized MNMs as MNMs could show some additional efficiency when interacting with an external magnetic field. The effect of an external magnetic field in various applications with the suitable selection of MNM and surface functionalization needs to look at in detail for possible practical applications.

ACKNOWLEDGMENTS

T.A. acknowledges Dr. Justin Joseyphus for his guidance and support. T.A. also acknowledges ANID-SA 77210070 and Universidad de Atacama for the financial support.

REFERENCES

[1] R.J. Joseyphus, A. Narayanasamy, N. Sivakumar, M. Guyot, R. Krishnan, N. Ponpandian, K. Chattopadhyay, Mechanochemical decomposition of Gd3Fe5O12 garnet phase, J. Magn. Magn. Mater. 272–276 (2004) 2257–2259. https://doi.org/10.1016/J. JMMM.2003.12.573.

[2] R.J.Joseyphus,A.Narayanasamy,D.Prabhu,L.K.Varga,B.Jeyadevan,C.N.Chinnasamy, K. Tohji, N. Ponpandian, Dipolar and exchange couplings in Nd2Fe14B/α-Fe ribbons, Phys. Status Solidi. 1 (2004) 3489–3494. https://doi.org/10.1002/PSSC.200405488.

[3] R.J. Joseyphus, A. Narayanasamy, N. Sivakumar, M. Guyot, R. Krishnan, N. Ponpandian, K. Chattopadhyay, Mechanochemical decomposition of Gd{sub 3}Fe{sub 5}O{sub 12} garnet phase, J. Magn. Magn. Mater. 272–276 (2004) 2257–2259. https:// doi.org/10.1016/J.JMMM.2003.12.573.

[4] R.J. Joseyphus, A. Narayanasamy, L.K. Varga, B. Jeyadevan, Studies on the exchange and dipolar couplings in Nd2Fe14B/α-Fe, Zeitschrift Fuer Met. Res. Adv. Tech. 99 (2008) 70–74. https://doi.org/10.3139/146.101597/MACHINEREADABLECITATION/RIS.

[5] T. Arun, M. Vairavel, S. Gokul Raj, R. Justin Joseyphus, Crystallization kinetics of Nd-substituted yttrium iron garnet prepared through sol-gel auto-combustion method, Ceram. Int. 38 (2012) 2369–2373. https://doi.org/10.1016/j.ceramint.2011.10.090.

[6] K. Prakash, R.J. Joseyphus, Magnetic Nanoparticle Flow Characteristics in a Microchannel for Drug Delivery Applications, AIP Conf. Proc. 1347 (2011) 27. https:// doi.org/10.1063/1.3601779.

[7] T. Arun, S.K. Verma, P.K. Panda, R.J. Joseyphus, E. Jha, A. Akbari-Fakhrabadi, P. Sengupta, D.K.K. Ray, V.S.S. Benitha, K. Jeyasubramanyan, P.V. V. Satyam, Facile synthesized novel hybrid graphene oxide/cobalt ferrite magnetic nanoparticles based surface coating material inhibit bacterial secretion pathway for antibacterial effect, Mater. Sci. Eng. C. 104 (2019) 109932. https://doi.org/10.1016/j.msec.2019.109932.

[8] A. Thirumurugan, A. Akbari-Fakhrabadi, R.J. Joseyphus, Surface Modification of Highly Magnetic Nanoparticles for Water Treatment to Remove Radioactive Toxins, in: Springer, Cham, 2020: pp. 31–54. https://doi.org/10.1007/978-3-030-16427-0_2.

[9] P. Karipoth, R.J. Joseyphus, Influence of annealing parameters on the magnetic properties of CoPt nanoparticles, Sci. Adv. Mater. 6 (2014) 1792–1798. https://doi. org/10.1166/SAM.2014.1943.

[10] P. Karipoth, R.J. Joseyphus, Evolution of High Coercivity in CoPt Nanoparticles Through Nitrogen Assisted Annealing, J. Supercond. Nov. Magn. 27 (2014) 2123–2130. https://doi.org/10.1007/S10948-014-2564-6/FIGURES/8.

[11] P. Karipoth, A. Thirumurugan, S. Velaga, J.-M.M. Greneche, R. Justin Joseyphus, Magnetic properties of FeCo alloy nanoparticles synthesized through instant chemical reduction, J. Appl. Phys. 120 (2016) 123906. https://doi.org/10.1063/1.4962637.

[12] P. Rajesh, S. Sellaiyan, A. Uedono, T. Arun, R.J. Joseyphus, Positron Annihilation Studies on Chemically Synthesized FeCo Alloy, Sci. Reports 2018 81. 8 (2018) 1–9. https://doi.org/10.1038/s41598-018-27949-2.

[13] P. Rajesh, J.-M.M. Greneche, G.A. Jacob, T. Arun, R.J. Joseyphus, Exchange Bias in Chemically Reduced FeCo Alloy Nanostructures, Phys. Status Solidi. 216 (2019) 1900051. https://onlinelibrary.wiley.com/doi/full/10.1002/pssa.201900051 (accessed July 18, 2022).

[14] P. Karipoth, R. Justin Joseyphus, Enhanced coercivity in non-equiatomic CoPt-Cu nanoparticles, J. Magn. Magn. Mater. 471 (2019) 475–481. https://doi.org/10.1016/J.JMMM.2018.09.081.

[15] G.A. Jacob, S. Sellaiyan, A. Uedono, R.J. Joseyphus, Magnetic properties of metastable bcc phase in Fe64Ni36 alloy synthesized through polyol process, Appl. Phys. A Mater. Sci. Process. 126 (2020) 1–7. https://doi.org/10.1007/S00339-020-3292-3/FIGURES/7.

[16] R. Ponraj, A. Thirumurugan, G.A. Jacob, K.S. Sivaranjani, R.J. Joseyphus, Morphology and magnetic properties of FeCo alloy synthesized through polyol process, Appl. Nanosci. 2019 102. 10 (2019) 477–483. https://doi.org/10.1007/S13204-019-01128-9.

[17] G.A. Jacob, R.J. Joseyphus, Enhanced Curie Temperature and Critical Exponents of Fe-Substituted NiCu Alloy, Phys. Status Solidi. 218 (2021) 2100050. https://doi.org/10.1002/PSSA.202100050.

[18] J.S. Anandhi, R.J. Joseyphus, Insights on the Heating Characteristics of Mn and Co Ferrites, Int. J. Thermophys. 42 (2021) 1–11. https://doi.org/10.1007/S10765-020-02782-W/FIGURES/3.

[19] G.A. Jacob, S.P.S. Prabhakaran, G. Swaminathan, R.J. Joseyphus, Thermal kinetic analysis of mustard biomass with equiatomic iron–nickel catalyst and its predictive modeling, Chemosphere. 286 (2022) 131901. https://doi.org/10.1016/J.CHEMOSPHERE.2021.131901.

[20] T. Arun, K. Prakash, R. Justin Joseyphus, Synthesis and magnetic properties of prussian blue modified Fe nanoparticles, J. Magn. Magn. Mater. 345 (2013) 100–105. https://doi.org/10.1016/J.JMMM.2013.05.058.

[21] K. Prakash, S. Nallamuthu, R.J. Joseyphus, Synthesis and Properties of Gold Coated Magnetic Nanoparticles, Front. Opt. 2012/Laser Sci. XXVIII (2012), Pap. FTu3A.42. (2012) FTu3A.42. https://doi.org/10.1364/FIO.2012.FTU3A.42.

[22] T. Arun, R. Justin Joseyphus, Prussian blue modified Fe3O4 nanoparticles for Cs detoxification, J. Mater. Sci. 49 (2014) 7014–7022. https://doi.org/10.1007/s10853-014-8406-x.

[23] P. Karipoth, R.J. Joseyphus, Magnetic properties of interacting CoPt nanoparticles synthesized through polyol process, Mater. Chem. Phys. 154 (2015) 53–59. https://doi.org/10.1016/J.MATCHEMPHYS.2015.01.044.

[24] K.S. Sivaranjani, G. Antilen Jacob, R. Justin Joseyphus, Coercivity and exchange bias in size reduced iron obtained through chemical reduction, J. Magn. Magn. Mater. 513 (2020) 167228. https://doi.org/10.1016/J.JMMM.2020.167228.

[25] R.J. Joseyphus, A. Narayanasamy, A.K. Nigam, R. Krishnan, Effect of mechanical milling on the magnetic properties of garnets, J. Magn. Magn. Mater. 296 (2006) 57–64. https://doi.org/10.1016/J.JMMM.2005.04.018.

[26] R.J. Joseyphus, D. Kodama, T. Matsumoto, Y. Sato, B. Jeyadevan, K. Tohji, Role of polyol in the synthesis of Fe particles, J. Magn. Magn. Mater. 310 (2007) 2393–2395. https://doi.org/10.1016/J.JMMM.2006.10.1132.

[27] R.J. Joseyphus, T. Matsumoto, H. Takahashi, D. Kodama, K. Tohji, B. Jeyadevan, Designed synthesis of cobalt and its alloys by polyol process, J. Solid State Chem. 180 (2007) 3008–3018. https://doi.org/10.1016/J.JSSC.2007.07.024.

[28] R.J. Joseyphus, K. Shinoda, Y. Sato, K. Tohji, B. Jeyadevan, Composition controlled synthesis of fcc-FePt nanoparticles using a modified polyol process, J. Mater. Sci. 43 (2008) 2402–2406. https://doi.org/10.1007/S10853-007-1951-9/FIGURES/6.

[29] R.J. Joseyphus, K. Shinoda, D. Kodama, B. Jeyadevan, Size controlled Fe nanoparticles through polyol process and their magnetic properties, Mater. Chem. Phys. 123 (2010) 487–493. https://doi.org/10.1016/j.matchemphys.2010.05.001.

[30] J. Shebha Anandhi, G. Antilen Jacob, R. Justin Joseyphus, Factors affecting the heating efficiency of Mn-doped Fe3O4 nanoparticles, J. Magn. Magn. Mater. 512 (2020) 166992. https://doi.org/10.1016/J.JMMM.2020.166992.

[31] R. Justin Joseyphus, A. Narayanasamy, K. Shinoda, B. Jeyadevan, K. Tohji, Synthesis and magnetic properties of the size-controlled Mn–Zn ferrite nanoparticles by oxidation method, J. Phys. Chem. Solids. 67 (2006) 1510–1517. https://doi.org/10.1016/J.JPCS.2005.11.015.

[32] T. Arun, K. Prakash, R. Kuppusamy, R.J. Joseyphus, Magnetic properties of prussian blue modified Fe3O4 nanocubes, J. Phys. Chem. Solids. 74 (2013) 1761–1768. https://doi.org/10.1016/j.jpcs.2013.07.005.

[33] T. Arun, R. Justin Joseyphus, Prussian blue modified FePt nanoparticles for the electrochemical reduction of H_2O_2, Ionics (Kiel). 22 (2016) 877–883. https://doi.org/10.1007/s11581-015-1617-6.

[34] P. Karipoth, A. Thirumurugan, R. Justin Joseyphus, Synthesis and magnetic properties of flower-like FeCo particles through a one pot polyol process, J. Colloid Interface Sci. 404 (2013). https://doi.org/10.1016/j.jcis.2013.04.041.

[35] J.S. Anandhi, T. Arun, R.J. Joseyphus, Role of magnetic anisotropy on the heating mechanism of Co-doped Fe3O4 nanoparticles, Phys. B Condens. Matter. 598 (2020) 412429. https://doi.org/10.1016/j.physb.2020.412429.

2 Surface Characterization Techniques

T. V. K. Karthik
Autonomous Hidalgo State University

M. Pérez-González
Universidad Autónoma del Estado de Hidalgo

M. Morales-Luna
Tecnológico de Monterrey

H. Gómez-Pozos and A. G. Hernández
Universidad Autónoma del Estado de Hidalgo

CONTENTS

DOI: 10.1201/9781003335580-2

15

2.1   INTRODUCTION TO SURFACE CHARACTERIZATION TECHNIQUES

Nanotechnology has been developing recently with a great interest among scientific community, and in fact, its advances allowed the obtaining of new synthetic materials in a wide range of applications like medical, technological, domestic, industrial, biological, and agricultural, among others. In order to understand the properties of the emerging nanomaterials, surface characterization techniques are important to understand these materials (Gandhi 2022).

On one hand, surfaces become relevant since their interaction with the environment, especially the surface properties of nanomaterials deviate considerably from the bulk properties owing to a difference in physical structure and chemistry (Klein et al. 2008). On the other hand, magnetic nanomaterials have received considerable attention in the last two decades due to their usage for controlled drug delivery in cancer treatments (Gul et al. 2019). A big challenge in the synthesis and characterization of magnetic nanomaterials like iron oxide has been the capping of nanoparticles, since some of them are highly reactive and prone to oxidation affecting their magnetic behavior. The complexity of surface properties of nanomaterials leads to the translation of surface characterization techniques to study their properties (Gul et al. 2019; Araujo et al. 2020).

Besides, different types of synthesis for magnetic nanomaterials have been developed in order to obtain modified properties, as well as other techniques like doping or the synthesis of composites (Bertolucci et al. 2015). Due to this, surface characterizations should be reliable techniques for obtaining information of the structural, chemical, morphological, magnetic, and electrical properties of the novel nanomaterials which emerge day by day.

In this work, we intend to show novel surface characterization techniques for magnetic nanomaterials. Furthermore, techniques are explained in detail and their physical principles are mentioned for a better understanding. The advantages and limitations of the techniques are present in this chapter and the requirements for sample preparation are described.

2.2   CHEMICAL CHARACTERIZATIONS

2.2.1   Definition of Chemical Characterizations

Chemical characterizations such as Raman spectroscopy or X-ray photoelectron spectroscopy (XPS) provide chemical structure information of the elements or compounds in a sample. Despite the two techniques that are able to give the same information in appearance, there are several substantial differences among them. Besides, XPS can quantify the amount of each element in a sample without the necessity of an external reference, but it cannot measure some elements such as hydrogen.

Furthermore, through infrared and Raman spectroscopies, the composition of a sample can be inferred not by its elements but by the chemical compounds in the material. For instance, in typical infrared or Raman spectra, as a fingerprint, several bands related to the bonds or vibration modes of molecules appear. Thus, the researcher is allowed to identify which chemical compounds (and the elements forming the compounds) are in the material. In this section, several chemical characterizations and the involved mechanisms will be presented.

2.2.2 IR SPECTROSCOPY

Materials science tries to describe different interactions like temperature, mechanical deformation, and irradiation with the matter to mention some of them. This section will focus on the interaction between irradiation and matter. It is well known that there is a range in the radiation spectrum that energetically could excite electrons and/or molecules to energy levels greater than the basal energy state in which the electrons or molecules are found, *e.g.*, the ultraviolet region. However, there is a region of the radiation spectrum that energetically does not have the ability to generate these changes, but what it can produce is a change in the vibrational energy of the molecule. This region goes from 750 to 1×10^6 nm; which, in turn, is divided into three regions: the near (750–2,500 nm), middle (2,500–14,900 nm), and far (14,900–1×10^6 nm) (Skoog, Holler, and Crouch 2021). It is important to mention at this point that the region of greatest interest in the characterization of materials is the mid-IR region. To study this region, the emission as well as the absorption and reflection spectra are analyzed and used for analyses either qualitative or quantitative. On the other hand, the near-IR region is often useful for the quantitative determination of certain species, such as water, carbon dioxide, sulfur, and low molecular weight hydrocarbons (Skoog, Holler, and Crouch 2021). Unlike the mid-IR, the interaction that is used is the diffuse reflectance spectra of samples that can be in liquid or solid form without previous treatment or in gas absorption studies. As regards the far infrared region, it is useful for determining the structures of inorganic and organometallic species by analyzing absorption measurements (Skoog, Holler, and Crouch 2021).

2.2.2.1 Mid-IR Spectroscopy

The basic questions that mid-IR spectroscopy can answer are as follows: What molecules are present in the sample? Can two samples be identical or is there a difference between them? And finally, do you know what are the concentrations of molecules in this sample? The first can be answered by locating and associating the positions of the peaks in the infrared spectrum, which are correlated with the molecular structure. The second question can be answered by correlating an unknown spectrum with a reference spectrum and seeing how well the positions, heights, and widths of the peaks in the two curves matched. So, IR spectra will always have significant differences if the samples are different. Finally, this characterization method allows knowing the concentrations of various molecules by measuring the spectra of samples of known concentration and then using Beer's law to prepare a calibration line that relates absorbance to concentration (Skoog, Holler, and Crouch 2021).

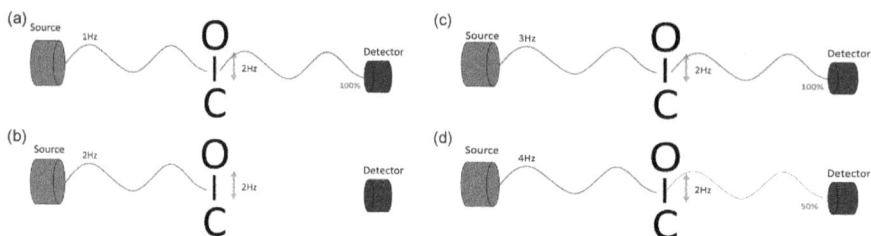

FIGURE 2.1 Schematic representation of a carbon and oxygen bond: (a) a contraction or elongation of the bond with a frequency value of 2 Hz and (b) a vibrational in plane y–z of the bond with a frequency value of 4 Hz.

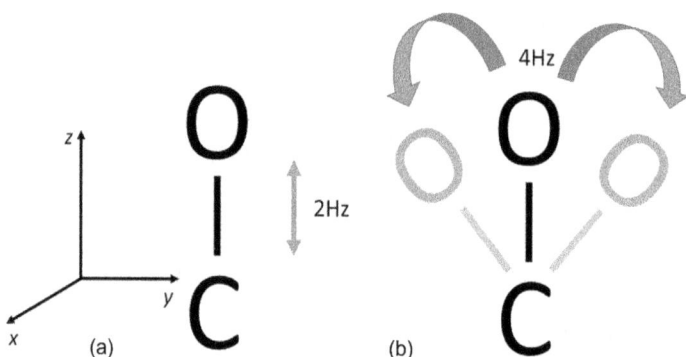

FIGURE 2.2 Schematic representation of the signal–matter interaction: (a) 1 Hz frequency, (b) 2 Hz frequency, (c) 3 Hz frequency, and (d) 4 Hz frequency. In all the cases, the frequency value is emitted by a source and the signal interacts with a carbon and oxygen bond.

The basic idea of how IR spectroscopy works is described by the following example: Let us consider a molecule that is formed by a bond between a carbon and an oxygen atom as shown in Figure 2.1. This bond could be vibrating, in contraction or elongation (with a frequency of 2 Hz) with respect to the "z" axis, as shown in Figure 2.1a. On the other hand, it could also be oscillating like a pendulum in the "y–z" plane, with a frequency of 4 Hz, see Figure 2.1b. With these values of natural vibration of the bond as a basis, now let us suppose a radiation with a frequency of 1 Hz as an incidence signal. As shown in Figure 2.2a, the signal when it interacts with the molecule does not match with the natural vibration frequency of the molecule, this signal passes throughout without interacting and reaches the detector 100% of the signal. But if the source modifies the signal and generated a signal with a frequency of 2 Hz (Figure 2.2b), when this signal interacts with the molecule since this frequency match with the frequency of elongation or contraction, the signal will be absorbed by the molecule and the detector will not receive the signal (the molecule absorbed the radiation). Now, let us consider that the signal is generated with a 3 Hz frequency, the interaction behavior could be the same as the signal at 1 Hz, and it will arrive at the detector without any modification (Figure 2.2c). With a 4 Hz signal, it could happen that it is totally absorbed because this molecule has a vibration with this frequency; however, it can happen that the signal could be attenuated, and a percentage of the signal can pass through the molecule; let's say 50% less of the signal (see Figure 2.2d). If the signal is not totally absorbed by the molecule, it might be due to various interactions between the signal and the molecule.

We can observe from Figure 2.2 that the absorption of radiation by the molecule is reflected in the IR spectrum, and the changes in transmittance are mainly associated with a vibration type between the different elements. Based on this, as we have already described, IR spectroscopy has been one of the most developed characterization techniques to know the type of molecule or compound that is being formed in the material under study. From these studies; two regions have been found, the region that goes from 4,000 to 1,500 cm^{-1} is called the region of fundamental bands or functional groups, and the region from 1,500 to 200 cm^{-1} is called fingerprints. The last is the one with the most valuable information to be able to distinguish the elements that compound the molecules. On the other hand, vibrations have been classified into two families (Mitsuo Tasumi 2014): (i) Stretching, longitudinal, or tension (v) where the distance between the atoms that form the union is the one that varies during a vibration, and, in turn, this is classified as symmetric or asymmetric stretching (see Figure 2.3a and b), respectively.

(ii) Deformation or flexion (δ): It is associated with a difference in the angle formed between atoms due to the vibration caused by the interaction of the radiation with the molecule. This is divided into in-plane swing, in-plane scissoring, out-of-plane flapping, and out-of-plane twisting (see Figure 2.4).

Finally, it has been found that hydrogen stretching occurs in the region of 3,600–2,500 cm^{-1}, where it can be found that the absorption is related with the vibrational stretching of hydrogen atoms bonded to carbon, oxygen, and nitrogen. On the other hand, triple bonds have been found in the region between 2,300 and 2,000 cm^{-1}, and in this so-called region, the absorption is related to vibrational stretching vibration of the triple bonds. In the double bond (1,900 – 1,550 cm^{-1}), the absorption process is correlated with the vibrational stretching of the carbon–oxygen, carbon–nitrogen, and carbon–carbon double bonds. Lastly, the hydrogen bending is between 1,600 and 1,250 cm^{-1}; in this case, the absorption is commonly due to the bending vibrations of the hydrogen atoms bonded to carbon and nitrogen (Skoog, Holler, and Crouch 2021).

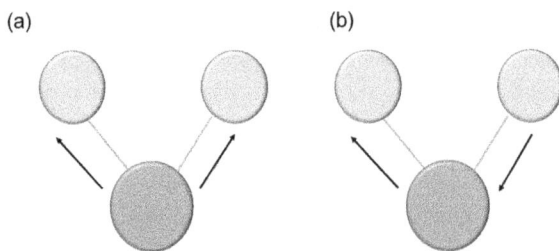

(a)   (b)

FIGURE 2.3 Schematic representation of (a) symmetric stretching and (b) asymmetric stretching. The arrow represents the movement of the atoms of the compound.

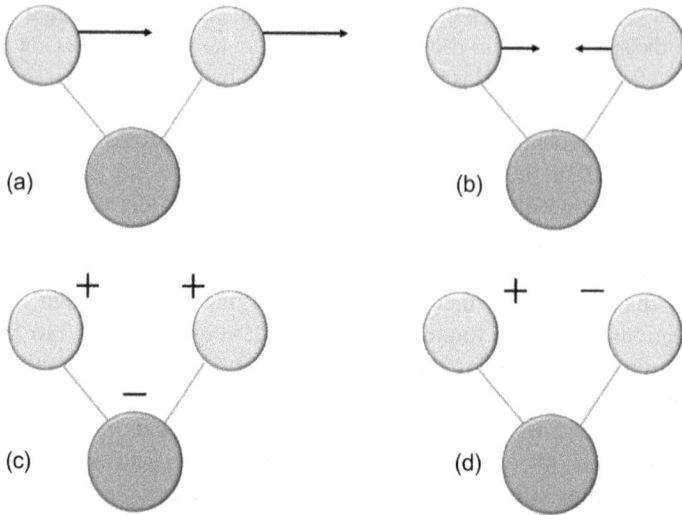

FIGURE 2.4 Diagram of (a) in-plane swing, (b) in-plane scissoring, (c) out-of-plane flapping, and (d) out-of-plane twisting. The arrow represents the movement of the atoms of the compound, and the plus and minus signs represent the movement out of the plane or enter to the plane, respectively.

2.2.2.2 Mathematical Analysis

The location of these absorption bands in the IR spectrum can be predicted by a mathematical expression that represents the vibrational motion of two atoms that form the chemical bond (Peter Larkin 2017). Characteristics of elongation or contraction vibrations are usually represented with a mechanical model that consists of two massive bodies joined by a spring, as shown in Figure 2.5. The deformation; x, produced by one of these masses along the spring axis, produces a vibration called simple harmonic motion.

To obtain the mathematical relation that describes the motion of this system, we should consider the vibration of only one mass; m_1, which will be attached to a spring that is suspended, as shown in Figure 2.6.

Now, if the mass deforms the spring a distance "x" from equilibrium point, the spring will apply a force contrary to the movement this force is considered as a restoring force F, which is proportional at displacement x, the relationship between these two variables is well described by Hooke's law; $F = -k\,x$. Being k the spring constant, the minus sign in the equation indicates a force that opposes the applied force by m_1.

To describe the movement of m_1, as a function of time "t", it is necessary to remember Newton's second law that establishes $F = m \times a$, where for our case $m = m_1$, and the acceleration "a" according to the fundamentals of classical mechanics is as follows:

$$a = \frac{d^2x}{dt^2} \tag{2.1}$$

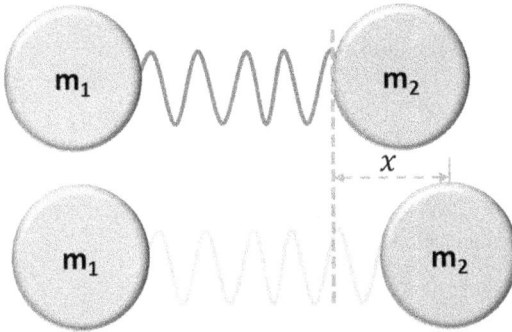

FIGURE 2.5 Schematic representation of motion of a chemical bond of two atoms.

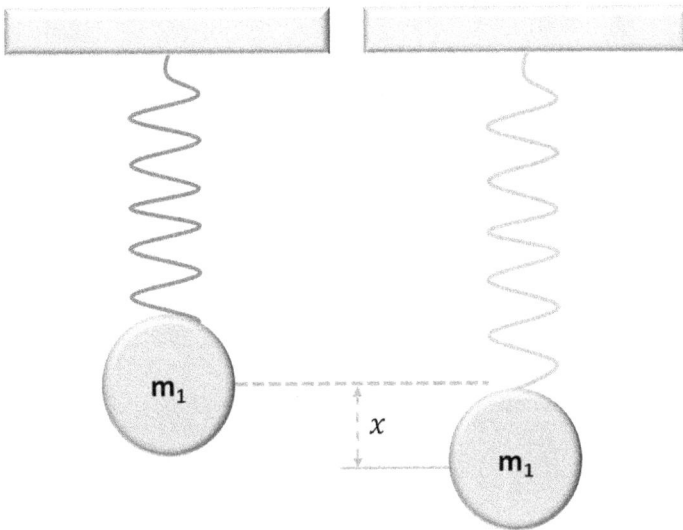

FIGURE 2.6 Schematic representation of motion of an atom.

For the case of the time description of the mass attached to a spring, it can match Hooke's law and Newton's second law to obtain the following equation:

$$m_1 \frac{d^2x}{dt^2} = -k\ x \tag{2.2}$$

To solve this second-degree differential equation, we can propose a function for "x" whose second derivative is equal to the original function. This condition is met by the Sine function. Therefore, the instantaneous shift of m_1 at time t can be expressed as follows:

$$x = A\ Sin\ (\omega t) \tag{2.3}$$

where ω is the angular frequency. Then, by applying a second derivative to the displacement function, we can obtain the following equation:

$$-m_1\, A\, \omega^2 Sin(\omega t) = -k\, A\, Sin\,(\omega t) \tag{2.4}$$

By simplifying and remembering that the angular frequency is related to the linear frequency as $\omega = 2\,\pi\,\nu$, we obtain the relationship for the natural frequency of oscillation for the system as follows:

$$\nu = \frac{1}{2\,\pi}\sqrt{\frac{k}{m_1}} \tag{2.5}$$

Now, we can rewrite this relationship as a function of wavelength, looking back that $\nu = c/\lambda$, so we can rewrite equation 2.1 as follows:

$$\bar{\nu} = \frac{1}{2\,\pi\,c}\sqrt{\frac{k}{m_1}} \tag{2.6}$$

This relationship can be modified in order to explain the conduct of a system, which includes two masses m_1 and m_2 joined with springs. For this case, is necessary to replace the mass m_1 with a reduced mass μ, which more interested readers can develop as an exercise.

$$\mu = \frac{m_1 m_2}{m_1 + m_2} \tag{2.7}$$

With this relation, then equation 2.2 can be rewritten as follows:

$$\bar{\nu} = \frac{1}{2\,\pi\,c}\sqrt{\frac{k}{\mu}} \tag{2.8}$$

From this equation, now we can predict theoretically the appearance of the absorption bands present in the IR spectrum. For these calculations, the k values that have been found are between 3×10^2 and 8×10^2 N/m for most single bonds. However, the value of 5×10^2 N/m can be taken as a reasonable mean value. In the case of double bonds, it has been found that the value of k is 1×10^3 N/m, and finally, for the triple bond, a value of 1.5×10^3 N/m is reported (Skoog, Holler, and Crouch 2021).

2.2.2.3 Fourier Transform (FTIR) Spectroscopy

As we have seen in the previous subsection, IR spectroscopy provides information on the type of bond that molecules have. However, the IR technique has been improving over time in the data acquisition to indicate or find the presence of certain chemical bonds. On the other side, it has been developed an alternative to IR spectroscopy to increase the quality of the signal, this alternative is called Fourier Transform IR spectroscopy (FTIR), which essentially works in the same way as IR spectroscopy,

Infrared source

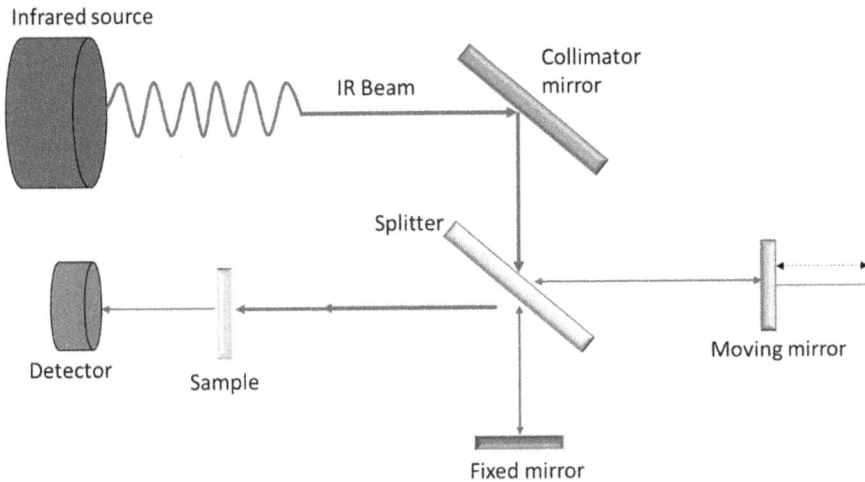

FIGURE 2.7 Schematic representation of Michelson's interferometer.

notwithstanding, the main modification is the optical arrangement that FTIR systems possess (Brian C. Smith 2011).

The optical arrangement is known as the Michelson interferometer, which, as is seen in Figure 3.7, this optical arrangement consists of a source of IR radiation, this generated signal goes to a collimating mirror where it directs the signal towards a divider, where the signal is split into two beams, one that goes to a fixed mirror and another part of the signal goes towards a moving mirror, the purpose of this moving mirror is that the signal that comes from the source has a different optical path than the signal that goes to the fixed mirror.

2.2.3 RAMAN SPECTROSCOPY

Raman spectroscopy turns out to be a complementary characterization to IR or FTIR. Raman also allows us to visualize the type of chemical bond that a molecule will present, through the vibration of the elements of which the molecule is composed. However, the subtle difference it has with IR spectroscopy is the type of interaction between the radiation (in this case it is a monochromatic light source) and matter. As we will see in this section, the Raman system will measure the type of dispersion that the molecule has when it interacts with specific radiation.

Based on the dispersion phenomenon, two types of dispersion can be studied: elastic and inelastic dispersions. The elastic dispersion case was a phenomenon studied by the British physicist John W. Strutt, also known as Lord Rayleigh, when he tried to describe the phenomenon of light polarization (Young 1981). Additionally, to these studies, he also contributed to the understanding of the process of light scattering in the atmosphere which is related to the bluish color in the sky. This process, as its name suggests, indicates that the radiation that interacts with the matter does not undergo any modification after the interaction, in other words, the radiation that enters the molecule is equal to the radiation after the interaction (Tyndall 1869). In

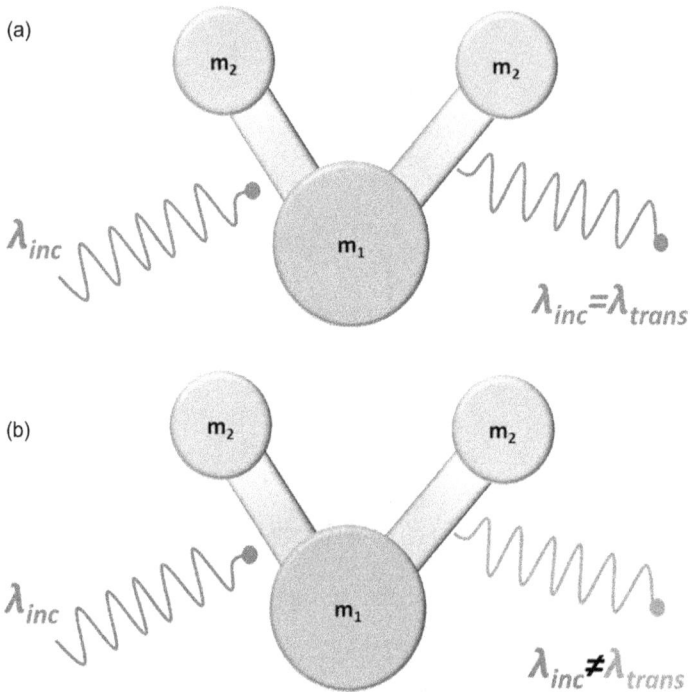

FIGURE 2.8 Schematic representation of (a) Rayleigh scattering process or elastic dispersion and (b) inelastic dispersion.

Raman, this process is known as Rayleigh-type scattering; see Figure 2.8a. On the other hand, when the dispersion type is inelastic, a significant change can be perceived in the input and output signals, and in this case, the output signal will be different from the input signal. This difference can be visualized as a red shift or a blue shift, radiation gain or loss, respectively, as shown in Figure 2.8b. The Raman effect is an inelastic light-scattering phenomenon in which two different types of processes can occur, the first and most common process is that the photon can lose energy during the interaction with the molecule (Stokes). The second process could be that the photon can gain energy after the interaction with the molecule (Anti-Stokes). These two processes will be discussed below in this section. The interaction can happen by distortion or polarization of the electron cloud that surrounds the nucleus to create a short lifetime that is called a virtual state. To generate this virtual state, it is necessary to use a high-intensity monochromatic beam, such as a laser (Vandenabeele 2013). The generated virtual state is not stable, so if a photon is excited to the virtual state, it tends to recombine rapidly. Become obvious that the energy of these states also depends on the excitation wavelength source. So, at room temperature, most molecules are at the lowest energy level (Skoog, Holler, and Crouch 2021). Among these, the most likely process that could occur is the Rayleigh scattering, where most of the photons that are scattered do not present a change in energy and therefore the incident photons on the material return to the same energy state, *that is*, when most

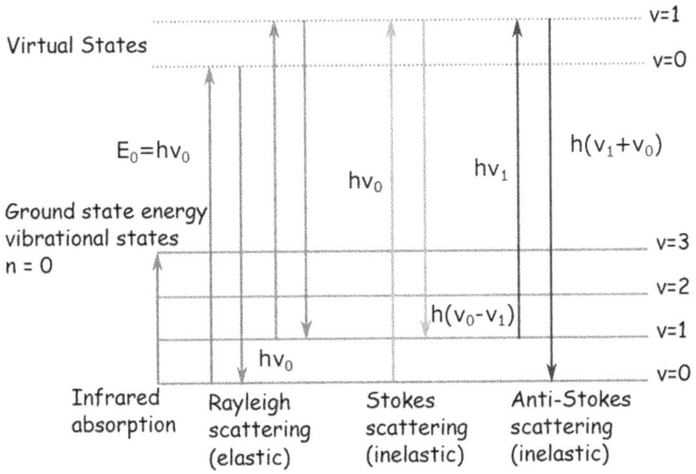

FIGURE 2.9 Raman scattering process, Rayleigh, Stokes, and anti-Stokes types.

photons return to their ground state, radiation with the same wavelength as the incident radiation is emitted. Therefore, changing the photon energy becomes a direct measure of the phonon energy.

Stokes-type scattering occurs when the energy of the reemitted photon is lesser than the incident photon, and that is why in Figure 2.9, the energy relationship is represented as a subtraction between the natural oscillation frequency ν_0 and the emission frequency, ν_1. This indicates a loss of energy of the photon. The process in which the energy of the reemitted photon has higher energy than the energy of the incident photon is called anti-Stokes-type scattering, here the energy difference is represented as a sum of frequencies between ν_0 and ν_1, and now here the photon increase his energy.

As is typical in Raman spectroscopy, it is necessary to indicate the differences between the intensities of these three types of interaction (Smith and Dent 2019). It is well known that the relative intensities will depend on the temperature and the occupation of the different states of the system. For the case of Stokes and anti-Stokes scattering, we should emphasize that anti-Stokes scattering compared to Stokes scattering has a much lower probability of occurring because there are few energy transitions where there is an energy gain. In addition, for Rayleigh scattering there is no change; as was mentioned this scattering is elastic, so there is no Raman shift. However, in the anti-Stokes case, the shift is toward minor wavelengths. Finally, it is important to highlight that the intensities corresponding to the Rayleigh and Stokes processes are much higher than the intensities that the anti-Stokes scattering presents, as previously mentioned.

As a context of this characterization technique, it has its beginnings in the 1920s, when the physicist Chandrasekhara Venkata Raman published a series of investigations on molecular diffraction, but it was not until 1928 that he published the results of this effect (Raman effect) (Gardiner and Graves 1989). And, in addition to the vibrational information that the molecule has, this technique has also been used to obtain crystallinity information, detect doping and/or matrix defects, and distinguish from heat treatments.

2.2.4   X-Ray Photoelectron Spectroscopy

Surfaces are very important since many properties of a material depend on the interactions between its surface and the surrounding environment. Among the techniques of analysis of surfaces, XPS has demonstrated their capability to provide value information of a surface related to the quantitative and qualitative chemical bonding, elemental composition, oxidation states, and so on. In this section, we will discuss the powerful of XPS through several examples.

The origins of XPS can be traced to the photoelectric effect noticed by Hertz and later described by Einstein in 1905. During the photoelectric process, when photons of specific energies given by hv impact with a material, the energy of the light is transferred to the atoms of the material. If the work function of the material is lower than the energy of the photons, then electrons with certain kinetic energy, E_k, will be ejected from the surface. The equation that relates the energies of photons and electrons is (Heide 2012)

$$BE = hv - E_k - \phi_{spectrometer} \qquad (2.9)$$

where BE is the binding energy of the electron, while $\phi_{\text{spectrometer}}$ is the work function of the spectrometer. An XPS spectrum is plotted as a function of the number of electrons detected per unit time versus BE. Since each element produces a unique set of XPS peaks at characteristic BE values, it allows the direct identification of the elements existing on the surface of the material being analyzed.

2.2.4.1   XPS Acquisition Modes and Some Examples

There are several acquisition modes for XPS, for instance, *conventional mode*. The XPS spectra allow identification of elemental composition of a surface. Detection limit for the atomic concentration is close to 0.05 at.%. Likewise, oxidation states of the elements can be studied. When an XPS measurement is performed, several analyses can be done to obtain the survey, high-resolution or valence band spectra, and the Auger peaks, among others.

In Figure 2.10, a survey spectrum of a Fe_3O_4 powder sample is shown, as seen there are several peaks at different binding energies. However, some of the most important are labeled according to the chemical composition of the sample (Fe 2p, O 1s, and C 1s). Iron and oxygen combine to form Fe oxides, in this case, Fe_3O_4. The presence of carbon is attributed to the interaction between the sample and air from the environment (Watts and Wolstenholme 2019). Despite the evident contamination of the surface of the sample, the C 1s peak is useful to calibrate the XPS signals since C 1s is assigned to adventitious hydrocarbon placed between 284.6 and 285.0 eV, which is important for interpretations of good results and will be discussed later (Moulder et al. 1992). Survey spectra measurements are useful in order to study the presence of chemical elements on the first surface layers (less than 12 nm). However, conventional XPS equipment performs this measurement with a step energy of 1 eV.

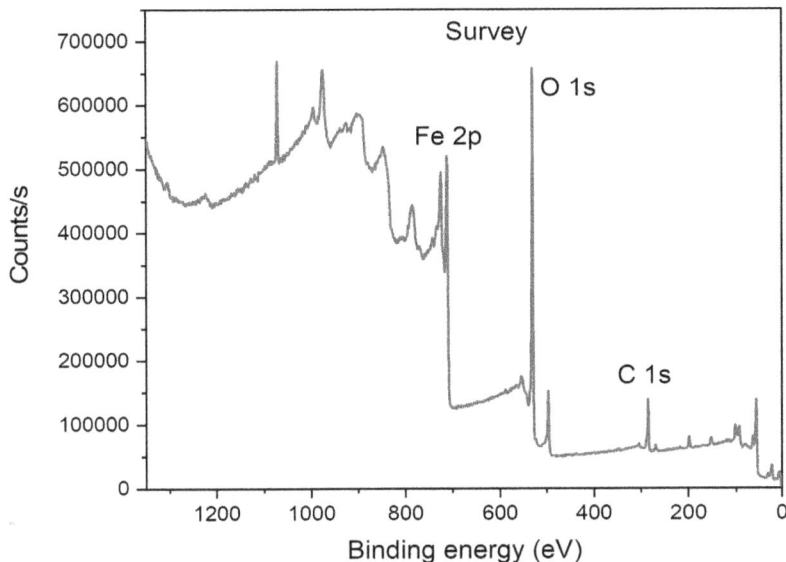

FIGURE 2.10 Survey of a Fe_3O_4 sample.

If some elements are of interest, there is another analysis mode that is able to measure step energies lower than 1 eV. It is known as high-resolution (HR-XPS) or core-level mode. HR-XPS is useful to investigate in more detail the bonds between elements and the surface chemical composition of a sample.

In Figure 2.11, the core-level spectra of (a) Fe 2p, (b) Co 2p, and (c) Ni 2p states are shown. As seen, 2p states are presented as a doublet, *that is*, two peaks appear. It should be mentioned that these doublets are formed since spin–orbit splitting arises from quantum mechanical considerations. The $2J+1$ degeneracy should be considered, where J is the total angular momentum quantum operator, which satisfies $J=L+S$, being L the orbital- and S the spin-quantum angular momentum operators (Morales-Luna et al. 2019).

As known, the associated quantum numbers (principal n, orbital l, and spin s) are constrained since l is an integer number (0, 1, 2, …), for electrons $s=\frac{1}{2}$, and allowed values of j are $|l-s| \leq j \leq |l+s|$. Then, for 2p states such as Fe 2p, Ni 2p, and Co 2p, the peaks are labeled as Fe $2p_{3/2}$ and Fe $2p_{1/2}$, Ni $2p_{3/2}$ and Ni $2p_{1/2}$, and Co $2p_{3/2}$ and Co $2p_{1/2}$, respectively. From the inspection of the $2p_{3/2}$ and $2p_{1/2}$ contributions of each element it is noted that the $2p_{3/2}$ contributions have higher area (or intensity) than the $2p_{1/2}$ photoelectron lines, this fact is explained since the areas (or intensities) of the peaks are connected by the $2J+1$ degeneracies as follows Area (M $2p_{1/2}$)/Area (M $2p_{3/2}$) $= [2(1/2)+1]/ [2(3/2)+1] = 1/2$, where M is the element analyzed. The same relation can be used for the peak intensities for most of the elements. However, it has been reported that some elements, such as Ti 2p peaks (as seen in Figure 2.12), do not satisfy that equation for intensities since the intensity of the Ti $2p_{1/2}$ peak is not the half of that displayed by the Ti $2p_{3/2}$ peak due to the Coster–Kronig effect. Moreover, while Ti $2p_{1/2}$ does not reach the expected intensity, its Full-Wide at Half-Maximum

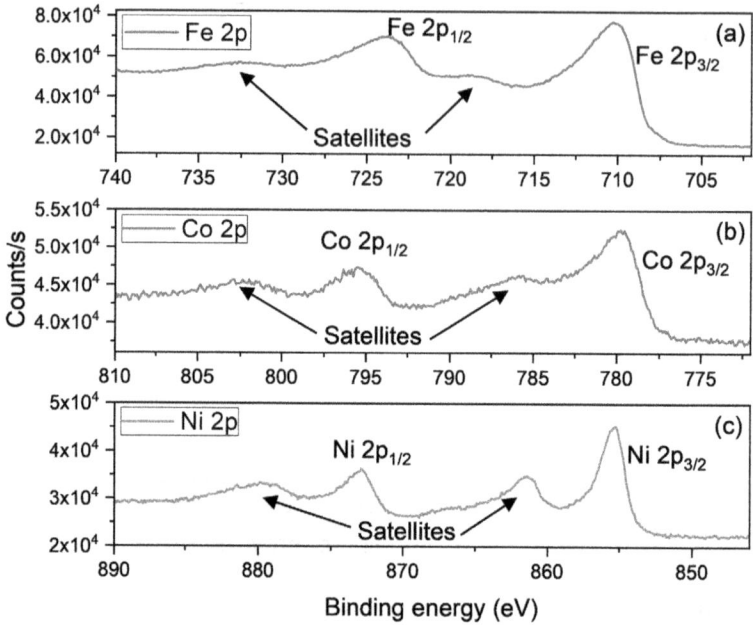

FIGURE 2.11 XPS core-level spectra of (a) Fe 2p, (b) Co 2p, and (c) Ni 2p states.

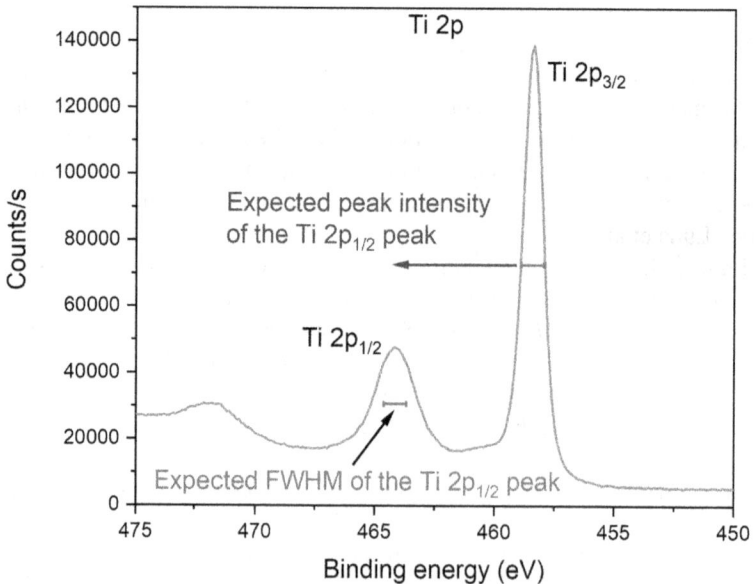

FIGURE 2.12 Ti 2p spin–orbit doublet of a thin film exhibiting the Coster–Kronig effect.

TABLE 2.1
XPS Area Ratios and Examples of Typical XPS Peaks Found in Reported Literature

Subshell (*l* Quantum Number)	States (*l* − *s*, *l* + *s*) with *s* = ½	$\dfrac{Area\,(peak\,1)}{Area\,(peak\,2)}$	Examples of XPS Peaks Typically Found in the Literature
s (*l* = 0)	$1s_{1/2}$ (usually denoted by 1s)	-	C1s N 1s O 1s
p (*l* = 1)	$2p_{1/2}$ and $2p_{3/2}$	1/2	Fe 2p Co 2p Ni 2p Ti 2p Zn 2p
d (*l* = 2)	$3d_{3/2}$ and $3d_{5/2}$	2/3	Mo 3d Ag 3d As 3d In 3d Pd 3d
f (*l* = 3)	$4f_{5/2}$ and $4f_{7/2}$	3/4	W 4f Pt 4f Au 4f Pb 4f

(FWHM) is higher than expected, which should be similar to that exhibited by the Ti $2p_{3/2}$ peak (Chambers et al. 2018).

Returning to the peak area ratios of an XPS doublet, Area (peak 1)/Area (peak 2), that relation can be extended to other states. For instance, for 3d states, the spin–orbit values are given by $3d_{3/2}$ and $3d_{5/2}$, while for 4f states, the peaks are labeled $4f_{7/2}$ and $4f_{5/2}$. In Table 2.1, the peak area ratios are shown for different subshells reported in the literature (Granada-Ramírez et al. 2022; Eguía-Eguía et al. 2021; Kloprogge and Wood 2020; Pérez et al. 2021).

With respect to the calibration of XPS signals, the most common method consists of using the C 1s peak attributed to adventitious hydrocarbon. This method is as follows: first, determine the position of the C 1s signal, and then shift at higher (or lower) binding energy until the C 1s peak matches the position of 284.6 eV. However, several references use other positions: for instance, 284.8 or even 285 eV (Moulder et al. 1992). Finally, shift the peaks of other elements in the same amount that is used for C 1s. The process is shown in Figure 2.13, where the C 1s peak was shifted an amount of ΔE (Figure 2.13a), and then the peaks that correspond to Pt 4f and Fe 2p were shifted in the same energy ΔE, as shown in Figure 2.13b and c.

It should be noticed in Figure 2.13 that the signal height is higher at higher energies (left side of the curves). That effect in the spectra arises due to several factors being the inelastic scattering of electrons as the main source (Heide 2012; Watts and Wolstenholme 2019).

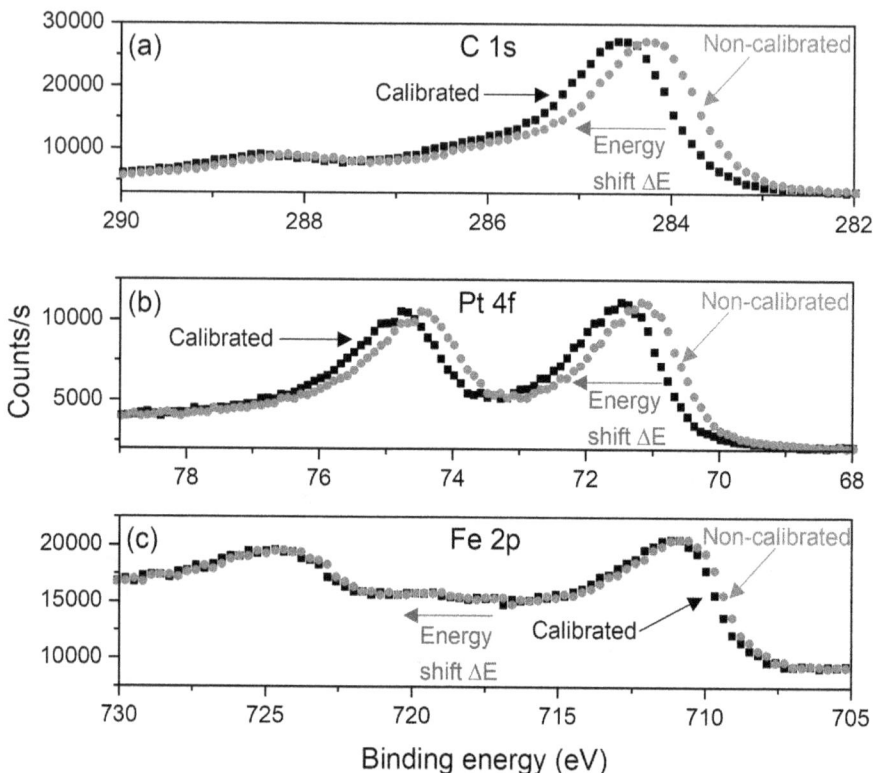

FIGURE 2.13   Calibration of (a) C 1s, (b) Pt 4f, and (c) Fe 2p peaks.

For quantification of the elemental composition of a sample, background signal must be removed. It has been pointed out that the background subtraction process is an open line of research since traditional methods do not give an exact amount of each element in a sample. However, at least three background subtraction routines are widely used: Linear, Shirley, and Tougaard (Heide 2012; Watts and Wolstenholme 2019). As seen in Figure 2.14, the background of Fe 2p peak has been adjusted using the Linear, Shirley, Tougaard, and "Shirley-type" or Smart routines. In general, for peaks with a behavior similar to that exhibited by the Fe 2p doublet, the linear background is not recommended to use since it does not discriminate between the XPS and the background signals.

It has been reported that peak areas (and consequently, the estimation of the elemental composition) varied as much as 50% if the integration points are slightly moved (Watts and Wolstenholme 2019). The Tougaard background has a theoretical basis that considers the scattering of electrons in solids. However, at the same time, several parameters must be adjusted to perform the analysis, thus limiting its practical application. The Shirley routine has no theoretical basis, but considering an S-shaped background has been demonstrated to be very effective to quantify the elemental composition of most of the samples with acceptable accuracy. On the other hand, the Smart background implemented in the Avantage software (Thermo Fisher

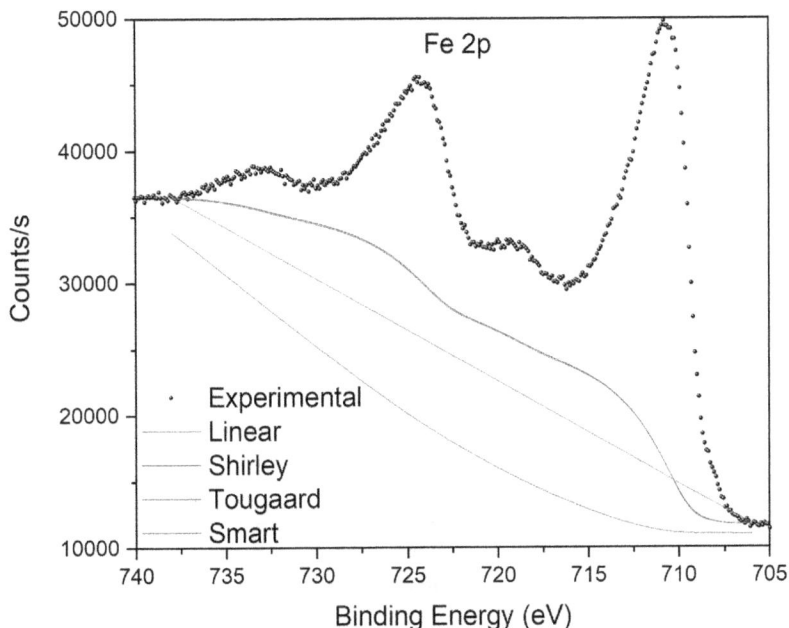

FIGURE 2.14 Background subtraction lines for the Fe 2p doublet.

TABLE 2.2
Estimated Peak Areas for the Fe 2p Doublet
(Figure 3.14) Using Different Subtraction Routines

Subtraction routine	Start BE (eV)	End BE (eV)	Area $\dfrac{Counts}{s} * eV$
Linear	738	706	36.0
Tougaard	738	706	51.8
Shirley	738	706	26.8
Smart	738	706	27.0

Scientific) is an "improved" Shirley routine with the constraint that the background is always below the experimental data. For the Fe 2p doublet presented in Figure 2.4, the peak area was estimated and presented in Table 2.2, considering the same region but different subtraction routines, as the Shirley and Smart backgrounds provide almost the same areas, while Linear and Tougaard backgrounds give completely different values.

Some elements display only one peak in appearance, for instance, the Se 3d signal. However, if a deconvolution process is carried out, the peak may be resolved in two components as shown in Figure 2.15, since two peaks are expected because of the spin–orbit splitting. When the principal and orbital (n and l) quantum numbers

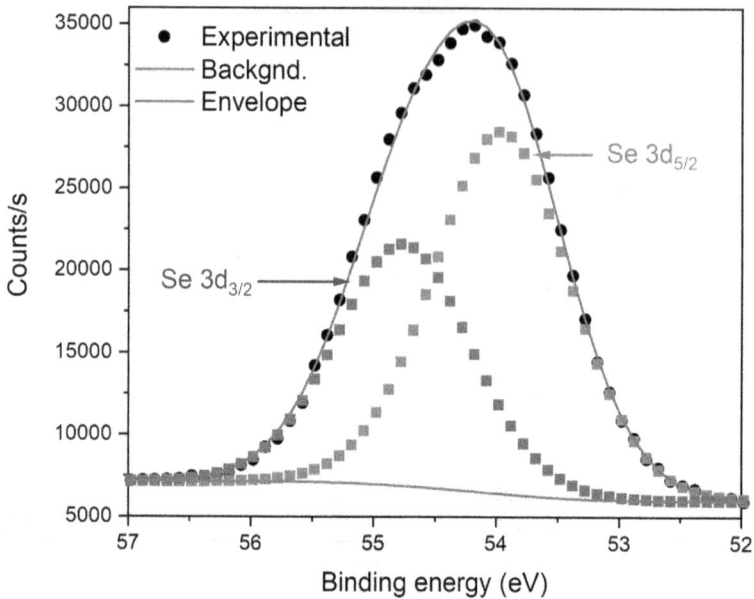

FIGURE 2.15 XPS spectra of Se 2d doublet. As seen, two components related to Se $3d_{5/2}$ and Se $3d_{3/2}$ emerge from the experimental data when a deconvolution process is done.

are fixed, the separation of the components is higher for higher atomic number of the atom (Watts and Wolstenholme 2019).

A very useful variant in XPS for thin-film analysis is the *depth profile mode*. By etching the surface of the material using an Ar+ ion gun, the elemental composition of a sample as a function of the etching time is registered. However, this is a destructive method since the ion/cluster bombardment can change the initial oxidation state of chemical elements in the studied surface. This can be observed in Figure 2.16: a cupper thin film was etched for 40 seconds, and measurement was performed each 10 seconds, obtaining a spectrum that provides information of the state of the sample at that time. It should be mentioned that the ion bombardment can change the original oxidation states of a sample. Then, it is recommended to analyze the sample as received without any cleaning process or ion etching. In summary, the XPS analysis can provide much information of the surface of a sample and even from deeper regions through a depth profile analysis.

2.2.5 PREPARATION OF SAMPLES FOR CHEMICAL CHARACTERIZATIONS

Raman and IR spectroscopies do not require a special preparation of the samples. These techniques usually are performed at room temperature and are not destructive.

For XPS measurements, the preparation of samples in thin-film form is quite simple, since no cleaning process is recommended because any change of the surface will affect the analysis. For that reason, samples are mounted on the sample holder as received and then introduced to the load lock to degas the film before analysis. For

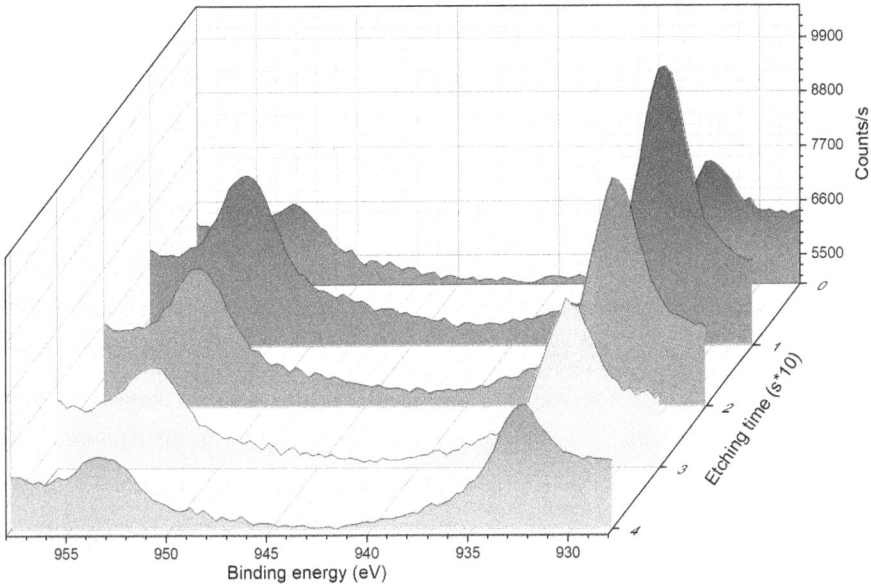

FIGURE 2.16   XPS depth profile spectra of Cu 2p core level from a copper sample.

mounting the sample on the holder, the laboratory technician usually uses gloves, face mask, and tweezers to avoid surface contamination of thin films. However, when samples are in powder form, there are several ways to mount the sample. For instance, a double-side carbon tape is set on the holder, and then a small amount of powder is deposited and then pressed on the carbon tape using a clean aluminum foil. Another option is to dissolve the powder in water or alcohol, depending on the characteristics of the sample, and then some drops of liquid are deposited on a "substrate" (e.g., clean glass, quartz, or aluminum). When the liquid has been evaporated, the substrate with the powder sample impregnated can be introduced inside the XPS load lock chamber, and subsequently to the analysis chamber for measurement.

2.3   MORPHOLOGICAL CHARACTERIZATION

2.3.1   Definition of Morphological Characterization

The morphological characterizations allow obtaining the shape, topography and grain size of the surfaces by different microscopies, which have evolved in time. In 1993, the scientists Ruska and Knoll built the first scanning electron microscopy (SEM), called transmission electron microscopic (TEM) (Erni 2015), in which the emitted electron passes through very thin layers. In 1938, Von Ardenne incorporated a scanning coil to the TEM using 23 kV voltage, improving image resolution from 50 to 100 nm, and magnifying the image by 8,000 times its size (von Ardenne 1938). In 1942, Zworykin, Hillier, and Snyder presented an SEM using secondary electron uptake to obtain the image contrast (Schmitt 2014). In 1952, Oatley developed an electrostatic lens with a voltage of 40 kV for the electron source (Oatley,

Nixon, and Pease 1966). The SEM images are generated by the emission of electrons, and these images are generated in a scale of grays and coat the areas from surfaces. These surfaces range from nanometer to micrometer scale. The images can be enlarged from 300,000 to 1,000,000 of its size without losing definition. Another important technique is energy dispersion spectroscopy (EDS), which works together with an SEM to supply information either qualitative or quantitative from the material that is under study, such as the composition of elements. The equipment consists of a system that contains a chamber at a certain pressure, in which samples with a size of up to $200 \, mm^2$ and height of $80 \, mm$ can be analyzed. Samples require minimal preparation for SEM characterization, and measurements can be performed for small samples as long as they fit in the vacuum chamber without altering their physical state. Samples should be mounted on a holder and coated with a thin film of heavy metal elements to permit spatial dispersion of electrical charges on the sample surface. This allows a better image production with higher definition.

2.3.2 SCANNING PROBE MICROSCOPY

One of the important features of SPM is that any type of interaction between the tip and the sample can be measured accordingly to the interaction measured with the tip-sample distance. During a typical SPM measurement, the tip performs a sweep over the surface of samples creating and resolving images related to the surface variations as a function of position. Generally, images are represented in the micrograph like color contrasts. This type of microscopy has its beginnings in 1998 (Binnig et al. 1998), with the invention of the scanning tunneling microscope, which is an instrument for obtaining images of surfaces at the atomic level. The first successful experiment with a tunneling microscope was carried out by Gerd Binnig and Heinrich Rohrer (Binnig et al. 1998; Binnig et al. 1983).

2.3.3 SCANNING TUNNELING MICROSCOPY

As previously described, this microscopy was developed by Binnig and Roher at the IBM laboratories (Switzerland) for this contribution they received the Nobel Prize in Physics in 1986 (Binnig and Rohrer 1987). This type of microscope generates surface images under a very important principle in quantum mechanics, which is the tunnel effect. To describe this phenomenon, we should consider a scenario where ideal surfaces of a metal and a semiconductor are separated by vacuum or any insulator (see Figure 2.17). In Figure 2.17, E_1 and E_2 represent the energy of the electrons in each surface, E_{b1} and E_{b2} depict the energy barrier in the position where the insulator is in contact with the metals, and d is the thickness of the insulator.

In the classical regime, the electrons of the metallic material cannot be transferred from one surface to another due to the presence of the insulator due to the presence of the energy barrier and which the electrons cannot cross. However, an alternative to promote the migration of electrons is the application of a voltage between the two plates, with value $V = E_1 - E_2$. This produces a modification in the barrier making the electrons show the best mobility to cross that energy barrier. So, if these electrons

(a)

E_{b1} E_{b2}

E_1 d E_2

(b)

E_{b1}

E_1

$V = E_1 - E_2$ *Electron Tunneling* E_{b2}

d E_2

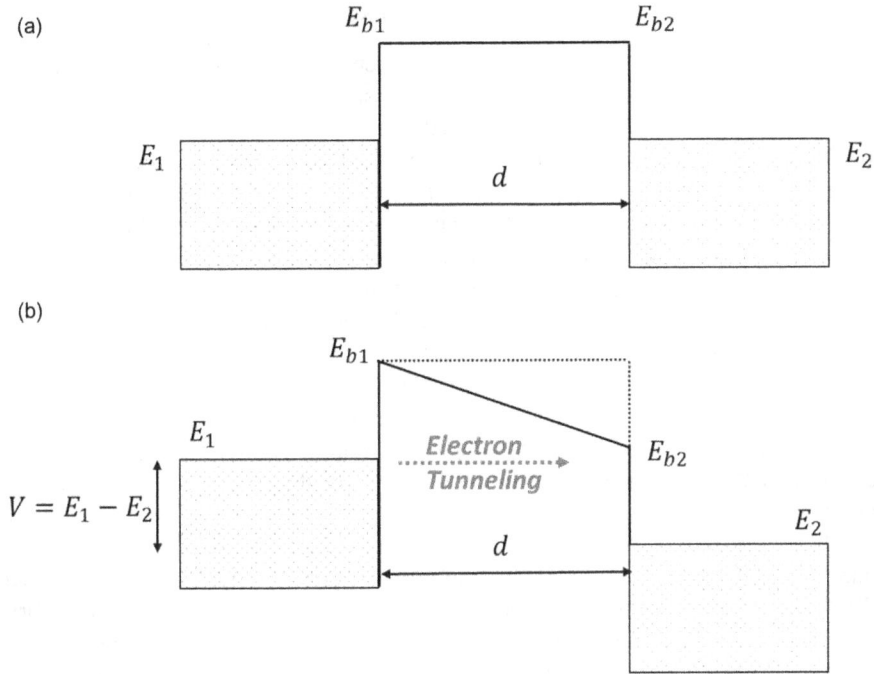

FIGURE 2.17 Schematic representation of the energy profile between two metal plates separated by an insulator: (a) without applying the potential and (b) an instant after the potential application. The red dotted line indicates the direction of propagation of electrons, and this is the phenomenon of tunneling.

have enough energy to move through the barrier, this effect is usually known as the tunneling effect (Hawkes and Spence 2019).

The tunneling effect for STM could be defined as the electron that is on the surface of the material after applying the voltage presents a non-zero probability of escape from the surface. Therefore, if the tip moves closer to a metallic material just before contact, the electrons from the tip will have a non-zero probability of jumping from the tip to the sample through the vacuum (Kochanski 1989).

As described above, in this characterization a sharp and conductive tip is used, and voltage is applied between the tip and the sample. This voltage modifies the energy of the electrons that are near the surface of the tip and the surface of the sample. So, these electrons can jump from the tip to the sample making the tunneling (tunneling current). In other words, to generate a surface image of some samples it is necessary to measure the change in the tunneling current. This current can increase if the tip and the sample are remarkably close but this current decreases exponentially as the tip and the sample separate.

In STM, there are usually two types of measurement (Leng 2009): constant voltage mode, in which the position of the tip is kept unchanged during the scanning of the sample surface. The second is the constant current mode, in which the distance between the tip and the sample surface is kept in constant adjustment.

2.3.4 SCANNING ELECTRON MICROSCOPY (SEM)

SEM has a source of primary electrons that reaches the surface of the sample, thus releasing small amounts of electrons known as secondary electrons, which were in the orbits of their atoms and were expelled from it. The electrons emitted from the surface called backscattered electrons, are originally primary electrons, when they penetrate the surface are deviated from their initial trajectory, these electrons don't lose neither speed nor energy, this deviation is denominated as elastic scattering, if the deviation is greater than 90°, the electron will leave the sample, these electrons generate contrast in the images, which means that we have areas with different chemical compositions, the elements that are heavier or have a high atomic number will appear brighter.

With the help of the condenser, objective lens, and scan coils, sample focus can be achieved by the electron beam. A scintillation detector is used to collect both secondary and backscattered electrons, with a positive voltage on the detector screen, both electrons will be collected (Worsfold et al. 2019). The diagram of SEM function can be observed in Figure 2.18.

There exist three types of SEM, conventional SEM, ambient SEM, and low vacuum SEM (Khursheed 2010; Clubb et al. 2004; Daniatos 1981). In a conventional SEM, the electron beam interaction with surface occurs in a vacuum of 10^{-6} Torr. However, in ambient SEM, this interaction occurs from 0.2 to 20 Torr. And the third SEM is the low vacuum, which is similar to a conventional SEM but adapted to operate under pressures from 0.2 to 2 Torr.

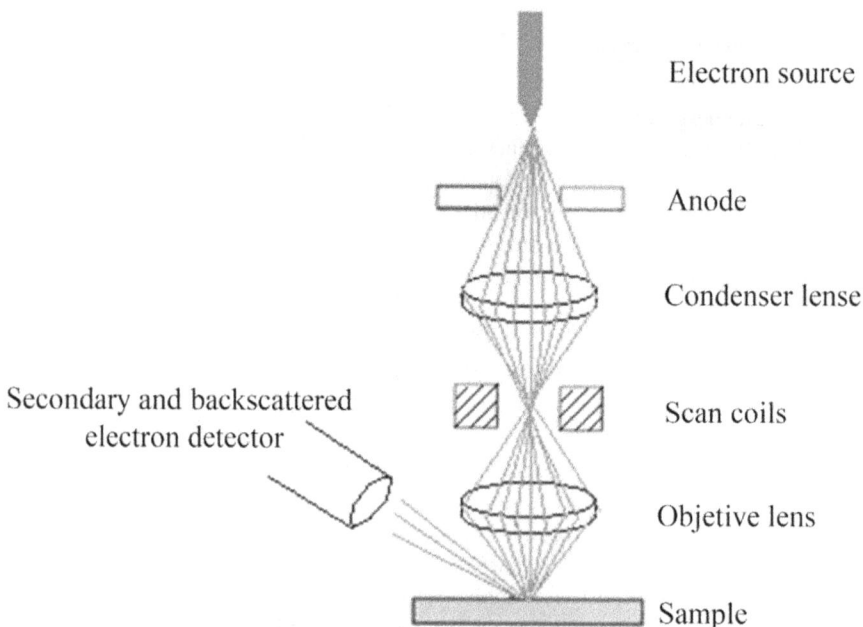

FIGURE 2.18 Schematic representation of SEM.

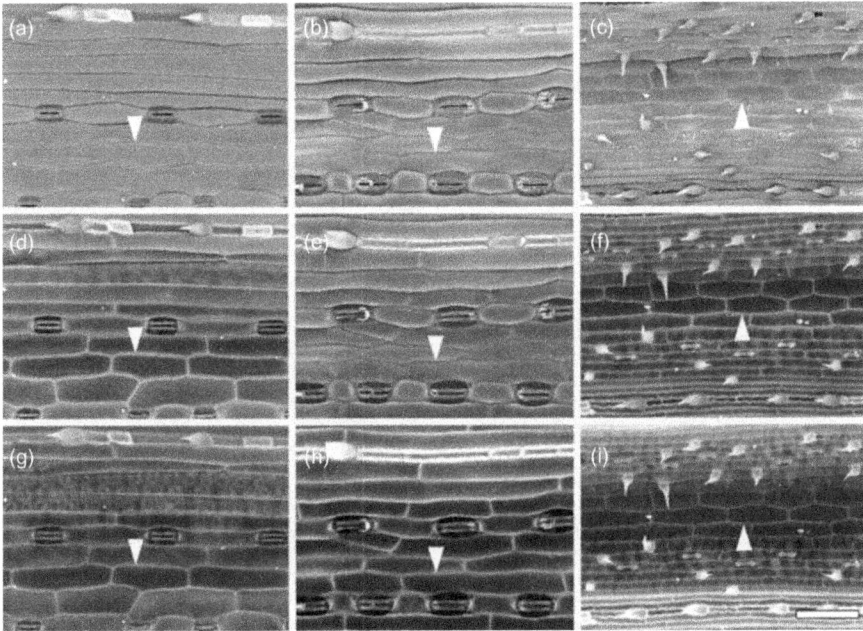

FIGURE 2.19 Effect of accelerating voltage on BSE imaging of cell wall outlines in critical point dried barley (a, d, g), wheat (b, e, h), and *Brachypodium distachyon* (c, f, i) leaves (Talbot and White 2013).

The acceleration voltage used in the electron source is proportional to the quality of the image. When the acceleration voltage is less than 5 kV, the image is not clear. However, between 15 and 30 kV, it causes the electrons to penetrate under the surface, and the generated image gets a higher quality. Figure 2.19a–c shows different cell wall outlines with respect to change in SEM accelerating voltage.

2.3.5 ENERGY DISPERSIVE X-RAY SPECTROSCOPY (EDS)

The spectroscopy of energy dispersive X-ray is an analytical technique that allows the chemical characterization of materials. This is done through an emitter of electrons that are directed to the material of study, and these electrons interact with the electrons of the internal orbits of the atoms that form the material, expelling them from it.

The expelled electrons are called secondary electrons and are fundamental in the SEM and the one that expelled it is called primary electron. However, secondary electrons leave behind empty energy states in the internal orbits of their atoms, causing those electrons from the outermost orbits of the atom, jump to occupy those internal energy states. This rearrangement of electrons, occupying energy states of internal orbits, causes photons to be emitted, and this is the basis of X-ray energy dispersion spectrometry. The effect described above is illustrated in Figure 2.20.

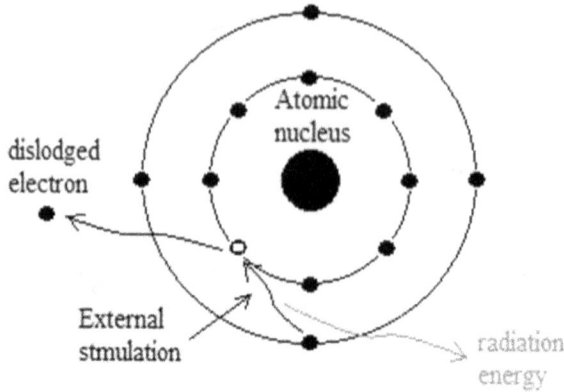

FIGURE 2.20 Emission of X-rays through the rearrangement of electrons in the orbits of the atom.

The emission spectrum produced by the transitions of electrons between the orbits of the atoms will give the atomic weight of the analyzed material or element based on the energy difference between the E_{photon} orbitals, by using the Rydberg formula:

$$Z = \sqrt{\frac{E_{photon}}{R\left(\dfrac{1}{n_{internal}^2} - \dfrac{1}{n_{external}^2}\right)}} \qquad (2.10)$$

where Z is the atomic number of the element under study, E_{photon} is the photon energy, R is the Rydberg constant, $R = 1.097 \times 10^7 \, \text{m}^{-1}$, $n_{internal}$ refers to internal orbits, and $n_{external}$ refers to external orbits. Therefore, the location of energy peaks tells us of what elements the sample is made, and the magnitude of these peaks will tell us the amount of these elements in the sample.

Most of the EDS systems are interfaced with SEM (Girão, Caputo, and Ferro 2017), where they use the same electron beam source to excite X-rays from the sample under study and the detector used for obtaining the information about X-rays is a Si or Li detector. An X-ray emitted by the sample produces a photoelectron at the detector. The photoelectron in turn dissociates to form electron–hole pair, and the pairs give the information such as amplitude of the generated voltage pulse which is proportional to the incident photon energy. All the received information is processed within a range of voltages (energies) and is finally transferred to a graph, like the one shown in Figure 2.21.

2.4 ELECTRICAL CHARACTERIZATIONS

2.4.1 DEFINITION OF ELECTRICAL CHARACTERIZATIONS

The physical properties that change when a material is subjected to an electric field are referred to as electrical properties (Araujo et al. 2020; von Ardenne 1938). Many metals are good conductors of electricity, and some ferromagnetic metals may also

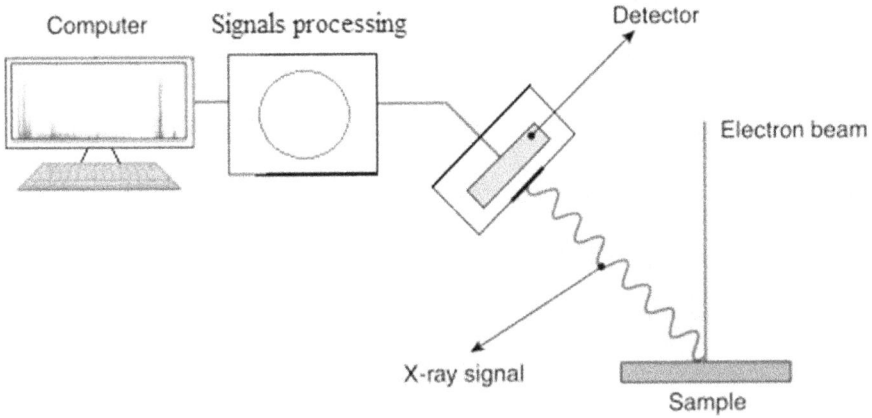

FIGURE 2.21 Schematic representation of EDS.

exhibit electrical properties, although with low electrical conductivity. It should be noted that magnetic and electrical properties are not identical and are not always related.

Due to this, magnetic materials are suitable for electrical appliances and micro electromechanical system (MEMS), etc. (Aslam et al. 2021). Some of the electrical properties are dielectric strength, surface resistance, resistivity, electrical conductivity, temperature coefficient of resistance, mobility, band gap, carrier concentration, carrier lifetime, capacitance with respect to voltage, impedance, among others (Batoo et al. 2013; Bertolucci et al. 2015; Binnig et al. 1983).

Magnetic nanoparticles are made of different elements possessing magnetic properties such as iron, nickel, gadolinium, to mention some of them and their chemical compounds. In general ferrite nanoparticles are the basic and most explored magnetic nanoparticles for different applications (Binnig et al. 1983; Binnig et al. 1987). In this section, electrical characterization techniques will be discussed by referring ferrite nanoparticles. The performance of magnetic nanoparticles also relies on different electrical parameters like mobility, resistivity, carrier density, etc. Most common techniques to measure previously mentioned electrical properties are Hall effect, IV measurements, and surface resistance measurements (Smith 2011).

2.4.2 HALL EFFECT

This measurement consists of the application of magnetic field to a current carrying conductor in order to determine different electrical parameters. This method is used to differentiate between electrons and holes which are the charge carriers and provides information about the density of electrons or holes depending on the semiconductor.

Hall effect is the phenomenon where the voltage generates in a sample or surface when a magnetic field is applied perpendicularly to a current carrying conductor. The generated voltage is called "Hall voltage (V_H)" (Chien and Westgate 2013).

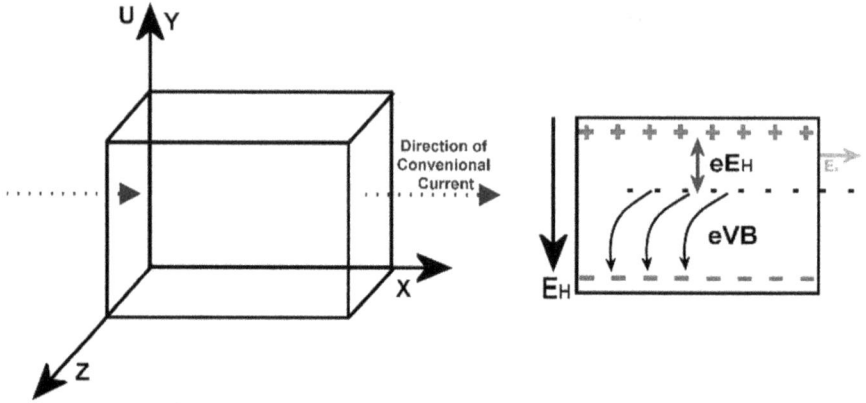

FIGURE 2.22   Hall field generation phenomena.

Therefore, if we consider a thin-film-type conductor in the electric field E, then it produces a current I and causes a force of a magnitude eE to act on the charge carriers. Subsequently, in the presence of magnetic field, the magnetic force is proportional to the magnetic field strength (B_z), and charge (e) and velocity (V_x) act on the charge carriers. If the magnetic force is at right angle to the direction of B_z and V_x, each charge is moved toward each side of the conductor resulting in the formation of an electrical charge on the surface of the thin-film conductor as the charge carriers reach its surface. This further gives rise to a transverse field called Hall field (E_H). This phenomenon can be seen in Figure 2.22.

Therefore, equations (2.11) and (2.12) represent the Hall field and Hall coefficient (R_H), respectively, at equilibrium (Karplus and Luttinger 1954).

$$E_H = V_x * B_z \tag{2.11}$$

$$R_H = E_H / (J_x * B_z) \tag{2.12}$$

where J_x is the current density of the specimen utilized in the experiment and is determined by equation (2.13).

$$J_x = (n * e) / (V_x) \tag{2.13}$$

Figure 2.23 shows a typical experimental setup for obtaining Hall constant (R_H) and Hall voltage (V_H). The Hall constant or coefficient is found by measuring Hall voltage V_H, which generates the Hall field E_H (CHein and Westgate 2013; Karplus and Luttinger 1954). If V_H is taken across the thickness "t" of the specimen, then

$$V_H = (E_H * t) = (R_H * J_x * B_z * t) \tag{2.14}$$

Where $J_x = I_x / (b * t)$ and $b =$ width of the specimen, as shown in Figure 2.23. Therefore,

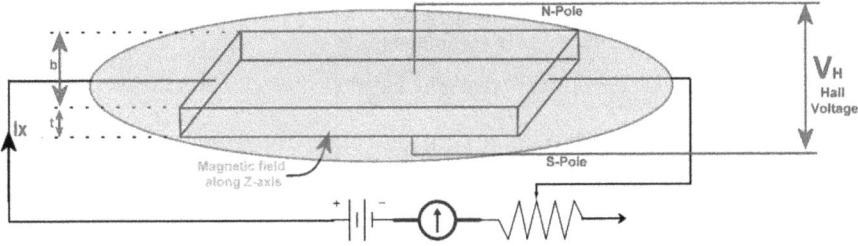

FIGURE 2.23 Experimental setup for obtaining Hall constant.

$$R_H = (V_H * b) / (I_x * B_z) \qquad (2.15)$$

The additional parameters that can be obtained by Hall measurement techniques are mobility, electrical conductivity electrons concentration, and holes concentration from equations (2.16), (2.17), (2.18), and (2.19), respectively. This means that Hall coefficient is negative for n-type semiconductor and positive for p-type semiconductor.

$$\text{Mobility } (\mu) = (R_H * J_x) / E_x = (\sigma * R_H) \qquad (2.16)$$

$$\text{Electrical conductivity } (\sigma) = J_x / E_x \qquad (2.17)$$

$$\text{Electron concentration} = -1 / (e * R_H) \qquad (2.18)$$

$$\text{Holes concentration} = 1 / (e * R_H) \qquad (2.19)$$

Specifically, Hall constant is always negative for n-type semiconductors and is positive for p-type semiconductors due to the type of mobility charge carriers. It has been reported by different authors (Dawn et al. 2022; Lan et al. 2022; Dawn et al. 2021) that the utilization of experimental setup as shown in Figure 2.2 obtains different electrical properties of magnetic nanoparticles, especially for ferrites and their composites.

2.4.3 SURFACE RESISTANCE

One of the most important electrical parameters that affect directly in practical applications of the magnetic nanoparticles, or any planar materials is surface resistance. For example, gas sensors for detecting reducing gases like carbon monoxide and propane utilizes samples with higher surface resistance because of easy adsorption and desorption process. Whereas materials with lower surface resistance makes the material to be photo active in UV-visible range, making them transparent conductive oxides. Therefore, surface resistance plays key role in many practical applications and is dependent on the operation temperature. In general, as the temperature increases the surface resistance of the material decreases due to the movement of charge carriers after gaining the thermal energy. The measurement techniques for

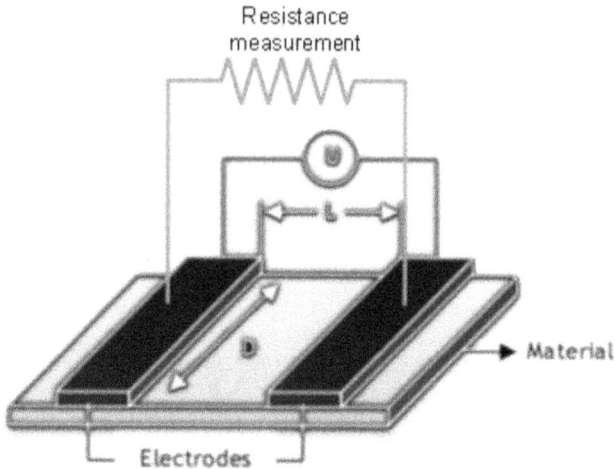

FIGURE 2.24   Configuration for measuring surface resistance of a planar material.

obtaining surface resistance vary slightly at room temperature and at higher temperatures. In general, in research laboratories the surface resistivity is measured by two-probe methods and for industrial device level measurements four-probe method is preferred.

Also, it is important to distinguish the surface resistance (Rs) from the surface resistivity (ρs). The surface resistance (Rs) is a measure between two points on the surface with electrodes, obtained in ohms (Ω) and is the ratio of voltage applied to the current flowing between those two measuring points, as shown in Figure 2.24. However, the ρs is dependent on the dimensions of the material like length and width, which is measured in ($\Omega/\square$). Therefore, ρs is defined as the ratio of DC voltage drop per unit length (L) to the surface current (Is) per unit width (D). Unlike surface resistance, the surface resistivity does not depend on the electrode configuration, but it depends on the operation temperature, which activates the material to conduct (Loss, Salvado, and Pinho 2014).

The surface conduction in magnetic nanoparticles, especially ferrites, is majorly due to the interchange between the electron and hole. Charge carriers rise due to different oxidation states of the ferrites at the surface (Shinde 2021). It is a common mechanism that the surface resistivity occurs due to the exchange of electrons in ferrites generally between Fe^{+2} and Fe^{+3}. However, the electrical and magnetic properties of ferrite are strongly dependent upon the particle or grain size and grain boundaries. It has been reported that the increase in density of grain boundaries in ferrite system also increases the resistivity (Panda et al. 2016). The resistivity of the magnetic nanomaterials is calculated by the Arrhenius relation as mentioned in equation (2.20):

$$\rho = \rho_o e^{\left(\frac{E_a}{k_B T}\right)}$$

(2.20)

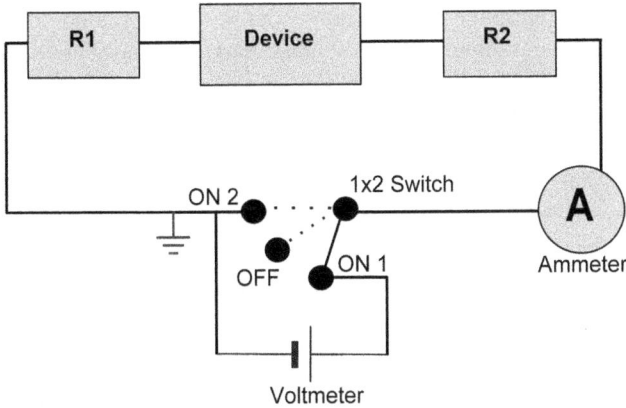

FIGURE 2.25 Experimental setup for obtaining IV characteristics.

where ρ is the resistivity, ρ_o is the exponential constant, E_a is the activation energy, k_B the Boltzmann constant, and T is the temperature in Kelvin (Shinde 2021; Panda et al. 2016; Loss, Salvado, and Pinho 2014; Panda, Muduli, and Behera 2015).

2.4.4 I–V Measurements

Utilizing current–voltage (I–V) measurements, it is possible to obtain the performance characteristics of the devices such as leakage current, switching current, static current, and DC resistivity. All the I–V measurements can be obtained utilizing two-probe method with Keithley Multimeters. A typical static I–V measurement setup is shown in Figure 2.25, which includes two external resistors R1 (between 0.5 and 1 MΩ) and R2 (between 0.2 and 0.6 MΩ), a switch, a DC power supply, and an ammeter A. Current and voltage values are obtained in parallel by voltmeter and ammeter, respectively, in two switch positions ON1 and ON2. ON1 is in connection with the DC power supply and ON2 is in short circuit. The results obtained are further plotted and analyzed in order to obtain the electrical parameters (Venkatramani et al. 2009 ; Zakir et al. 2021; Aslam et al. 2021; Saqib et al. 2019).

By utilizing the I–V curves, it is possible to obtain the resistivity (ρ) of the specimen and is calculated from equation (2.21):

$$\rho = (R*A) / L \tag{2.21}$$

where R is the resistance, "$A = \pi r^2$" is known as the area of the specimen, and L is the thickness of the specimen.

Also, it is possible to confirm the behavior of the device whether ohmic or non-ohmic by observing and interpreting the variation in the current with respect to the voltage applied. An ohmic behavior is observed when voltage is directly or linearly proportional to current, which is observed at lower voltages, that is, V < 1 V and is given in equation (2.22).

$$V \alpha I^{\alpha} \tag{2.22}$$

When $\alpha=1$, ohmic behavior is observed, and as α increases, a non-ohmic behavior is observed. In the non-ohmic region, voltage is exponentially proportional to current and is observed at higher voltages, that is, $V > 1\,V$ (Singh and Chauhan 2019; Venkatramani et al. 2009). It has been reported (Devi et al. 2011; Batoo et al. 2013) that non-ohmic region in magnetic nanomaterials occurs due to the space charge effects and is termed as space charge-limited current region, which is used to understand the conduction mechanism in nanoferrites (Singh and Chauhan 2019).

2.4.5 PREPARATION OF SAMPLES FOR ELECTRICAL CHARACTERIZATIONS

The electrical characterization is used in almost all research centers and is an important part of the investigation of the physical properties of materials. The traditional way of measuring electrical resistivity is the four-point method, which is used for thin films with a rectangular shape. However, it cannot be used with irregular shapes. The Van der Pauw technique solves this problem and improves the reliability of the measurements. The technique is nondestructive, and only small amounts of metal are placed on the surface to be measured. The technique is more precise, and we only require knowing the thickness of the sample. The technique allows us to measure the electrical resistivities of up to $10^9\ \Omega$-cm. The Van der Pauw technique requires four contacts on the surface of the sample to be measured (van der Pauw 1991), as shown in Figure 2.26.

We made eight measurements combining the four contacts, that is, electrical current is passed from contacts 1 to 2 and the voltage between contacts 3 and 4 is measured, then when the direction of this electrical current is reversed, the current will flow from the contact 2 to 1, and also the voltage between contacts 3 and 4 is measured and so on. The contact points can be cleaned with acetone and then with methanol. Later, they will be immersed in dilute HCl, and leaving the contacts to drain will preserve the tips of the tips. The Van der Pauw equation assumes small contacts of the order of 10 microns.

The cleansing of silicon wafer is standardly done as follows (Kern 1990):

The wafers are immersed in deionized water. If there are visible contaminants, a preliminary cleansing in piranha solution is required. The wafers are rinsed with

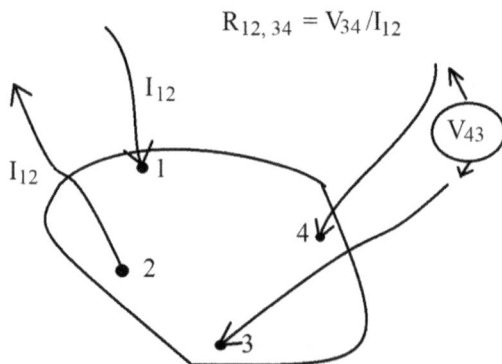

FIGURE 2.26 Implementation of electrical resistivity measurement in semiconductor thin films by the Van der Pauw method.

deionized water between each step. The containers used for cleansing are made from fused silica or fused quartz, and also the chemicals products utilized must be of electronic grade or CMOS grade to avoid contamination of the wafers.

1. SC-1 cleansing, where SC means standard cleansing, is done as follows: five parts of deionized water, one part of ammonia water, 29% weight of NH, and one part of aqueous solution of HO, 30%. Cleansing is done between 75°C and 80°C for 10 minutes. The mixture removes organic waste. This treatment forms a thin layer of silicon dioxide of approximately 1 nm on the silicon, with a certain degree of iron contamination, this will be removed in subsequent steps.
2. The second is an optional step, to remove the thin layer of dioxide and small ionic contaminants formed on the surface. This is done in a 1:100 or 1:50 solution of aqueous HF at 25°C for about 15 seconds. This step is performed within ultra-high-purity materials and ultra-clean containers, since the silicon surface without surface oxide is very reactive.
3. The third and last step called SC-2 is carried out with a solution of the following proportions: six parts of deionized water, one part of aqueous HCl of 37% in weight, one part of aqueous solution of HO, 30%, at a cleaning temperature between 75°C and 80°C. The substrates were ultrasonicated in the above mentioned SC-2 solution for 10 min. The process removes traces of ionic metal contaminants. In the end, there's a passive thin layer sitting on the surface of the wafer.
4. And finally, a rinse with water is carried out and drying is carried out with nitrogen.

2.5 SUMMARY AND PERSPECTIVES

In this work, we have described the most common surface characterization techniques for magnetic nanomaterials. By the usage of these techniques, it is possible to obtain accurate information regarding chemical, morphological, and electrical properties. The chemical characterizations described in this work include IR, Raman, and XPS, which provide information regarding the chemical composition, crystalline structure, and chemical bonding between the atoms that built up the material. Furthermore, the morphological characterizations provide information related to the surface shape, size, aspect ratio, and topography. The discussed techniques were SMP, STM, and SEM-EDS. By this group of microscopies, it is possible to obtain the size of grains/particles as their aspect ratio.

Finally, the Hall effect, surface resistance, and IV measurements provide information related to changes in electrical properties like surface resistance or behavior of the relation current–voltage. These properties are important especially for materials applications and device fabrication.

Although there are various characterization techniques, which are not possible to cover in this work, we have focused on the characterization techniques most used in the manufacture of devices in semiconductors applied to manufacture devices such as sensors, catalysts, and magnetic nanomaterials.

REFERENCES

Araujo, Jefferson F.D.F., Tahir, Soudabeh Arsalani, Fernando L. Freire, Gino Mariotto, Marco Cremona, Leonardo A.F. Mendoza, et al. 2020. "Novel Scanning Magnetic Microscopy Method for the Characterization of Magnetic Nanoparticles." *Journal of Magnetism and Magnetic Materials* 499 (April): 166300. doi:10.1016/J.JMMM.2019.166300.

Ardenne, M. Von. 1938. *The Scanning Electron Microscope: Practical Construction.* 19th ed. Z. Phys.

Ardenne, Manfred von. 1938. "Das Elektronen-Rastermikroskop." *Zeitschrift Für Physik* 109 (9): 553–572. doi:10.1007/BF01341584.

Aslam, Asma, Abdul Razzaq, S. Naz, Nasir Amin, Muhammad Imran Arshad, M. Ajaz Un Nabi, Abid Nawaz, et al. 2021. "Impact of Lanthanum-Doping on the Physical and Electrical Properties of Cobalt Ferrites." *Journal of Superconductivity and Novel Magnetism* 34 (7): 1855–1864. doi:10.1007/s10948-021-05802-4.

Batoo, Khalid Mujasam, Feroz Ahmed Mir, M. S. Abd El-sadek, Md Shahabuddin, and Niyaz Ahmed. 2013. "Extraordinary High Dielectric Constant, Electrical and Magnetic Properties of Ferrite Nanoparticles at Room Temperature." *Journal of Nanoparticle Research* 15 (11). doi:10.1007/s11051-013-2067-6.

Bertolucci, Elisa, Anna Maria Raspolli Galletti, Claudia Antonetti, Mirko Marracci, Bernardo Tellini, Fabio Piccinelli, and Ciro Visone. 2015. "Chemical and Magnetic Properties Characterization of Magnetic Nanoparticles." *Conference Record - IEEE Instrumentation and Measurement Technology Conference* 2015 (July). Institute of Electrical and Electronics Engineers Inc.: 1492–1496. doi:10.1109/I2MTC.2015.7151498.

Binnig, G., H. Rohrer, Ch Gerber, and E. Weibel. 1983. "7×7 Reconstruction on Si(111) Resolved in Real Space." *Physical Review Letters* 50 (2). American Physical Society: 120. doi:10.1103/PhysRevLett.50.120.

Binnig, G, H Rohrer, Ch Gerber, and E Weibel. 1998. "Tunneling through a Controllable Vacuum Gap." *Applied Physics Letters* 40 (2): 178. doi:10.1063/1.92999.

Binnig, Gerd, and Heinrich Rohrer. 1987. "Scanning Tunneling Microscopy—From Birth to Adolescence." *Reviews of Modern Physics* 59 (3). American Physical Society: 615. doi:10.1103/RevModPhys.59.615.

Brian C. Smith. 2011. *Fundamentals of Fourier Transform Infrared Spectroscopy.* Edited by CRC. 2nd ed. https://books.google.com.mx/books?id=LR9HkK2cP_0C&printsec=frontcover &dq=B.C.+Smith, +Fundamentals+of+Fourier+Transform+Infrared+Spectroscopy& hl=en&sa=X&ved=2ahUKEwic5vGR3sn7AhUjLUQIHfdKAsUQ6AF6BAgMEAI#v =onepage&q&f=false.

Chambers, S. A., Y. Du, Z. Zhu, J. Wang, M. J. Wahila, L. F.J. Piper, A. Prakash, et al. 2018. "Interconversion of Intrinsic Defects in SrTi O3(001)." *Physical Review B* 97 (24). American Physical Society: 245204. doi:10.1103/PHYSREVB.97.245204/FIGURES/ 13/MEDIUM.

Chien, CL, and CR Westgate. 2013. *The Hall Effect and Its Applications.* 1st ed. Springer Science + Business Media. https://books.google.com.mx/books?hl=en&lr=&id=Tq rVBwAAQBAJ&oi=fnd&pg=PA1&dq=hall+effect&ots=W6xmKL_Ucx&sig=JJqAr 5paqzwoNDwDyuLAAyzdiOg&redir_esc=y#v=onepage&q=hall effect&f=false.

Clubb, F J Jr, N Underhill, M R Coscio, S Sedlik, C B McClay, and L M Buja. 2004. "Low Vacuum Scanning Electron Microscopy (Lvsem) For Integrated Microscopic Evaluation of Implantable Devices." *Asaio Journal* 50 (2). ASAIO: 140. https://journals.lww. com/asaiojournal/Fulltext/2004/03000/LOW_VACUUM_SCANNING_ELECTRON_ MICROSCOPY__LVSEM_.120.aspx.

Daniatos, G. D. 1981. "Design and Construction of an Atmospheric or Environmental SEM (Part 1)." *Scanning* 4 (1): 9–20. doi:10.1002/SCA.4950040102.

Dawn, R., M. Zzaman, R. R. Bharadwaj, C. Kiran, R. Shahid, V. K. Verma, S. K. Sahoo, K. Amemiya, and V. R. Singh. 2021. "Direct Evidence to Control the Magnetization in Fe3O4 Thin Films by N2 Ion Implantation: A Soft X-Ray Magnetic Circular Dichroism Study." *Journal of Sol-Gel Science and Technology* 99 (3): 461–468. doi:10.1007/S10971-021-05606-X.

Dawn, R., M. Zzaman, F. Faizal, C. Kiran, A. Kumari, R. Shahid, C. Panatarani, et al. 2022. "Origin of Magnetization in Silica-Coated Fe3O4 Nanoparticles Revealed by Soft X-Ray Magnetic Circular Dichroism." *Brazilian Journal of Physics* 52 (3). doi:10.1007/S13538-022-01102-X.

Devi, P. Indra, N. Rajkumar, B. Renganathan, D. Sastikumar, and K. Ramachandran. 2011. "Ethanol Gas Sensing of Mn-Doped CoFe2O4 Nanoparticles." *IEEE Sensors Journal* 11 (6): 1395–1402. doi:10.1109/JSEN.2010.2093881.

Eguía-Eguía, Sandra I., Lorenzo Gildo-Ortiz, Mario Pérez-González, Sergio A. Tomas, Jesús A. Arenas-Alatorre, and Jaime Santoyo-Salazar. 2021. "Magnetic Domains Orientation in (Fe3O4/γ-Fe2O3) Nanoparticles Coated by Gadolinium-Diethylenetriaminepen taacetic Acid (Gd3+-DTPA)." *Nano Express* 2 (2). IOP Publishing: 020019. doi:10.1088/2632–959X/AC0107.

Erni, Rolf. 2015. *Aberration-Corrected Imaging in Transmission Electron Microscopy : An Introduction.*

Gandhi, Ashish Chhaganlal. 2022. "Synthesis and Characterization of Functional Magnetic Nanomaterials." *Coatings 2022, Vol. 12, Page 857* 12 (6). Multidisciplinary Digital Publishing Institute: 857. doi:10.3390/COATINGS12060857.

Gardiner, Derek J., and Pierre R. Graves. 1989. *Practical Raman Spectroscopy.* Edited by Derek J. Gardiner and Pierre R. Graves. doi:10.1007/978-3-642-74040-4.

Girão, Ana Violeta, Gianvito Caputo, and Marta C. Ferro. 2017. "Chapter 6- Application of Scanning Electron Microscopy–Energy Dispersive X-Ray Spectroscopy (SEM-EDS)." *Comprehensive Analytical Chemistry* 75: 153–168. doi:10.1016/BS.COAC.2016.10.002.

Granada-Ramírez, D. A., A. Pulzara-Mora, C. A. Pulzara-Mora, A. Pardo-Sierra, J. A. Cardona-Bedoya, M. Pérez-González, S. A. Tomás, S. Gallardo-Hernández, and J. G. Mendoza-Álvarez. 2022. "Study of the Surface Chemistry, Surface Morphology, Optical, and Structural Properties of InGaN Thin Films Deposited by RF Magnetron Sputtering." *Applied Surface Science* 586 (June): 152795. doi:10.1016/J.APSUSC.2022.152795.

Gul, Saima, Sher Bahadar Khan, Inayat Ur Rehman, Murad Ali Khan, and M. I. Khan. 2019. "A Comprehensive Review of Magnetic Nanomaterials Modern Day Theranostics." *Frontiers in Materials* 6 (July). Frontiers Media S.A.: 179. doi:10.3389/FMATS.2019.00179/BIBTEX.

Hawkes, Peter W., and John C. H. Spence. 2019. *Springer Handbook of Microscopy.* Edited by Peter W. Hawkes and John C. H. Spence. Springer Handbooks. Cham: Springer International Publishing. doi:10.1007/978-3-030-00069-1.

Heide, Paul van der. 2012. *X-Ray Photoelectron Spectroscopy: An Introduction to Principles and Practices.* Vol. 123. John Wiley & Sons, Ltd. https://www.wiley.com/en-us/X-ray+Photoelectron+Spectroscopy%3A+An+introduction+to+Principles+and+Practices-p-9781118062531.

John Tyndall. 1869. "IV. On the Blue Colour of the Sky, the Polarization of Skylight, and on the Polarization of Light by Cloudy Matter Generally." In *Proceedings of the Royal Society of London*, 17:223–233. The Royal Society London. doi:10.1098/RSPL.1868.0033.

Karplus, Robert, and J. M. Luttinger. 1954. "Hall Effect in Ferromagnetics." *Physical Review* 95 (5). American Physical Society: 1154. doi:10.1103/PhysRev.95.1154.

Kern, Werner. 1990. "Evolution of Silicon Wafer Cleaning Technology." *Proceedings - The Electrochemical Society* 90 (9). Publ by Electrochemical Soc Inc: 3–19. doi:10.1149/1.2086825/XML.

Khursheed, Anjam. 2010. "Scanning Electron Microscope Optics and Spectrometers." *Scanning Electron Microscope Optics and Spectrometers*, January. World Scientific Publishing Co., 1–403. doi:10.1142/7094.

Klein, K. L., A. V. Melechko, T. E. McKnight, S. T. Retterer, P. D. Rack, J. D. Fowlkes, D. C. Joy, and M. L. Simpson. 2008. "Surface Characterization and Functionalization of Carbon Nanofibers." *Journal of Applied Physics* 103 (6): 061301. doi:10.1063/1.2840049.

Kloprogge, J. Theo., and Barry J. Wood. 2020. "Handbook of Mineral Spectroscopy. Volume 1, X-Ray Photoelectron Spectra." Elsevier.

Kochanski, Greg P. 1989. "Nonlinear Alternating-Current Tunneling Microscopy." *Physical Review Letters* 62 (19). American Physical Society: 2285. doi:10.1103/PhysRevLett. 62.2285.

Lan, Feifei, Rui Zhou, Ziyue Qian, Yuansha Chen, and Liming Xie. 2022. "Chemical Vapor Deposition of Ferrimagnetic Fe3O4 Thin Films." *Crystals* 12 (4). MDPI. doi:10.3390/CRYST12040485.

Loss, Caroline, Rita Salvado, and Pedro Pinho. 2014. "Developing Sustainable Communication Interfaces Through Fashion Design," no. June.

Mitsuo Tasumi. 2014. *Introduction to Experimental Infrared Spectroscopy: Fundamentals and and Practical Methods*. Edited by John Wiley & Sons. Wiley. https://books.google.com.mx/boo ks?hl=en&lr=&id=lh3iBQAAQBAJ&oi=fnd&pg=PA153&dq=M.+Tasumi, +Introduct ion+to+Experimental+Infrared+Spectroscopy:+Fundamentals+and+Practical+Methods &ots=oH-_6PMcMb&sig=BHGKJp5QsLnF821W9BOVMc2TAXs#v=onepage&q=M. Tasumi%2C Introd.

Morales-Luna, M., M. A. Arvizu, M. Pérez-González, and S. A. Tomás. 2019. "Effect of a CdSe Layer on the Thermo- A Nd Photochromic Properties of MoO3 Thin Films Deposited by Physical Vapor Deposition." *Journal of Physical Chemistry C* 123 (28): 17083–17091. doi:10.1021/ACS.JPCC.9B02895/SUPPL_FILE/JP9B02895_SI_001.PDF.

Moulder, J., W. Stickle, W. Sobol, and K. D. Bomben. 1992. *Handbook of X-Ray Photoelectron Spectroscopy*.

Oatley, C. W., W. C. Nixon, and R. F. W. Pease. 1966. "Scanning Electron Microscopy." *Advances in Electronics and Electron Physics* 21 (C): 181–247. doi:10.1016/S0065-2539(08)61010-0.

Panda, R K, R Muduli, and D Behera. 2015. "Electric and Magnetic Properties of Bi Substituted Cobalt Ferrite Nanoparticles : Evolution of Grain Effect". *Journal of Alloys and Compounds* 634: 239–245.

Panda, R K, R Muduli, G Jayarao, D Sanyal, and D Behera. 2016. "Effect of Cr 3 þ Substitution on Electric and Magnetic Properties of Cobalt Ferrite Nanoparticles." *Journal of Alloys and Compounds* 669: 19–28. doi:10.1016/j.jallcom.2016.01.256.

van der PAUW, L. J. 1991. "A Method of Measuring Specific Resistivity and Hall Effect of Discs of Arbitrary Shape." *Semiconductor Devices: Pioneering Papers*, March: 174–182. doi:10.1142/9789814503464_0017.

Pérez, Joel Jiménez, Alicia Bracamontes Cruz, José Luis Jiménez Pérez, Zormy Nacary Correa Pacheco, Mario Pérez González, and Alfredo Cruz Orea. 2021. "Application of XPS and PA Techniques in the Study of Lime Used in the Talavera House from the Historical Center of Mexico City." *Superficies y Vacío* 34 (August): 210701–210701. doi:10.47566/2021_SYV34_1-210701.

Peter Larkin. 2017. *Infrared and Raman Spectroscopy: Principles and Spectral Interpretation*. Edited by Elsevier. 2nd ed. https://books.google.com.mx/books?id=bMgpDwAAQBAJ& printsec=frontcover&dq=P.+Larkin, +Infrared+and+Raman+Spectroscopy:+Princi ples+and+Spectral+Interpretation&hl=en&sa=X&ved=2ahUKEwiNmryA3sn7AhX sHEQIHV1GAoEQ6AF6BAgFEAI#v=onepage&q=P. Larkin%2C Infrared and R.

Peter Vandenabeele. 2013. *Practical Raman Spectroscopy: An Introduction*. 1st ed. Wiley. https://books.google.com.mx/books?hl=en&lr=&id=HjFi5eOUYvgC&oi=fnd&pg =PT5&dq=P.+Vandenabeele, +Practical+Raman+Spectroscopy:+An+Introduction,

+1st+edition, +ed., +Chichester, +UK:+Wiley, +2013&ots=m4gvvupsS5&sig=DGq ucz6Fgcx0yMHys3Eh-5jcq5U#v=onepage&q=P. Vanden.

Saqib, H., S. Rahman, Resta Susilo, Bin Chen, and Ning Dai. 2019. "Structural, Vibrational, Electrical, and Magnetic Properties of Mixed Spinel Ferrites Mg1-XZnxFe2O4 Nanoparticles Prepared by Co-Precipitation." *AIP Advances,* 9: 055306. doi:10.1063/1.5093221.

Schmitt, Robert. 2014. "Scanning Electron Microscope." *CIRP Encyclopedia of Production Engineering.* 1085–1089. doi:10.1007/978-3-642-20617-7_6595.

Shinde, A B. 2021. "Structural and Electrical Properties of Cobalt Ferrite Nanoparticles," no. 4: 64–67.

Singh, Anshika, and Pratima Chauhan. 2019. "Structural, Electrical and Optical Properties of Mn0.2Co0.8Fe2O4nano Ferrites." *Materials Today: Proceedings* 46 (xxxx): 6264–6269. doi:10.1016/j.matpr.2020.04.878.

Skoog, Douglas A., F. James Holler, and Stanley R. Crouch. 2021. *Principles of Instrumental Analysis.* Edited by CENGAG. 7th ed. https://books.google.com.mx/books?hl=es&lr=&id=D13EDQAAQBAJ&oi=fnd&pg=PP1&dq=Douglas+A.+Skoog,+F.+James+Holler,+Stanley+R.+Crouch,+Principles+of+Instrumental+Analysis,+7th+edition,+CENGAGE.&ots=DLInBCRsfp&sig=AT-L-l3dwH3kibbjht3iOqZxyL8&redir_esc=y#v=onep.

Smith, Ewen, and Geoffrey Dent. 2019. *Modern Raman Spectroscopy: A Practical Approach.* John Wiley & Sons, Ltd. https://books.google.com.mx/books?hl=en&lr=&id=WsqFDwAAQBAJ&oi=fnd&pg=PP9&dq=E.+Smith,+G.+Dent,+Modern+Raman+Spectroscopy+-+A+Practical+Approach,+Chichester,+UK:+Wiley,+2005&ots=rMaUoXKcuJ&sig=3AMx44yklybFLQsnmU_sSzigMfU#v=onepage&q&f=false.

Talbot, Mark J., and Rosemary G. White. 2013. "Cell Surface and Cell Outline Imaging in Plant Tissues Using the Backscattered Electron Detector in a Variable Pressure Scanning Electron Microscope." *Plant Methods* 9 (1): 1–16. doi:10.1186/1746-4811-9-40/FIGURES/11.

Venkatramani, Ravindra, De Yu Zang, Choon Oh, James Grote, and David Beratan. 2009. "Photoconductivity and Current-Voltage Characteristics of Thin DNA Films: Experiments and Modeling." *Nanobiosystems: Processing, Characterization, and Applications II* 7403 (March 2015): 74030B. doi:10.1117/12.831024.

Watts, John F., and John Wolstenholme. 2019. *An Introduction to Surface Analysis by XPS and AES.* Wiley. doi:10.1002/9781119417651.

Worsfold, Paul, Alan Townshend, Colin F. Poole, and Manuel Miró. 2019. *Encyclopedia of Analytical Science.* Elsevier. https://books.google.com.mx/books?hl=en&lr=&id=Nu2SDwAAQBAJ&oi=fnd&pg=PP1&dq=Encyclopedia+of+Analytical+Science:+Atomic+Absorption+Spectrometry:+Fundamentals,+Instrumentation+and+Capabilities&ots=D2ath7jW5V&sig=5u5DD2giGDPRgNQ1E80OTHObdgw#v=onepage&q=Ency.

Yang Leng. 2009. *Materials Characterization: Introduction to Microscopic and Spectroscopic Methods.* John Wiley & Sons, Ltd. https://books.google.com.mx/books?hl=en&lr=&id=Oku4IBjiUKAC&oi=fnd&pg=PR5&dq=Y.+Leng,+Materials+Characterization:+Introduction+to+Microscopic+and+Spectroscopic+Methods,+Wiley-VCH,+Germany,+2013&ots=AvdAQK6zAI&sig=SNgbVlh9LZ_92_xFBUJUKmXmSuM#v=onepage&q&f=.

Young, Andrew T. 1981. "Rayleigh Scattering." *Applied Optics, Vol. 20, Issue 4, Pp. 533–535* 20 (4). Optica Publishing Group: 533–535. doi:10.1364/AO.20.000533.

Zakir, Ruqayya, Sadia Sagar Iqbal, Atta Ur Rehman, Sumaira Nosheen, Tasawer Shahzad Ahmad, Nimra Ehsan, and Fawad Inam. 2021. "Spectral, Electrical, and Dielectric Characterization of Ce-Doped Co-Mg-Cd Spinel Nano-Ferrites Synthesized by the Sol-Gel Auto Combustion Method." *Ceramics International* 47 (20): 28575–28583. doi:10.1016/j.ceramint.2021.07.016.

3 Core–Shell Magnetic Nanostructures

S. Jasmine Jecintha Kay and N. Chidhambaram
Rajah Serfoji Government College (Autonomous)

Arun Thirumurugan
Sede Vallenar, Universidad de Atacama

S. Gobalakrishnan
Noorul Islam Centre for Higher Education
Deemed to be University

CONTENTS

DOI: 10.1201/9781003335580-3

3.1   INTRODUCTION

Core–shell nanostructures are distinct from regular nanoparticles in several attributes, based on the structure, functionalities, and materials used to create them. Core–shell nanostructures fall under the category of hybrid nanostructures, which are made up of two or more constituent elements but are only joined chemically rather than by molecular bridges at the boundaries (Zhang et al. 2009). Due to their special qualities, which are not accessible from either the core or the shell but are only feasible when the core–shell structure occurs, these nanomaterials have received a lot of attention in recent days (Kalska-Szostko, Wykowska, and Satuła 2015). The physical characteristics of the core and the shell can be modified depending on the need for applications. A multi-shell nanostructure with a lot of shells surrounding a central nanoparticle is also conceivable. By altering the core–shell dimensions and the properties of the material of the core and shell, it is conceivable to adopt the magnetic properties of core–shell nanostructures. Elements like iron (Fe) and cobalt (Co) and their chemical derivatives (magnetite (Fe_3O_4), iron oxides, and maghemite (γ-Fe_2O_3)) are recurrently cast off in magnetic core–shell nanostructures. When an external magnetic field is present, the core–shell aligns its magnetic moments with the external magnetic flux density. An extensive contact between the core and shell nanostructures results in an effective exchange coupling in such structures (Zeng et al. 2004). This causes cooperative magnetic switching between the shell and the core, allowing the characteristics of core–shell nanostructures to be tuned as a result. This magnetic response draws the magnetic core–shell nanostructures toward the direction of the applied magnetics gradient. Due to these magnetic properties, the core–shell nanostructures are beneficial in fields like data storage, molecular and cellular separation, spintronics, magnetic resonance imaging (MRI), drug delivery system, and cancer therapy (Tamer et al. 2010). Magnetic core–shell nanostructures have received a great deal of attention recently because of their enhanced design and adaptability in many fields of study, including biological sciences. The core–shell nanostructures consist of a thin shell composed of whichever material is suitable for the application surrounding a highly magnetic core. Magnetic core–shell nanostructures have been researched for their potential use in-vivo imaging, drug transport, and therapeutic hyperthermia, in addition to the many methods for their manufacturing. This chapter mainly focuses on the core–shell nanostructures, synthesis, tailoring of their properties, and applications of core–shell magnetic nanostructures.

3.2   CLASSIFICATION OF CORE–SHELL NANOSTRUCTURES

Contrary to simple nanoparticles, which are often formed of a single material, core–shell nanostructures are typically made of two or more components. The core–shell type nanostructures are generally defined as having an inner core constituent and an outer shell component. Figure 3.1 illustrates the different types of core–shell nanostructures. They are (i) concentric spherical core–shell nanostructures (simple spherical core and shell of different materials), (ii) hexagonal core–shell nanostructures (the core of hexagonal shape and coated with shell coatings), (iii) solitary shell

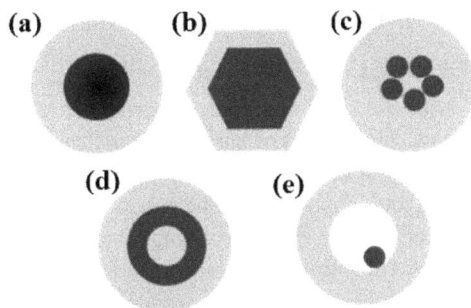

FIGURE 3.1 Different forms of core–shell nanostructures. Adapted with permission from Reference (Ghosh Chaudhuri and Paria 2012), Copyright (2012), American Chemical Society.

material covered onto numerous core materials, (iv) alternative concentric metal shell coating onto dielectric core materials, and (v) hollow shell material with a movable core inside (Ghosh Chaudhuri and Paria 2012). The ultimate usage and application have a noteworthy influence on the choice of shell material for the core–shell nanoparticle. The two or more different materials that constitute the core of most of the particles can be used to characterize core–shell nanostructures. It might be impossible to articulate all the conceivable permutations of the large diversity of substances and elements found on Earth. This makes it appear more acceptable to divide the ingredients that make core/shell nanostructures into organic and inorganic compounds. These can be constructed from a wide range of distinctive combinations that function well together, including organic/organic, inorganic/organic, inorganic/inorganic, and organic/inorganic materials.

3.2.1 ORGANICS

The organic constituents are associated with polymers, carbonaceous constituents, sugars, etc. to progress bio and cyto affinity, particularly for cases utilizing the human body (Liu et al. 2010). For the utmost part, polymerization processes are cast off in organic material preparations to generate the organic core, shell, or both. The fundamental method of polymerization entails the addition of numerous units of proper monomeric substrates with vital functions to produce 3D network-like assemblies. Organic shells can be created in situ, even though organic cores are frequently synthesized first. Surfactants, polyelectrolytes, and other surface modifiers can indeed be implemented to enhance the coating's effectiveness and stability.

3.2.2 INORGANICS

The inorganic materials can be further divided into silica-based and metal-based subcategories. They encompass all metals, metallic alloys, inorganic materials, metal salts, etc. The wide category of "metal-based core–shell nanostructures" requires further sub-classification and can be separated into three categories: metallic nanoparticles, metal oxides, and metal salts, in that order.

3.2.3 Combinations of Core–Shell Nanostructures

3.2.3.1 Organic–Organic Core–Shell Nanoparticles

Under this category, both the shell and core of particles are made of polymers or other organic components. These sorts of particles, also denoted as "smart particles," have a wide variety of uses in several works, including catalysis, chemical separation, biosensing, drug delivery, and biomaterials. The benefits of having a polymer coating on yet another polymer include the ability to change the material's physical features, such as durability or glass transition temperature (Si and Yang 2011). The capacity to synthesize polymers with certain features, such as biocompatibility, size, structure, and functionality, has been a key factor in the creation of core/shell materials made of polymers. The glass transition temperature (T_g) of the polymer/polymer type core/shell nanostructures is taken into consideration when selecting them for a given application. The difference in T_g across a polymer–polymer boundary may depend on the border width, the stark difference between bulk T_g values, network connectivity, and relative Debye–Waller factors. Because polymers are in what is referred to as a "glassy state" below this temperature, the T_g is a crucial property of a polymer. The mechanical properties of the polymers shift and that of a glassy (fragile) to a rubbery (stretchy) substance when the material's temperature crosses the glass transition point. While a low T_g shell material enhances film-forming capacity, a high T_g core particle influences enhancing mechanical firmness. Pekarek et al. (1996) described how a polymer–polymer functionalization was used to create a core/shell polymer system, which included a core of one polymer and a covering of another (shell).

3.2.3.2 Inorganic–Inorganic Core–Shell Nanostructures

The utmost substantial class of core/shell nanostructures is composed of inorganic/inorganic particles. Photonic bioimaging, bio labeling, optical devices, catalysis, quantum dots, and improvements in semiconductor effectiveness are mostly just a few applications for these kinds of particles. These types of core/shell particles can be roughly divided into two groups based on the composition of the shell material: those that contain silica and those made of any other inorganic substance. Additionally, it is clear that silica, metallic, metal salts, or other inorganic compounds typically make up the cores and shells of the many kinds of inorganic/inorganic nanoparticles (Ghosh Chaudhuri and Paria 2012). The silica covering as a core particle has several benefits. The following are the silica coating's primary advantages over other inorganic or organic coatings: It lowers bulk conductivity and improves the core particles' dispersion resilience. Moreover, silica is perhaps the most unreactive substance that is commercially available; it can cover the inner surface without interfering with the ongoing redox process.

3.2.3.3 Inorganic–Organic Core–Shell Nanostructures

Inorganic compounds, a metallurgical mixture, metallic salts, silica, polymers, or indeed any densely packed organic material form up the core of inorganic–organic nanostructured materials with an organic shell (Chiozzi and Rossi 2020). The organic layer over the inorganic material has numerous benefits. One of the main features of the organic layer is the metal core's increased oxidation stability that prevents the metal core's surface atoms from oxidizing to metal oxide in a typical environment. They also have improved bioactivity for use in biological applications. The particles have often been encapsulated in a suspension medium for a variety of applications, and the attractive and repulsive forces between both the particles play a major role in the durability of the resulting colloidal dispersion. Short-range isotropic attractions, Van der Waals' forces, steric repulsion, and electrostatic repulsion are the four types of interacting forces. The electrostatic and steric repulsion forces can be managed by relying on the synthesizing medium, which prevents the nanoparticles from aggregating. Based on their material properties, the core particles can be roughly separated into two classes: magnetic-organic and nonmagnetic-organic core–shell nanostructures.

3.2.3.4 Organic–Inorganic Core–Shell Nanostructures

Since organic–inorganic core–shell nanostructures combine exceptional qualities that none of the separate components possessed, hybrid materials made of organic and inorganic key components are attracting more and more attention. A metal oxide covering on organic material, in particular, is advantageous because it increases the material's overall effectiveness, barrier properties, thermal and colloidal stability, and damage tolerance (Chen et al. 2010). These particles can also increase the fragility of inorganic particles while simultaneously exhibiting polymeric features such as outstanding optical qualities, flexibility, and durability. These nanostructures have recently generated a lot of consideration due to their wide range of applications in a variety of material science domains, such as paints, ferrofluids, catalysis, microelectronics, and bioengineering. The core of this particular class of core–shell nanostructures is mainly made up of polymers (polystyrene, polyethylene oxide, polyurethane, polyvinyl benzyl chloride, polyvinyl pyrrolidone, dextrose, surfactant) and different copolymers (acrylonitrile butadiene styrene, poly styrene-acrylic acid, and styrene methyl methacrylate). Additionally, a variety of substances, including metallic materials, inorganic materials, metal chalcogenides, and silica, can be used to create the shell.

3.2.4 GENERAL MECHANISM FOR THE SYNTHESIS OF CORE–SHELL NANOSTRUCTURES

The general mechanism for the synthesis of core–shell nanostructures was put forth by Cozzoli et al. (Carbone and Cozzoli 2010; Casavola et al. 2008). They described the technicalities of general mechanisms for the synthesis of core–shell nanostructures as follows: Typically, two-stage mechanisms are used to form core–shell nanostructures, with the first step including the production and purification of the core nanostructure and the second step including the growth of the shell. The first step includes directly nucleating and growing the shell material onto already-formed nanocrystal cores. The most popular technique for this method is known be SILAR (successive ionic layer adsorption and reaction) approach, which includes the inclusion of diluted solutions containing shell precursors or alternating infusions of cation and anion. The further process includes the formation of a crystalline shell through annealing of an amorphous or discontinuous shell during its deposition, and the seeds are primed for growth, which is followed by polymerization. Subsequently through a redox reaction, sacrificed shell material replaces the core's outer layer. The preceding process results in the Kirkendall effect, which arises when two different materials are positioned near one another and diffusion is permitted between them. This effect results in the formation of hollow shell nanostructures. The restraints on the growth and formation of the nuclei, phase separation accompanying annealing, diffusion, and flocculation of solid-state reactions end in the formation of core–shell nanostructures.

3.3 SYNTHESIS OF MAGNETIC CORE–SHELL NANOSTRUCTURES

The fabrication of magnetic nanomaterials and their subsequent coating with suitable organic or inorganic components, depending on the application, are the two primary phases in the production of magnetic core–shell nanostructures. The usability and ultimate utilization of core–shell nanostructures heavily influence the choice of shell materials. Figure 3.2 shows an innovative method for fabricating various core–shell nanoparticles made of carbon and magnetite. It involves the thermal decarboxylation of iron acetyl sulfonyl acetate and the solvothermal reaction of glucose in nonaqueous solutions (Shen et al. 2013). As protective coatings, silica, other kinds of metallic and nonmetal oxides, polymeric materials, and bioactive compounds are frequently used. Thus, the development of magnetic nanoparticles is the most critical problem herein (core). The following four techniques can be used to create core–shell nanostructures.

3.3.1 MICROEMULSION METHOD

The thermodynamically constant isotropic solution, at which the micro-domain of any of that or even both liquids is stabilized by an interfacial surfactant coating, is a so-called microemulsion. Microdroplets of the aqueous phase, bound by a monolayer of surfactant molecules in the continuous hydrocarbon phase, are disseminated in water-in-oil microemulsions. The size of the reverse micelle was determined by the molar ratio of water to surfactant. When two identical water-in-oil microemulsions with the preferred reactants are combined, the outcome is a continual collision and break, followed by a subsequent break, which eventually forms a precipitate in the micelles. When a solvent, such as acetone or ethanol, is added to the microemulsions, a precipitate is created that can be removed by filtration or centrifugation. A microemulsion was used in this concept as a nanoreactor to create nanoparticles. Vogt et al. (2010) described how monodispersed, solitary magnetite core–SiO_2 shell nanoparticles with variable shell thickness around 5–13 nm using a painstakingly perfected inverse microemulsion technique. The synthesized superparamagnetic core of iron oxide nanoparticles with SiO_2 shell showed a superparamagnetic nature with a high enough magnetization, making them suitable for biomedical applications.

3.3.2 THERMAL DECOMPOSITION METHOD

The notion of high-quality semiconductor nanocrystalline fabrication and oxide-nanoparticle production utilizing thermal decomposition technology in a nonaqueous

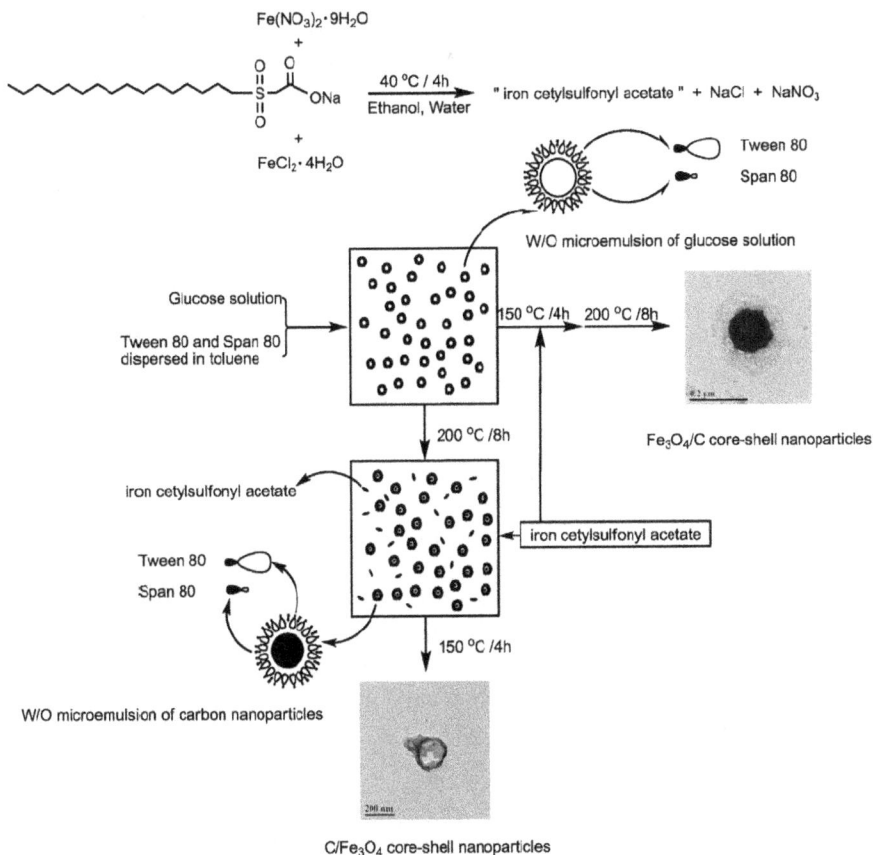

FIGURE 3.2 Synthesis of the magnetic core and carbon shell and vice versa. Adapted with permission from Reference (Shen et al. 2013), Copyright (2013), Elsevier.

medium led to the formation of magnetic particles having appropriate form and size (Murray, Norris, and Bawendi 1993). Monodisperse magnetic nanocrystals with smaller sizes are produced when organometallic compounds thermally decompose in highly flammable organic solvents with stabilizing surfactants (Sun et al. 2004). To minimize manipulation between phases and simplify synthesis control, Maria Eugenia F. Brollo et al. created a remarkable single-step methodology. In this instance, just before the creation of iron oxide, the Ag seeds are generated in the same reaction mixture. On a silver colloid produced in a certain reaction medium by the introduction of $AgNO_3$ salt, the Fe_3O_4 precursor thermally decomposes resulting in Ag@Fe_3O_4 core–shell nanostructures (Brollo et al. 2015).

3.3.3 COPRECIPITATION METHOD

In the coprecipitation approach, for the synthesis of core–shell nanoparticles, bases are often added to a combination of anhydrous Fe^{2+} or Fe^{3+} salts in the absence of air at a higher or ambient temperature depending on the iron oxide (Fe_3O_4 or γ-Fe_2O_3) nanoparticles needed. The coprecipitation technique is the name given to this process. The precursor salt and reactivity parameters, such as pH, temperature, and ionic strength of the reaction mechanism, have a significant effect on the formation and structure of the magnetic nanocomposites that were prepared by the coprecipitation process. The two-step coprecipitation procedure was used to fabricate the magnetic core–shell structures. The initial phase involved creating the magnetic core nanostructures depending on Fe_3O_4, $CoFe_2O_4$, $ZnCoFe_2O_4$, etc. based on magnetic materials. The identical process used to synthesize the core was employed in step two, with the addition of the shell precursors, to cover the created core with a shell layer (Darwish et al. 2020).

3.3.4 Hydrothermal Method

A solid–liquid–solution array including metal $C_{18}H_{31}O_2$ (solid), ethanol $C_{18}H_{31}O_2$ (liquid), and ethanol solution is used in the development under hydrothermal treatment. By using a hydrothermal carbonization procedure, Fe_3O_4-C magnetic core–organic/inorganic shell nanostructures with various shell thicknesses were created, and their physical, magnetic, and electromagnetic properties were studied by Omid Khani et al. (2016). Despite producing nanoparticles of exceptional quality, hydrothermal synthesis is a largely untapped technique for creating magnetic particles with different morphologies. On the other hand, coprecipitation and thermal decomposition processes are currently used on a wide scale to create magnetic core–shell nanostructures.

3.4 TAILORING THE PROPERTIES OF MAGNETIC CORE–SHELL NANOSTRUCTURES

Nano optimization is a frequently used method to develop the desired properties of nanostructures. This process is known as tailoring because the attributes are inextricably linked to the nanostructures and so are difficult to alter. Due to its significance in determining the scaling limitations of magnetic data, storage technology, and comprehending spin-dependent processes, nanoscale magnetism has received much interest. Core–shell nanostructures, wherein the magnetic core is covered with a casing of a nonferrous, or ferro–ferri-magnetic shell, antiferromagnetic, are an intriguing type of nanostructures, and the coating equipped is appropriate for the instance. Detailed research on magnetism, polarization, magnetization reversal processes, and interlayer couplings of the particles with various sizes and surface features would be possible in an array of monodisperse magnetic nanoparticles with regulated interparticle spacing. As the shell thickness varies, there can be two separate magnetic domains for core–shell magnetic nanoparticles: the exchange coupling domain, and the enhanced spin canting domains. The coupling relationship between the hard magnetic core and the soft magnetic shell has fascinated a lot of interest over the past years as a successful method of logically designing nanoparticles to control their magnetism. The modified magnetism of the core–shell nanostructures was studied by Moon et al. (2017) concerning the effects of shell thickness. They additionally shown are the effectiveness of magnetic features, increased energy products, and magnetic heating. Ultrathin-shelled nanoparticles can open up new possibilities for further customizing nanoscale magnetism in addition to the exchange coupling domain. The strong interaction between both the core and shell in a bimagnetic core/shell, where the core and shell are indeed firmly magnetic, results in effective exchange aligning and consequently cooperative magnetic switching, making it easier to create nano-structured magnetic materials with adjustable characteristics. The magnetization of these core–shell nanostructures may be customized by manipulating the core/shell diameters and by adjusting the material parameters of both the core and shell. Zeng et al. (2004) fabricated a 3.5 nm thickness of FePt core with MFe_2O_4 (M = Fe, Cd) shell and postulated that such systems might exhibit intriguing nano-magnetism as a result of the exchange coupling between both the core and the shell, and they might produce materials that are precisely suited for different nano-magnetic applications.

3.5 APPLICATIONS OF MAGNETIC CORE–SHELL NANOSTRUCTURES

3.5.1 BIOMEDICAL APPLICATIONS

The chemical nature of the core/shell nanostructures, which boosts their tendency to associate with medicines, sensors, ligands, etc., makes them primarily suitable for biological applications. Due to their superior biological compatibility with bulk material, these new nanostructures have been created (Sahoo and Labhasetwar 2003). Typically, the core substance is where core–shell nanostructures get their capacity to be compared. When compared with pure core particles, magnetic particles capped with a functional substance such as noble metals, semiconductors, or suitable oxide have improved electric, catalytic, magnetic, optical, and thermal capabilities. In general, the shell content of core–shell nanostructures is in charge of surface character-istics which alters the biological properties of the living systems which are in contact with the shell surfaces and the attachment of bioactive compounds due to the pres-ence of reactive monomers on the surfaces. The thickness of the shell can be adjusted to give sufficient contrast qualities as a contrast agent and binding of biomolecules for purposes such as targeted drug administration, specific binding, biosensing, etc. Additionally, magnetic core–shell nanostructures can alter or change the covering components after fabrication, providing customized control over surface properties, functional groups, and the physical size of the nanoparticles.

3.5.1.1 Drug Delivery Nanocarriers

Magnetic core–shell nanostructures are being intensively researched as the next level of tailored drug delivery nanocarriers due to their distinct physicochemical features and capacity to operate at the molecular and cellular levels of the living systems. Magnetic core–shell drug delivery nanocarriers are used to treat cancer. The main option for the treatment of tumors is magnetic core–shell nanomateri-als, particularly superparamagnetic iron oxide nanoparticles because they can detect and treat cancer cells when an external magnetic field is manipulated, increasing the restorative effects of the drug (Nguyen et al. 2016). These magnetic core–shell nanostructures are accurate even though many drug delivery systems have huge potential in chemotherapy. Figure 3.3 depicts the drug release response of magnetic material (hydrophobic Fe_3O_4) surface modified with $HSCH_2CH_2COOCH_3$ capsules—encapsulated with stimuli-responsive smart polymer (e dextran-g-poly-(NIPAAm-co-DMAAm)), designed for controlled release and magnetic drug targeting (Zhang and Misra 2007). Their superparamagnetic nature enables instant tumor treatment and malignant transformation monitoring with magnetic resonance

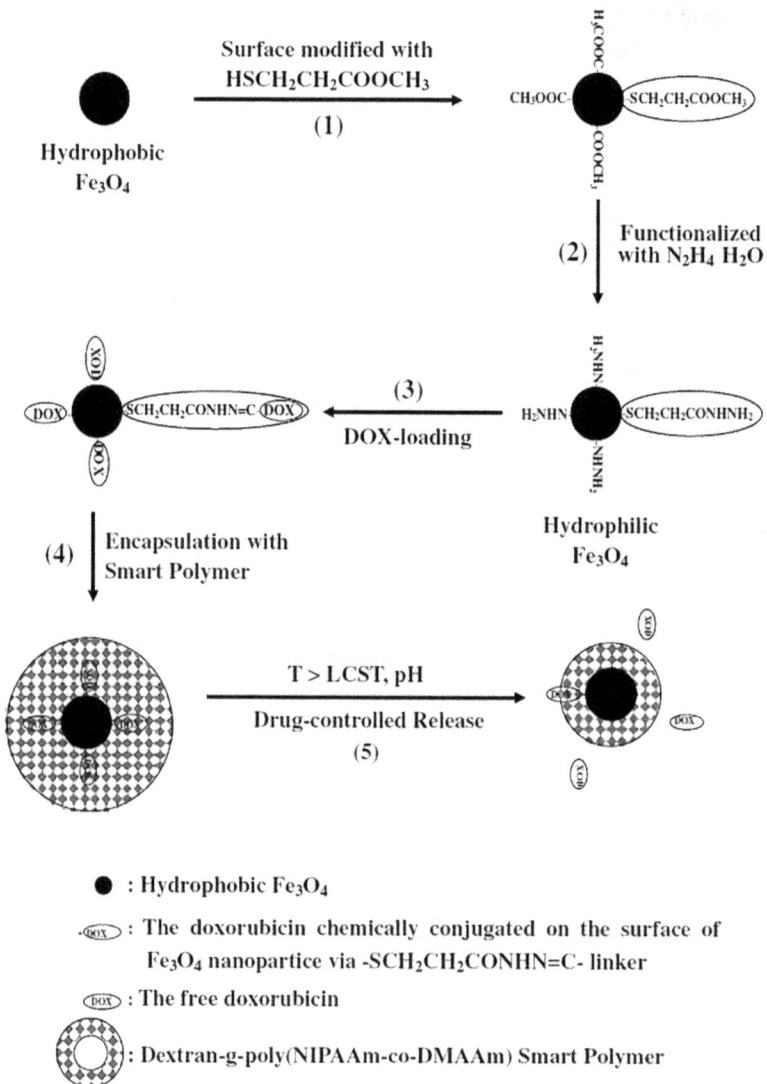

FIGURE 3.3 Schematic illustration of the step-by-step preparation of core–shell magnetic nanocarriers. Adapted with permission from Reference (Zhang and Misra 2007), Copyright (2007), Elsevier.

imaging (MRI). Surface functionalization has received a lot of attention as a key technique for overcoming the drawbacks of conventional superparamagnetic iron oxide nanoparticles. Superparamagnetic iron oxide nanoparticles were created with heparin polymer by Hoang Thi et al. (2019), who also claimed that the core–shell had the preferred size range and a comparatively high drug loading effectiveness without significantly altering their morphological characteristics, crystalline structure, or distinctive magnetic properties.

3.5.1.2 Magnetic Resonance Imaging (MRI)

Magnetic resonance imaging or nuclear MRI is a crucial non-invasive image processing technology that makes use of the magnetic properties of the numerous interacting magnetic nuclei that are prevalent within humans to examine internal biological components with precision. The overall magnetism of the magnetic nuclei within a person's body aligns with the orientation of a strong static magnetic field when they are sustained under it. These magnetic nuclei can collect this radiation and switch the spin to the opposite direction when excited by some other external electromagnetic field with the appropriate frequency which is known as the resonance frequency. As drug delivery carriers, "magnetic drug delivery" is the term for magnetic field-induced drug delivery systems (Mandal 2016). Injecting magnetic nanoparticles with drugs inside the living system and directing them to a target spot under the guidance of a magnetic field gradient is the principle behind magnetic drug delivery. These nanoparticles are kept at the treatment site until it is complete, at which point they are eliminated. This method allowed for the creation of high local concentrations of the targeted medicine, minimizing toxic effects and other unwanted concomitant effects on healthy cells in the remainder of the body.

3.5.1.3 Cancer Therapy

Figure 3.4 illustrates the core–shell nanostructures of magnetic core covered with polymer coating for cancer therapy, herein a suitable linker was used to decorate the external polymer coating encasing the anticancer agent with an imaging agent for improved imaging and/or a targeting ligand for active targeting (Mitra et al. 2015). In addition, the use of folic acid coupled $FePt@Fe_2O_3$-PEG nanomaterials facilitates effective targeting of tumor cells that express the folate receptor. To distribute the chemotherapeutic medication doxorubicin to these nanomaterials in a targeted manner for the destruction of cancer cells, hydrophobic adsorption is used. These $FePt@Fe_2O_3$-PEG nanomaterials have been used in-vivo MRI to produce tumor magnetic resonance contrasts that may either be used to start targeting the tumor or aggregate in a passive tumor. For synchronized magnetic resonance imaging as well as drug delivery, unique, monodisperse, and properly sized-regulated core–shell nanostructures with Fe_3O_4@mesoporous-silica with a size of less than 100nm were used (Kim et al. 2008).

3.5.2 Electrochemical Applications

Instruments like sensors can gauge a quantitative measurement and transform it into a signal that can be interpreted by a person or an instrument, whether analog or digital.

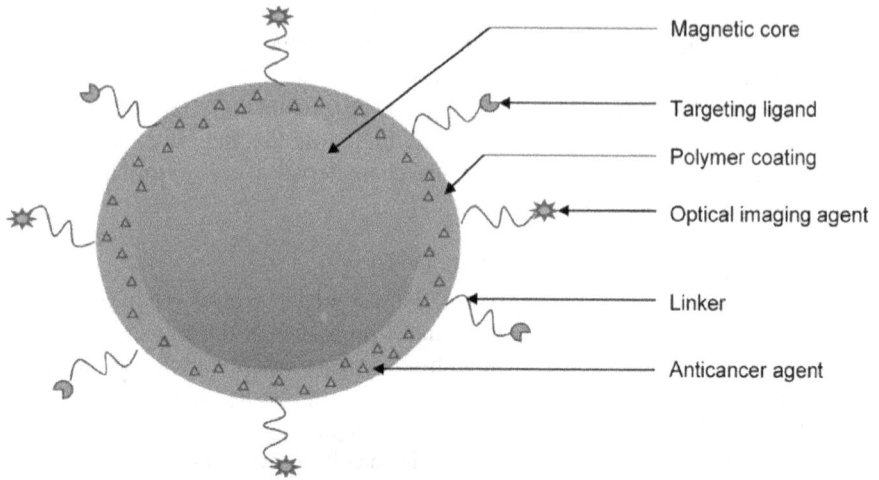

FIGURE 3.4 Schematic illustration of magnetic-polymer core–shell nanostructures. Adapted with permission from Reference (Mitra et al. 2015), Copyright (2015), Elsevier.

Magnetic core–shell nanostructures are utilized as sensors for the detection of damaged cells, glucose, DNA, cholesterol, RNA, etc. in in-vivo applications. A fluorescence-coated magnetic substance can be utilized as a sensor. Here, the magnetic coating works to transmit heat at that point via magnetic excitation, while the fluorescent dye monitors the position of the particle. As bioanalytical sensors, magnetic-based core–shell nanocomposites covered with any additional material, such as a fluorescent one, silica, a metallic, or a polymer, are employed. The identification of damaged DNA is carried out specifically using Fe/Fe_2O_3 core–shell nanostructures (Qiu et al. 2010). These particles have bioactive proteins including cytochrome, myoglobin, and hemoglobin linked to them to simulate the toxicity that occurs in living organisms.

3.5.3 SPINTRONICS

The embellished magnetic properties of bimagnetic core–shell nanostructures are highly favorable for the feasibility of application in spintronic devices, owing to the tunability of their magnetic properties, such as their ability to dominate the superparamagnetic edge and adaptability. It attributes to the probable advantages of magnetic bubble memory, data retrieval, curtail consumption of power, size reduction of devices, speedy performance, logic operations, etc. The ongoing drive for device component downsizing is a particularly significant driving force in spintronics (Hossain et al. 2018). Electronics that are based on spin are known as spintronics. Utilizing both the electron's charge and spin is the notion. The two orientations that electrons can spin are Spin-Up and Spin-Down, that is clockwise and anticlockwise orientations are measurable as indistinct magnetic energy. Khurshid et al. (2014) by examining the exchange bias in $Fe/\gamma\text{-}Fe_2O_3$ (ferro/ferri) core–shell nanostructures with a range of particle sizes (8–15 nm) illustrated that there is a critical particle size (10 nm) above where the interface spin consequence governs the exchange bias, but below which the surface spin effect becomes more significant. This finding clarified the cause of the noticed exchange bias effects and offered a method for adjusting exchange bias in core–shell nanostructures for spintronic applications.

3.5.4 MAGNETIC NANOPARTICLES FUNCTIONALIZATION STRATEGIES

By functionalizing the surface of nanomaterials, it is possible to produce strong binding and active sites for biomolecules and nanostructured materials. Magnetic nanoparticle functionalization of metals and polymers increases their effectiveness by making them reusable. A variety of nanomaterials, such as metal oxide nanoparticles, carbon-based nanomaterials, synthetic and natural polymers, antibodies, biomolecules, and zwitterionic materials, have all been used as functionalization agents. One important feature is that the surface hydrophobicity or hydrophilicity can be adjusted by precoating or functionalizing nanoparticles. Reza Eivazzadeh-Keihan et al. proposed the functionalization of magnetic nanoparticles decorated with metal ions such as iron, copper, nickel, gallium, cobalt, or zinc are powerful candidates for the separation and purification of histidine-tagged proteins. Figure 3.5 represents the schematic of magnetic nanoparticles decorated with diverse metals for the separation and purification of binding proteins. A commonly used tag for such protein purification is the histidine tag. In histidine, there are six or more sequential histidine residues. The binding of histidine-tagged proteins and subsequent purification in the suitable column is used to purify proteins for which there is no specific affinity column used (Eivazzadeh-Keihan et al. 2021). Junhua You et al. presented a report on amino-functionalized magnetic nanoparticles for water purification mainly based on three major mechanisms that consist of (i) impurities are attracted to the surface of the activated magnetic nanoparticles by chelation, electrostatic, or complexation; (ii) magnetic nanoparticles are removed magnetically from the solutions; and (iii) the magnetic nanoparticles are recovered again, and this happens efficiently when surface functionalization occurs (You et al. 2021). New reports in the identification and detachment of toxic heavy metal ions utilizing colorimetric or fluorometric systems were reported by Jong Hwa Jung et al. They also reported the preparation of a diverse array of functionalized magnetic nanoparticles and their implementations in multiple environmental fields. Due to the simplicity of covalently linking suitable functional organic molecules, magnetic nanoparticles like Fe_3O_4, Fe_3C, $Ni@SiO_2$, and $Fe_3O_4@SiO_2$ are ideal as supporting materials to manufacture functional hybrid nanomaterials via the sol-gel gluing approach (Jung, Lee, and Shinkai 2011) (Figure 3.5).

3.6 CONCLUSIONS AND FUTURE OUTLOOK

With the formation of a novel class of materials known as hybrid core–shell nanostructures, nanotechnology has ushered in a new era for achieving revolutionary applications. Constructing core–shell nanostructures with various surface-coating techniques has made it possible to attain the versatility of nanomaterials. Future research in this area will lead to significant advancements in the synthesis, characteristics, alteration, and usage of this new class of nanomaterials. The combinations of core and shell materials, the magnetic properties of core–shell nanostructure synthesis, and their various applications are discussed in this chapter. It is particularly advantageous because the

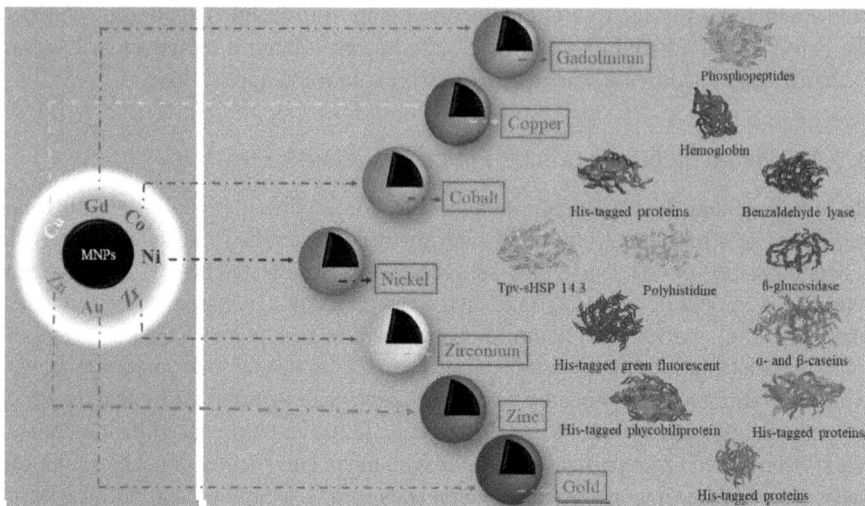

FIGURE 3.5 Schematic illustration of decorating various metals on magnetic nanoparticles for selective binding of proteins. Adapted with permission from Reference (Eivazzadeh-Keihan et al. 2021), Copyright (2021), Elsevier.

magnetic characteristics of core–shell materials are a class of nanostructures that may be manipulated using magnetic fields. In several biomedical applications, magnetic core–shell nanostructures have shown promising applications, including magnetic resonance imaging (MRI) data, assisting with tissue engineering, and facilitating medication delivery to hard-to-reach microinches. The use of these nanostructures is that it will bind the research establishment to the general public in the biomedical field. These aids can report new empathetic and indorse designs of core–shell magnetic nanostructures for providing novel routes and uses in the future.

REFERENCES

Brollo, Maria Eugênia F., RománLópez-Ruiz, Diego Muraca, Santiago J. A. Figueroa, Kleber R. Pirota, and Marcelo Knobel. 2015. "Compact Ag@Fe3O4 Core-Shell Nanoparticles by Means of Single-Step Thermal Decomposition Reaction." *Scientific Reports* 4 (1): 6839. doi:10.1038/srep06839.

Chen, Bo, Jianping Deng, Linyue Tong, and Wantai Yang. 2010. "Optically Active Helical Polyacetylene@silica Hybrid Organic–inorganic Core/Shell Nanoparticles: Preparation and Application for Enantioselective Crystallization." *Macromolecules* 43 (23): 9613–9619. doi:10.1021/ma102157e.

Chiozzi, Viola, and Filippo Rossi. 2020. "Inorganic–Organic Core/Shell Nanoparticles: Progress and Applications." *Nanoscale Advances* 2 (11): 5090–5105. doi:10.1039/D0NA00411A.

Darwish, Mohamed S. A., Hohyeon Kim, Hwangjae Lee, Chiseon Ryu, Jae Young Lee, and Jungwon Yoon. 2020. "Engineering Core-Shell Structures of Magnetic Ferrite Nanoparticles for High Hyperthermia Performance." *Nanomaterials* 10 (5): 991. doi:10.3390/nano10050991.

Carbone, Luigi, and P. Davide Cozzoli. 2010. "Colloidal Heterostructured Nanocrystals: Synthesis and Growth Mechanisms." *Nano Today* 5 (5): 449–493. doi: 10.1016/j.nantod.2010.08.006.

Casavola, Marianna, Raffaella Buonsanti, Gianvito Caputo, and Pantaleo Davide Cozzoli. 2008. "Colloidal Strategies for Preparing Oxide-Based Hybrid Nanocrystals." *European Journal of Inorganic Chemistry* 2008 (6): 837–854. doi:10.1002/ejic.200701047.

Eivazzadeh-Keihan, Reza, Hossein Bahreinizad, Zeinab Amiri, Hooman Aghamirza Moghim Aliabadi, Milad Salimi-Bani, Athar Nakisa, Farahnaz Davoodi, et al. 2021. "Functionalized Magnetic Nanoparticles for the Separation and Purification of Proteins and Peptides." *TrAC Trends in Analytical Chemistry* 141 (August): 116291. doi:10.1016/j.trac.2021.116291.

Ghosh Chaudhuri, Rajib, and Santanu Paria. 2012. "Core/Shell Nanoparticles: Classes, Properties, Synthesis Mechanisms, Characterization, and Applications." *Chemical Reviews* 112 (4): 2373–2433. doi:10.1021/cr100449n.

Hoang Thi, Thai, Diem-Huong Nguyen Tran, Long Bach, Hieu Vu-Quang, Duy Nguyen, Ki Park, and Dai Nguyen. 2019. "Functional Magnetic Core-Shell System-Based Iron Oxide Nanoparticle Coated with Biocompatible Copolymer for Anticancer Drug Delivery." *Pharmaceutics* 11 (3): 120. doi:10.3390/pharmaceutics11030120.

Hossain, Mohammad Delower, Robert A. Mayanovic, Ridwan Sakidja, Mourad Benamara, and Richard Wirth. 2018. "Magnetic Properties of Core–Shell Nanoparticles Possessing a Novel Fe(ii)-Chromia Phase: An Experimental and Theoretical Approach." *Nanoscale* 10 (4): 2138–2147. doi:10.1039/C7NR04770C.

Jung, Jong Hwa, Ji Ha Lee, and Seiji Shinkai. 2011. "Functionalized Magnetic Nanoparticles as Chemosensors and Adsorbents for Toxic Metal Ions in Environmental and Biological Fields." *Chemical Society Reviews* 40 (9): 4464. doi:10.1039/c1cs15051k.

Kalska-Szostko, B., U. Wykowska, and D. Satuła. 2015. "Magnetic Nanoparticles of Core–Shell Structure." *Colloids and Surfaces A: Physicochemical and Engineering Aspects* 481 (September): 527–536. doi:10.1016/j.colsurfa.2015.05.040.

Khani, Omid, Morteza Zargar Shoushtari, Mohammad Jazirehpour, and Mohammad Hossein Shams. 2016. "Effect of Carbon Shell Thickness on the Microwave Absorption of Magnetite-Carbon Core-Shell Nanoparticles." *Ceramics International* 42 (13): 14548–14556. doi:10.1016/j.ceramint.2016.06.069.

Khurshid, Hafsa, Manh-Huong Phan, Pritish Mukherjee, and Hariharan Srikanth. 2014. "Tuning Exchange Bias in Fe/γ-Fe$_2$O$_3$ Core-Shell Nanoparticles: Impacts of Interface and Surface Spins." *Applied Physics Letters* 104 (7): 072407. doi:10.1063/1.4865904.

Kim, Jaeyun, Hoe Suk Kim, Nohyun Lee, Taeho Kim, Hyoungsu Kim, Taekyung Yu, In Chan Song, Woo Kyung Moon, and Taeghwan Hyeon. 2008. "Multifunctional Uniform Nanoparticles Composed of a Magnetite Nanocrystal Core and a Mesoporous Silica Shell for Magnetic Resonance and Fluorescence Imaging and for Drug Delivery." *AngewandteChemie International Edition* 47 (44): 8438–8441. doi:10.1002/anie.200802469.

Liu, HongLing, Peng Hou, WengXing Zhang, and JunHua Wu. 2010. "Synthesis of Monosized Core–Shell Fe3O4/Au Multifunctional Nanoparticles by PVP-Assisted Nanoemulsion Process." *Colloids and Surfaces A: Physicochemical and Engineering Aspects* 356 (1–3): 21–27. doi:10.1016/j.colsurfa.2009.12.023.

Mandal, Samir. 2016. "Engineered Magnetic Core Shell Nanoprobes: Synthesis and Applications to Cancer Imaging and Therapeutics." *World Journal of Biological Chemistry* 7 (1): 158. doi:10.4331/wjbc.v7.i1.158.

Mitra, Ashim K., VibhutiAgrahari, Abhirup Mandal, Kishore Cholkar, Chandramouli Natarajan, Sujay Shah, Mary Joseph, et al. 2015. "Novel Delivery Approaches for Cancer Therapeutics." *Journal of Controlled Release* 219 (December): 248–268. doi:10.1016/j.jconrel.2015.09.067.

Moon, Seung Ho, Seung-Hyun Noh, Jae-Hyun Lee, Tae-Hyun Shin, Yongjun Lim, and Jinwoo Cheon. 2017. "Ultrathin Interface Regime of Core–Shell Magnetic Nanoparticles for Effective Magnetism Tailoring." *Nano Letters* 17 (2): 800–804. doi:10.1021/acs. nanolett.6b04016.

Murray, C. B., D. J. Norris, and M. G. Bawendi. 1993. "Synthesis and Characterization of Nearly Monodisperse CdE (E = Sulfur, Selenium, Tellurium) Semiconductor Nanocrystallites." *Journal of the American Chemical Society* 115 (19): 8706–8715. doi:10.1021/ja00072a025.

Nguyen, Dai Hai, Jung Seok Lee, Jong Hoon Choi, Kyung Min Park, Yunki Lee, and Ki Dong Park. 2016. "Hierarchical Self-Assembly of Magnetic Nanoclusters for Theranostics: Tunable Size, Enhanced Magnetic Resonance Imagability, and Controlled and Targeted Drug Delivery." *ActaBiomaterialia* 35 (April): 109–117. doi:10.1016/j. actbio.2016.02.020.

Pekarek, Kathleen J., Martinus J. Dyrud, Kathy Ferrer, Yong S. Jong, and Edith Mathiowitz. 1996. "In Vitro and in Vivo Degradation of Double-Walled Polymer Microspheres." *Journal of Controlled Release* 40 (3): 169–178. doi:10.1016/0168-3659(95)00176-X.

Qiu, Jian-Ding, Hua-Ping Peng, Ru-Ping Liang, and Xing-Hua Xia. 2010. "Facile Preparation of Magnetic Core–Shell Fe3O4@Au Nanoparticle/Myoglobin Biofilm for Direct Electrochemistry." *Biosensors and Bioelectronics* 25 (6): 1447–1453. doi:10.1016/j. bios.2009.10.043.

Sahoo, Sanjeeb K., and Vinod Labhasetwar. 2003. "Nanotech Approaches to Drug Delivery and Imaging." *Drug Discovery Today* 8 (24): 1112–1120. doi:10.1016/S1359-6446(03) 02903-9.

Shen, Kaihua, Jiwei Wang, Ying Li, Yanse Wang, and Yang Li. 2013. "Preparation of Magnetite Core–Shell Nanoparticles of Fe3O4 and Carbon with Aryl Sulfonyl Acetic Acid." *Materials Research Bulletin* 48 (11): 4655–4660. doi:10.1016/j. materresbull.2013.07.040.

Si, Jingyu, and Hua Yang. 2011. "Preparation and Characterization of Bio-Compatible Fe3O4@Polydopamine Spheres with Core/Shell Nanostructure." *Materials Chemistry and Physics* 128 (3): 519–524. doi:10.1016/j.matchemphys.2011.03.039.

Sun, Shouheng, Hao Zeng, David B. Robinson, Simone Raoux, Philip M. Rice, Shan X. Wang, and Guanxiong Li. 2004. "Monodisperse MFe$_2$ O$_4$ (M = Fe, Co, Mn) Nanoparticles." *Journal of the American Chemical Society* 126 (1): 273–279. doi:10.1021/ja0380852.

Tamer, Uğur, Yusuf Gündoğdu, İsmail HakkıBoyacı, and KadirPekmez. 2010. "Synthesis of Magnetic Core–Shell Fe3O4–Au Nanoparticle for Biomolecule Immobilization and Detection." *Journal of Nanoparticle Research* 12 (4): 1187–1196. doi:10.1007/s11051-009-9749-0.

Vogt, Carmen, Muhammet S. Toprak, Mamoun Muhammed, Sophie Laurent, Jean-Luc Bridot, and Robert N. Müller. 2010. "High Quality and Tuneable Silica Shell– Magnetic Core Nanoparticles." *Journal of Nanoparticle Research* 12 (4): 1137–1147. doi:10.1007/s11051-009-9661-7.

You, Junhua, Lu Wang, Yao Zhao, and Wanting Bao. 2021. "A Review of Amino-Functionalized Magnetic Nanoparticles for Water Treatment: Features and Prospects." *Journal of Cleaner Production* 281 (January): 124668. doi:10.1016/j.jclepro.2020.124668.

Zeng, Hao, Shouheng Sun, J. Li, Z. L. Wang, and J. P. Liu. 2004. "Tailoring Magnetic Properties of Core/shell Nanoparticles." *Applied Physics Letters* 85 (5): 792–794. doi:10.1063/1.1776632.

Zhang, J., and R. D. K. Misra. 2007. "Magnetic Drug-Targeting Carrier Encapsulated with Thermosensitive Smart Polymer: Core–Shell Nanoparticle Carrier and Drug Release Response." *ActaBiomaterialia* 3 (6): 838–850. doi:10.1016/j.actbio.2007.05.011.

Zhang, Jiatao, Yun Tang, Lin Weng, and Min Ouyang. 2009. "Versatile Strategy for Precisely Tailored Core@Shell Nanostructures with Single Shell Layer Accuracy: The Case of Metallic Shell." *Nano Letters* 9 (12): 4061–4065. doi:10.1021/nl902263h.

4 Functionalized Magnetic Nanoparticles for Biomedical Applications (Treatment, Imaging, and Separation and Detection Applications)

O. Icten
Hacettepe University

CONTENTS

4.1 INTRODUCTION

Compared with bulk materials, nanomaterials possessing magnetic properties have been widely used in various fields such as electronics, materials sciences, and biomedical sciences due to their many advantages. Generally, magnetic nanomaterials consist of an inorganic core with paramagnetic or superparamagnetic properties and a second organic or inorganic layer that provides stability, solubility, or differential properties (Lu et al. 2007; Yildiz 2016). In recent years, developments in both synthesis and characterization techniques have offered new opportunities to obtain magnetic nanoparticles with controlled magnetic properties (Peixoto et al. 2020). Especially magnetic nanoparticles are one of the most remarkable nanomaterials in the biomedical field because of their nontoxicity, biocompatibility, and remote control property by applying a magnetic field (Fraceto and de Araujo 2014; Gao et al. 2014). In the biomedical field, functional magnetic nanoparticles are generally utilized in treatment applications such as drug delivery, hyperthermia, neutron capture therapy

DOI: 10.1201/9781003335580-4

67

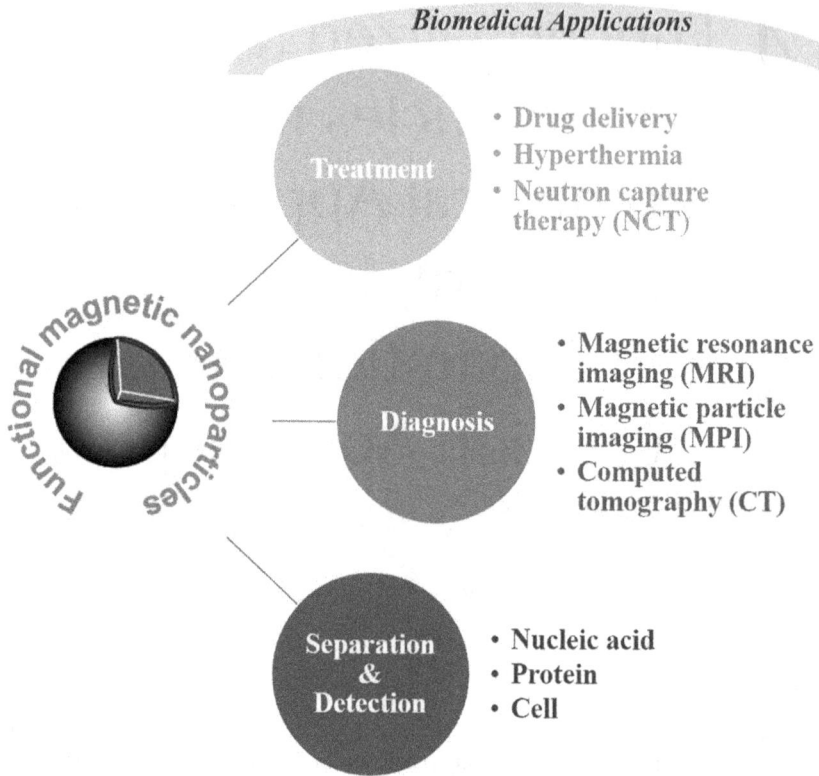

FIGURE 4.1 The biomedical applications of functional magnetic nanoparticles.

(NCT), imaging applications such as magnetic resonance imaging (MRI), magnetic particle imaging (MPI) and computed tomography (CT), and separation and detection applications for nucleic acids, proteins, and cells as depicted in Figure 4.1.

Magnetic nanoparticles can be grouped into pure metals, metal oxides, and magnetic nanocomposites. Cobalt, iron, nickel, titanium, and iron oxides or some ferrites such as $BaFe_{12}O_{19}$, $CoFe_2O_4$, and $MnFe_2O_4$ can be employed as magnetic nanoparticles in biomedical applications. However, cobalt and nickel, which are highly magnetic sensitive materials, are toxic and susceptible to oxidation, causing a barrier to their biomedical use (Cardoso et al. 2018; Majidi et al. 2016). On the contrary, iron oxide nanoparticles such as magnetite (Fe_3O_4) and maghemite (γ-Fe_2O_3) are the most studied and applied in the biomedical field owing to their biocompatibility, superparamagnetic behavior, and chemical stability at room temperature (Li et al. 2016; Majidi et al. 2016). Magnetite has an inverted spinel structure and ferromagnetic features. Thirty-two oxygen ions form a face-centered cubic structure, and half of the Fe^{3+} ions occupy the tetrahedral sites, and Fe^{2+} ions and half of the Fe^{3+} ions occupy the octahedral sites in the unit cell. The chemical formula of magnetite is $(Fe^{3+})_{tet.}(Fe^{2+}Fe^{3+})_{oct.}O_4$ (tet.: tetrahedral site; oct.: octahedral site). In the maghemite structure, 32 oxygen ions form a cubic close-packed structure, and Fe^{3+} ions are located in tetrahedral and octahedral sites in this structure.

Unlike magnetite structure, maghemite has a vacancy of Fe^{2+} ions in its structure. The chemical formula of maghemite is $0.75(Fe^{3+})_{tet.}(Fe^{3+}_{5/3}V_{1/3})_{oct.}O_4$ (tet.: tetrahedral site; V: Fe^{2+} vacancy; oct.: octahedral site) (Shabatina et al. 2020; Wu et al. 2015). Magnetic iron oxide nanoparticles, which can be synthesized by various methods such as hydro/solvothermal, coprecipitation, thermal decomposition, and sol-gel, can be obtained in spherical, nanorod, nanowire, and nanocube morphologies depending on the requirements of the application areas (Monteserín et al. 2021). However, instead of being used without modification, magnetic nanoparticles to be applied in the biomedical field should be functionalized to prevent aggregation in the body and provide colloidal stability, increase their biocompatibility, ensure transport to specific regions, and increase transport efficiency or gain different properties (Cardoso et al. 2018).

4.2 FUNCTIONALIZATION OF MAGNETIC NANOPARTICLES FOR BIOMEDICAL APPLICATIONS

As stated in the previous part, the synthesized nanoparticles should also be biocompatible in addition to desired shapes and sizes for biomedical applications. Functionalizing the surfaces of nanoparticles is a process that can be applied to reduce or eliminate their toxic side effects (Markides et al. 2012). In addition, the functionalization of magnetic nanoparticles with diverse structures can also be aimed at changing cell interactions and their distribution in the body, as well as making them biocompatible. The second or other layers on the surface of magnetic nanoparticles also provide some advantages, such as protecting the active agents carried into the body and increasing the delivery efficiency and also preventing their accumulation in the undesired region in the body (Cardoso et al. 2018; Ruiz et al. 2014). However, it should be noted that every modification or coating performed on the nanoparticle surface causes a decrease in magnetization, so optimizing the amount of modification or coating thickness depending on the magnetization behavior of magnetic nanoparticles will eliminate the negative situations that may occur (Duan and Li 2013). Studies have indicated that changes in the surface charges of nanoparticles with functionalization are essential for interactions with biological molecules. For example, iron oxide nanoparticles with a positive surface charge can aggregate and are more easily uptake by cells with a negative charge. Administration of high doses of positively charged nanoparticles can cause nonspecific accumulation in cells, resulting in a toxic effect (Liu et al. 2013). Conversely, nanoparticles with negative and neutral surface charges are appropriate for targeted imaging probes and drug delivery applications where long circulation times in the body are needed (Blanco et al. 2015; Cardoso et al. 2018).

Until now, many organic and inorganic structures have been employed to functionalize magnetic iron oxide nanoparticles. The most common organic structures for functionalization are polyethylene glycol (PEG) (Bloemen et al. 2012), polyvinylpyrrolidone (PVP) (Torresan et al. 2021), polydopamine (PDA) (Icten et al. 2022), chitosan (Song et al. 2015), and dextran, while inorganic structures for that are silica, Au, and Ag (Korolkov et al. 2020). However, modifications are not limited to these structures. Apart from these, diverse organic and inorganic materials can be utilized for surface functionalization. Therefore, many examples of these structures are given in Table 4.1 with application area, and also some of these studies are described in detail below.

TABLE 4.1

Example Studies for Functional Magnetic Nanoparticles

Functional Magnetic Nanoparticle*	Functionalization Agent	Application	Reference
$CuFeSe_2$-PEG-FA	PEG and folic acid (FA)	MRI and CT	Yan et al. (2021)
PEG/LA-CS@Fe_3O_4	Chitosan (CS), PEG and lactobionic acid (LA)	MRI	Song et al. (2015)
Biocompatible Fe-B	PVP	NCT Guided by MRI	Torresan et al. (2021)
PVP-$MnFe_2O_4$	PVP	Drug delivery	Wang et al. (2018)
γ-Fe_2O_3/PVP	PVP	MRI	Li et al. (2015)
PVP-SPIO (SPIO: superparamagnetic iron oxide)	PVP	MRI and tracking of stem cells	Reddy et al. (2009)
$MnFe_2O_4$@PDA-Au-BA	PDA, gold, and boric acid (BA)	A potential candidate for MRI, NCT, and photothermal therapy	Icten et al. (2022)
Fe_3O_4@PDA@SNA Fe_3O_4@PDA@SH-DNA	PDA and the spherical (SNA) or linear (SH-DNA) nucleic acids	DNA extraction and detection	Zandieh and Liu (2021)
Ti^{4+}-PDA@Fe_3O_4	Ti^{4+} ions and PDA	Separation of phosphorylated proteins	Xiangdong Ma and Jia (2016)
Fe_3O_4@CS	Chitosan (CS)	MRI	Nguyen et al. (2020)
Fe_3O_4/Chitosan	Chitosan	Drug delivery and MRI	Zhao et al. (2014)
Dextran-Fe_3O_4	Dextran	Hyperthermia	Linh et al. (2018)
MPs_Dex_22/8 (MPs: iron oxide magnetic particles)	Dextran (Dex) and a protein A mimetic ligand (22/8)	IgG purification	Santana et al. (2012)
DIO/^{64}Cu (DIO: dextran-coated iron oxide nanoparticles)	Dextran and ^{64}Cu	Dual-mode MR/PET imaging	Wong et al. (2012)
Water-based FePt	Mercaptoacetic acid and streptavidin-biotin pair	A potential candidate for detection of specific biomolecules	Chiang et al. (2007)
$NiFe_2O_4$@CA $NiFe_2O_4$@PAA	Citric acid (CA) and polyacrylic acid (PAA)	MPI	Irfan et al. (2021)
Fe_3O_4/Au-PEI-HA	Gold, poly polyethyleneimine (PEI) and hyaluronic acid (HA)	MRI and CT	Hu et al. (2015)

(Continued)

TABLE 4.1 (*Continued*)
Example Studies for Functional Magnetic Nanoparticles

Functional Magnetic Nanoparticle*	Functionalization Agent	Application	Reference
$Fe_3O_4@SiO_2@$ alginate/CQDs (CQDs: carbon quantum dots)	Silica, alginate, and CQDs	Drug delivery and a potential candidate for bioimaging	Molaei and Salimi (2022)
$Fe_3O_4@SiO_2$/MIPs (MIP: molecularly imprinted polymer)	Silica and 4-formylphenylboronic acid (FPBA)	Glycoprotein extraction	Guo et al. (2022)
$Fe_3O_4@SiO_2$/PVP	Silica and PVP	Drug delivery	Ehteshamzadeh et al. (2021)
$Fe_3O_4@SiO_2$-DFFPBA	Silica and 2,3-difluoro-4-formyl phenylboronic acid (DFFPBA)	Recognition and isolation of Glycoproteins	Wang et al. (2021)
Fe_3O_4/TEOS/TMSPM/ GMA/Carborane	Tetraethoxysilane (TEOS), 3-(trimethoxysilyl) propyl methacrylate (TMSPM), glycidyl methacrylate (GMA), and carborane	A potential candidate for BNCT	Korolkov et al. (2020)
Fe_3O_4/SiO$_2$/NH$_2$/CHO	Silica, 3-aminopropyltriethoxysilane for NH$_2$, and glutaraldehyde for CHO.	Immobilization of Protein A for disease diagnosis	Thanh et al. (2019)
Fluorescent PEGylated $(Mn_xFe_{1-x})Fe_2O_4@SiO_2$	Silica, fluorescein isothiocyanate, methoxy polyethylene glycol (PEG)	Fluorescence imaging, MRI, and hyperthermia	Sheng et al. (2018)
$Fe_3O_4@SiO_2@$ILs	Silica and imidazolium ionic liquid (IL)	Heme proteins isolation and purification	Liu et al. (2014)
$Fe_3O_4@$silica@chitosan	Silica and chitosan	DNA isolation	Tiwari et al. (2015)
MGCE/Fe_3O_4/Fe_2O_3@Au-DNA/BSA	Magnetic glassy carbon electrode (MGCE), gold, DNA aptamer, and bovine serum albumin (BSA)	VEGF$_{165}$ protein detection	Zhang et al. (2022)
Fe_3O_4-Au	Gold	MRI and CT	Keshtkar et al. (2020)
Fe_3O_4-Au	Gold	MRI and hyperthermia	Efremova et al. (2018)
$CoFe_2O_4@$Ag-HB5	Silver and HB5 aptamer	Cancer cell separation and detection	Vajhadin et al. (2022)

(Continued)

TABLE 4.1 (*Continued*)
Example Studies for Functional Magnetic Nanoparticles

Functional Magnetic Nanoparticle[*]	Functionalization Agent	Application	Reference
MNPs-Herceptin and GQNPs-Herceptin (MNPs: magnetic nanoparticles) (GQNPs: graphene quantum nanoparticles)	Herceptin	Cancer cell separation and detection	Digehsaraei et al. (2021)
Ab-IONPs (IONPs: iron oxide nanoparticles)	Anti-HER2 antibody (Ab)	Cancer cell separation	Saei et al. (2020)
Anti-D-dimer antibody-PSS-MA-GoldMag (GoldMag: Fe_3O_4 and Au nanoparticles)	Poly(4-styrenesulfonic acid-co-maleic acid) (PSS-MA) and anti-D-dimer antibody	D-dimer detection	Zhang et al. (2017)
Anti-CD3 mAb conjugated Fe_3O_4@Au	Gold and Anti-CD3 monoclonal antibody (mAb)	Cell separation	Cui et al. (2011)
RITC labeled MNPs (R-MNPs) (MNPs: Magnetic nanoparticles)	Rhodamine isothiocyanate (RITC)	Cell labeling	Perlstein et al. (2010)

[*]indicated as in the cited articles.

The surfaces of nanoparticles synthesized in an organic solvent environment are generally hydrophilic, which poses a significant obstacle to biomedical applications. In this respect, their surfaces should be modified with hydrophilic ligands for use in various biomedical applications (Huang et al. 2022). PEG is a synthetic, hydrophilic, and biocompatible polymer that is most widely used for surface modification (Icten 2021b). For instance, Bloemen et al. functionalized the superparamagnetic iron oxide nanoparticles with various trialkoxy silanes, which have PEG chains, carboxylic acid, amine, and thiol groups, to increase their colloidal stability and enable further modifications. Amine and thiol groups can act as anchor points for further functionalization with gold nanoparticles; PEG chains allow nanoparticles to be dispersed in aqueous media, and modifications with carboxylic acid and amine groups can be used for the bioconjugation of proteins. In this study, oleic acid-coated iron oxide nanoparticles with an average size of 9.3 nm were modified with these groups under an ultrasonic bath in a shorter time. The colloidal stability of functionalized magnetic nanoparticles in different environments such as various pHs, human plasma, and serum was investigated. The results obtained from this study revealed that especially PEG and carboxylic acid-modified iron oxide nanoparticles might hold promise for future in vivo experiments (Bloemen et al. 2012).

Another widely used coating agent polymer is PVP, which possesses biodegradable and nontoxic features. As an example study, Ehteshamzadeh et al. developed PVP-capped magnetite nanoparticles for doxorubicin (DOX) drug delivery. In this study, iron oxide nanoparticles with a 35 nm average size were first coated with tetraethyl orthosilicate (TEOS) and then with ((3-aminopropyl)triethoxysilane) APTES after they were synthesized by the coprecipitation method. Magnetic nanoparticles modified with silica were also coated with PVP to load the DOX and control release. The results of the analysis showed that the drug release increased in an acidic environment which was strong enough to break the bond between the DOX and PVP (Ehteshamzadeh et al. 2021). Another study in which PVP modification was performed was carried out by Li et al. In this study, oleic acid-coated iron oxide nanoparticles (γ-Fe_2O_3) were prepared by thermal decomposition. However, since these prepared nanoparticles were not soluble in aqueous media, they were coated with PVP to provide hydrophilic properties. PVP-coated γ-Fe_2O_3 nanoparticles exhibited the saturation magnetization of 57.7 emu/g at 295 K, which can be sufficient to provide a T2-weighted image on MRI (Li et al. 2015).

PDA, which has catechol and amine groups, is another exciting polymer in biomedical applications because it is easily coated, biocompatible, hydrophilic, and allows further modification. In addition, PDA molecules coated on the surface of nanoparticles also provide a reduction of some metals such as silver and gold to the surface. (Icten 2021a). Icten et al. coated multi-core manganese ferrite nanoparticles ($MnFe_2O_4$) with PDA to reduce gold nanoparticles and bind boron atoms to the surfaces of the nanoparticles, as shown in Figure 4.2. $MnFe_2O_4$

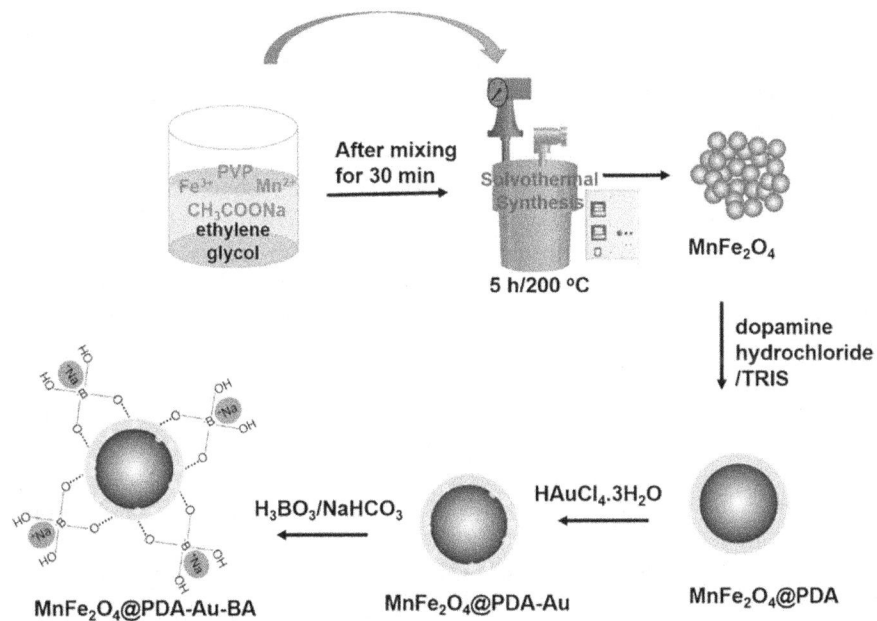

FIGURE 4.2 The preparation routes of $MnFe_2O_4$@PDA-Au-BA nanoparticles. Adapted from (Icten et al. 2022), ACS.

nanoparticles prepared by the solvothermal method possessed sizes of 10–15 nm, and multi-core nanoparticles with a total size of 100–200 nm were coated with PDA (thickness of 30–40 nm). After spherical gold nanoparticles with 20–30 nm dimensions were reduced on the surface of the PDA-coated $MnFe_2O_4$ nanoparticles, boric acid was attached via catechol groups of PDA. The final sample ($MnFe_2O_4$@PDA-Au-BA) indicated no significant cell death and may be a potential candidate for various biomedical applications, MRI, NCT, and photothermal therapy (Icten et al. 2022).

Unlike the iron oxide nanoparticles, Chiang et al. prepared superparamagnetic FePt nanoparticles with 2–3 nm size by chemical reduction method for biomedical applications such as detecting specific biomolecules. In this study, FePt nanoparticles were modified with mercaptoacetic acid to provide them dispersed in an aqueous environment. Further functionalization of mercaptoacetic acid-modified FePt nanoparticles was achieved with streptavidin-biotin pair, which is a biomolecule binding tool (Chiang et al. 2007).

Chitosan, a natural polysaccharide, is a pH-sensitive polymer and occurs of β-(1–4)-2-acetamido-d-glucose and β-(1–4)-2-amino-d-glucose units. Due to its low cost, controlled release of active drugs, biocompatibility, biodegradability, and nontoxicity, chitosan is utilized as a coating material in biomedical applications such as MRI and targeted drug delivery (Assa et al. 2017). The study related to developing biofunctionalized chitosan/Fe_3O_4 magnetic nanoparticles for MRI application performed by Song et al. can be an example of chitosan modification for magnetic nanoparticles. In this study, chitosan (CS) was firstly modified with PEG and lactobionic acid (LA), then Fe_3O_4 nanoparticles with approximately 14.1 nm in size, prepared by coprecipitation method, were encapsulated with PEG/LA-CS through self-assembly. These modifications for magnetic nanoparticles can contribute to improving their circulation time, water solubility, biocompatibility, and targeting of hepatocytes (via lactobionic acid) (Song et al. 2015).

Silica coating is widely preferred for modifying magnetic nanoparticles to be used in the biomedical field, as it provides biocompatibility and further functionalization (Kim et al. 2018). For instance, the combined modifications were performed on iron oxide nanoparticles before carborane immobilization for BNCT applications in a study performed by Korolko et al., as seen in Figure 4.3. Carboranes are boron-rich compounds used as agents in BNCT, but carriers such as iron oxide nanoparticles are needed to deliver an effective dose of boron isotopes to the tumor site since carboranes cannot be targeted directly to tumor sites. Therefore, some surface functionalizations are required beforehand as carboranes cannot be directly immobilized to the surfaces of iron oxide nanoparticles. Firstly, the iron oxide nanoparticles were coated with silica and 3-(trimethoxysilyl) propyl methacrylate (TMSPM) to induce C=C double bond formation, and then biocompatible glycidyl methacrylate (GMA) was attached to the C=C double bonds on the surface of iron oxide nanoparticles via graft polymerization. Lastly, Isopropyl-o-carborane was immobilized on the surface of the final functional nanoparticle through covalent bonding via the catalytic azide-alkyne cycloaddition (Korolkov et al. 2020).

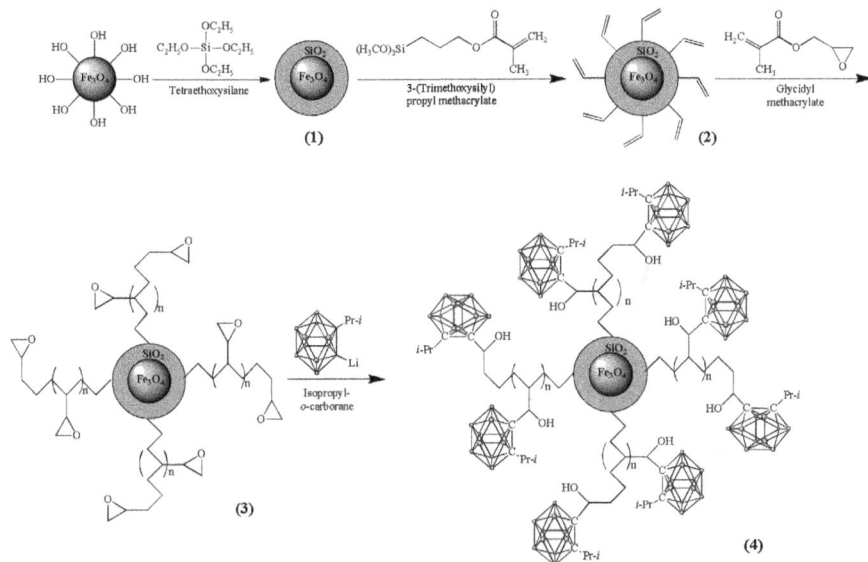

FIGURE 4.3 The synthesis scheme of carborane immobilized iron oxide nanoparticles for potential application of BNCT. Adapted with permission from (Korolkov et al. 2020), Copyright 2020, Elsevier.

Guo et al. developed a new imprinted material (Fe$_3$O$_4$@SiO$_2$/MIPs) for selective glycoprotein detection. Glycoproteins play an essential role in the determination of some diseases. Detection of low glycoprotein levels is challenging due to the complex sample structure and the presence of trace amounts in biological samples. However, this study indicated that the new imprinted material formed by the combination of molecularly imprinted polymer (MIP) and iron oxide nanoparticles had a high binding capacity and fast mass transfer rate toward the sample glycoprotein. As seen in this study, bare iron oxide nanorods could not be employed directly due to the formation of aggregates related to the high magnetization and the absence of active groups on the surface. Instead, the nanorods were coated with silica shells to improve their biocompatibility and hydrophilicity (Guo et al. 2022).

The surfaces of magnetic nanoparticles can also be modified with organic or inorganic dyes to visualize cells. In the sample study, Perlstein et al. prepared fluorescent-labeled magnetic nanoparticles by a novel method based on nucleation and controlled the growth of iron oxide layers on the iron oxide/gelatin-dye structure. Rhodamine isothiocyanate (RITC) as a fluorescent dye was covalently bound to gelatin via thiourea bonds, and then maghemite-type iron oxide (γ-Fe$_2$O$_3$) thin films were grown gelatin-RITC/iron oxide nuclei to obtain fluorescent magnetic nanoparticles. The reason for the covalent bonding of the RITC was to improve the probe property while maintaining the size and magnetic properties of nanoparticles. In addition, the leaving of the RITC from the structure could be prevented by covalent bonding. As a result, the fluorescent-labeled magnetic nanoparticles may be effective for labeling and imaging in biological applications (Perlstein et al. 2010).

4.3 BIOMEDICAL APPLICATIONS OF FUNCTIONALIZED MAGNETIC NANOPARTICLES

4.3.1 TREATMENT APPLICATIONS

One of the most effective methods to destroy cancerous cells is chemotherapy, in which anticancer drugs are applied to patients. Although these anticancer drugs are not selective for cancer cells and damage normal cells, they are still developed to kill cancerous cells. The other major problems with anticancer drugs are that they have low solubility in the biological environment, cannot be transported to the desired area, and have low treatment efficacy (Materón et al. 2021; Talegaonkar and Bhattacharyya 2019; Tran et al. 2020). However, such barriers can be reduced by using magnetic nanocarriers that enable the delivery of anticancer drugs to the desired region. For example, Fe_3O_4@SiO_2@alginate/carbon quantum dots (CQDs) nanohybrid material was synthesized in a newly published study by Molaei and Salimi for anticancer drug delivery. In order to ensure colloidal stability, the magnetic nanoparticles were modified with alginate, an anionic hydrophilic copolymer. In addition, alginate can be easily eliminated from the body due to its biodegradability after drug release. Finally, the hybrid material possessed the fluorescent emission ability by attaching CQDs to the surface of alginate-modified magnetic nanoparticles. In this way, the nanohybrid material can be used simultaneously in drug delivery and bioimaging applications. The drug loading efficiency of Fe_3O_4@SiO_2@alginate/CQDs was 81% and showed a higher drug release rate (38%) at pH 5.5 compared to pH 7.4 (25%) (Molaei and Salimi 2022). In another study performed by Avedian et al., a different anticancer drug, Erlotinib, was loaded into the magnetic carrier. After Fe_3O_4 nanoparticles were coated with mesoporous silica, Erlotinib was loaded, and folic acid conjugated poly (ethyleneimine) (PEI-FA) was modified to the functional nanoparticles. Similar to the previous study, Erlotinib release was higher at a pH of 5.5 (63%) than at a pH of 7.4 (33%). While the unloaded functional nanoparticles did not show significant toxic effects, Erlotinib-loaded magnetic nanoparticles inhibited the proliferation of HeLa cells (Avedian et al. 2018). Another study in which functional magnetic nanoparticles were used as drug carriers was carried out by Dhavale et al. In this study, Fe_3O_4 nanoparticles were synthesized with a coprecipitation method and directly modified with APTES to provide amino groups on the surface. Telmisartan, as a model anticancer drug, was loaded on the APTES-modified Fe_3O_4 nanoparticles. The loading of the drug into the functional magnetic nanoparticles was provided by the amide bond formed between the amino group of APTES and the carboxyl group of Telmisartan. The maximum drug loading ratio (85.6%) was obtained in the loading experiments possessing the APTES-modified Fe_3O_4 /drug ratio of 15/1. While the amide bond was stable at physiological pH, it degraded in acidic environments to allow controlled drug release. Telmisartan-loaded and APTES-modified Fe_3O_4 nanoparticles showed a dose-dependent cytotoxic effect on the PC-3 human prostate cancer cell line (Dhavale et al. 2018).

Hyperthermia is another cancer treatment method that utilizes magnetic nanoparticles and has recently attracted attention because it has fewer side effects than chemotherapy and radiotherapy. In this treatment method, magnetic nanoparticles are localized to the tumor regions, and these regions are heated with the externally applied magnetic field to magnetic nanoparticles. The death of cancer cells occurs when the temperature rises above 42°C–43°C depending on the heat generated. The heating efficiency can be defined by using the thermal dose and the specific absorption rate (SAR). In this respect, developing magnetic nanoparticles with low thermal dose and high SAR value is crucial for hyperthermia applications (Fatima et al. 2021; Sanad et al. 2021). As seen in the equation "SAR = absorbed power/mass of magnetic nanoparticles", the SAR value is defined as the efficiency of magnetic nanoparticles in converting magnetic energy into heat (Huang et al. 2012). In clinical applications, the applied magnetic field frequency and the optimum applied AC magnetic field are two of the most critical parameters affecting heating efficiency. However, increasing frequency and high field raise the application cost, so applying these values at optimum levels would be beneficial. In a study performed by Hergt and Dutz (2007), it was stated that the frequency and field amplitude values of 500 kHz and 10 kA m^{-1}, respectively, were sufficient for hyperthermia applications.

The study published by Daboin et al. can be presented as an exciting study for hyperthermia application. Manganese-cobalt ferrite nanoparticles ($Mn_{1-x}Co_xFe_2O_4$) were coated with SiO_2 and decorated with $Au@Fe_3O_4$ nanoparticles, as shown in Figure 4.4a. The magnetic properties of the nanocomposite ($Au@Fe_3O_4$-SiO_2-$Mn_{1-x}Co_xFe_2O_4$) were investigated using a vibrating sample magnetometer depending on the Mn^{2+} content. It was observed that the saturation magnetization values of nanocomposites decreased with the increase of Mn^{2+} contents due to the decrease in the numbers of Mn^{2+} ions in the octahedral site (Figure 4.4b). The temperature change curves in the water and SAR values of $Au@Fe_3O_4$ decorated and undecorated SiO_2-$Mn_{1-x}Co_xFe_2O_4$ nanocomposites are shown in Figure 4.4c–e. Hyperthermia experiments were performed at an alternating magnetic field of 5.5 mT (frequency of 454 20±kHz). When the temperature rise and SAR values of decorated and undecorated nanocomposites were examined, it was seen that the decorated nanocomposites possessed higher values. The high temperature and SAR values of decorated nanocomposites could be due to the contribution of $Au@Fe_3O_4$ nanoparticles to the total magnetization because SAR values are directly proportional to the nanoparticle's effective diameter and magnetization (Daboin et al. 2019). In addition to these studies, various functional magnetic nanoparticles such as $Fe_3O_4@YPO_4$:5Eu (Prasad et al. 2013), $Zn_xCo_{1-x}Fe_2O_4@MnFe_2O_4$ (Hammad et al. 2016), $Mn_{0.25}Fe_{2.75}O_4$-PEG (Saputra et al. 2019), $NaYF_4$:Er^{3+}/$Yb^{3+}@SiO_2@AuNP@Fe_3O_4$ (Soni et al. 2019), Fe_3O_4/RGO/PEG (RGO: reduced graphene oxide) (Alkhayal et al. 2021), and $Fe_3O_4@ZnO$ (Gupta et al. 2021) have been tried for hyperthermia applications in recent years.

Another method used to treat various tumors, such as brain, head, and neck, is neutron capture therapy (NCT). ^{10}B and ^{157}Gd isotopes are commonly used agents for NCT, as they possess high cross-section values, 3840 and 254,000 barn (barn: 10^{-24} cm^2), respectively (Icten 2021b). The treatment method using the ^{10}B isotope is entitled boron neutron capture therapy (BNCT), while the method using the ^{157}Gd isotope is called gadolinium neutron capture therapy (GdNCT). BNCT is a method based on the production of high-energy particles such as $^4He^{2+}$ (α-particle) and a$^7Li^{3+}$ ion formed as a result of the nuclear reaction between the ^{10}B isotope and thermal neutrons. As shown in the reaction (4.1), the energy particles formed by the nuclear

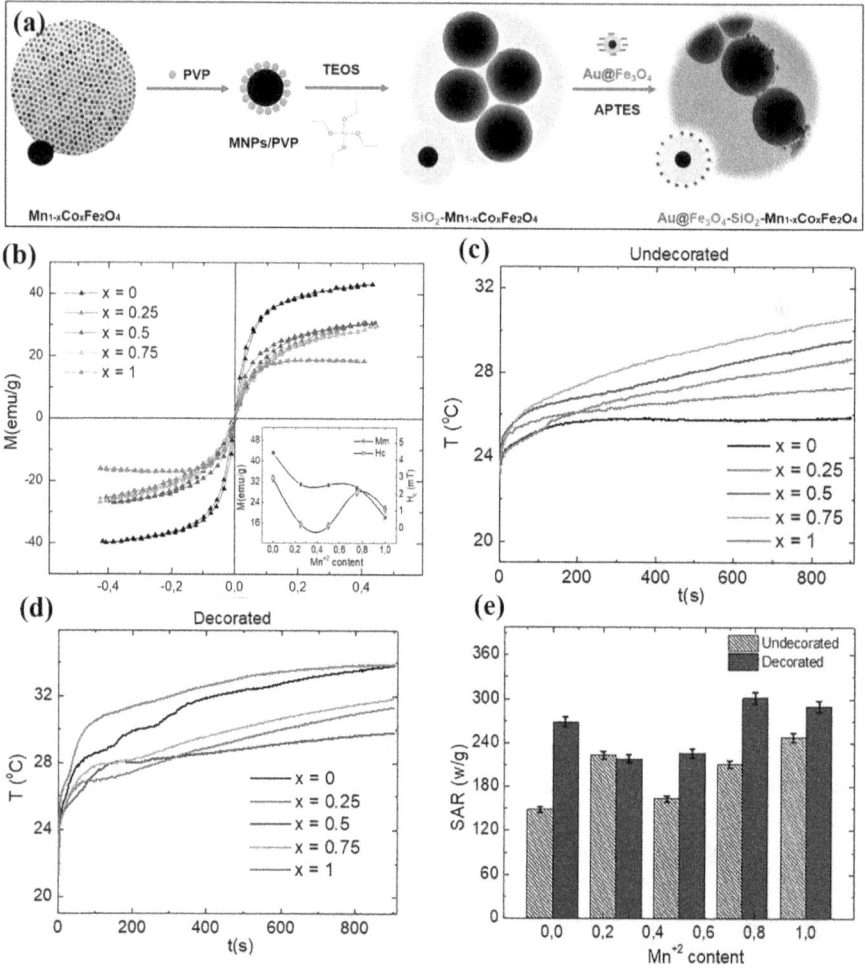

FIGURE 4.4 Schematic description of the synthesis of $Au@Fe_3O_4$-SiO_2-$Mn_{1-x}Co_xFe_2O_4$ nanocomposite (a), room-temperature magnetization curves of the nanocomposite according to Mn^{2+} contents (b), temperature change curves (c, d) and SAR values (d) of the undecorated and decorated nanocomposite. Adapted with permission from (Daboin et al. 2019), Copyright (2019), Elsevier.

reaction cause the destruction of cancerous cells because the particles transfer all energy to cancer cells, and their average path is smaller than the radius of a cell (4–9 μm). Therefore, they only destroy cancerous cells in the region where they are located without harming normal tissues (Icten 2021c). In the GdNCT method (reaction 4.2), after the ^{157}Gd isotope is irradiated with thermal neutrons, internal conversion, and Auger–Coster–Kronig (ACK) electrons are formed together with X-ray and photon emissions. ACK electrons have the highest linear energy transfer (LET) value (0.3 MeV/μm; LET: 0.2 MeV/μm for BNCT) and are the most effective in destroying cancer cells (De Stasio et al. 2005).

$$^{10}B + n_{th}\left(0,025\,eV\right) \rightarrow {}^{11}B^{*} \rightarrow {}^{4}He^{+2}$$
$$\left(1.78\,MeV\right) + {}^{7}Li^{+3}\left(1.01\,MeV\right) + 2.79\,MeV\left(\%6\right)$$
$$\rightarrow {}^{4}He^{+2}\left(1.47\,MeV\right) + {}^{7}Li^{+3}\left(0.84\,MeV\right) +$$
$$2.31\,MeV\left(\%94\right)$$

(4.1)

$$^{157}Gd + n_{thermal} \rightarrow \left[{}^{158}Gd\right] \rightarrow {}^{158}Gd + \gamma + 7.94\,MeV \rightarrow *$$

Internal conversion electrons *

(4.2)

Auger – Coster – Kronig (ACK) electrons

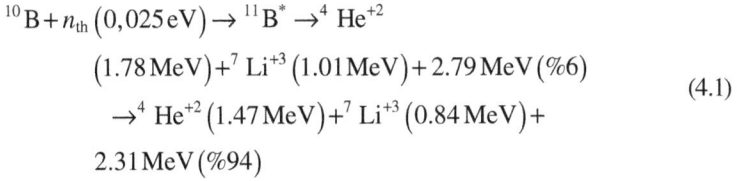

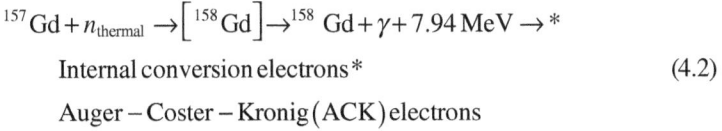

One of the essential requirements for both BNCT and GdNCT methods is to accumu-late a sufficient number of isotopes in cancer cells (25–35 $\mu g^{10}B$/g tumor for BNCT; 50–200 $\mu g^{157}Gd$/g tumor for GdNCT). The combination of NCT agents with mag-netic nanoparticles ensures sufficient accumulation of isotopes in cancerous cells by magnetic targeting (Hosmane 2011). Therefore, there is increasing interest in NCT agents containing magnetic nanoparticles (Icten et al. 2022).

Dukenbayev et al. synthesized the Fe_3O_4-carborane nanoparticles for BNCT application. After preparing Fe_3O_4 nanoparticles with a size of about 18.9 nm by the coprecipitation method, their surfaces were modified with an amino group using (3-aminopropyl)-trimethoxysilane (APTMS). The carborane borate containing 21 boron atoms was attached to the surface of nanoparticles through the ionic interaction between the amino group and borate. Incubation of the mouse embryonic fibroblasts (MEFs) cells with Fe_3O_4-carborane nanoparticles for 24 hours caused no toxic effect up to the concentration of 0.015 mg/mL. Additionally, Fe_3O_4 nanoparticles exhibited an IC_{50} value of 0.110 mg/mL while Fe_3O_4-carborane nanoparticles showed that of 0.405 mg/mL. It was concluded that the nanoparticles developed in this study have the potential to be used in BNCT (Dukenbayev et al. 2019). Unlike the iron oxide nanoparticles, Torresan et al. developed biocompatible Fe-B nanoparticles for MRI-guided BNCT applications. In this study, Fe-B nanoparticles were synthesized with the laser ablation in liquid (LAL) method, which is a straightforward, low-cost approach to preparing bimetallic nanoparticles. Additionally, Fe-B nanoparticles were coated with PVP, which is a biocompatible polymer, while the process was carried out in the presence of PVP solution in acetone. The final biocompatible Fe-B nanoparticles had a 1.0:0.4 atomic ratio for Fe/B, an average size of 26 nm, and a saturation magnetiza-tion of 70 emu/g Fe. Fe-B nanoparticles showed a nontoxic effect in the concentration range of 1–100 $\mu g\,mL^{-1}$ and degradation in the lysosomal environment, which con-tributes to their clearance via the liver-spleen-kidney pathway (Torresan et al. 2021).

Compared to BNCT applications, a few studies have been related to functional magnetic nanoparticles containing gadolinium for GdNCT applications (Eguía-Eguía et al. 2021; Gao et al. 2013; Pylypchuk et al. 2016; Zibert et al. 2022). One of the magnetic nanoparticles possessing the potential use for GdNCT applications is the $GdFeO_3$/Fe_3O_4/SiO_2 nanocomposite prepared by Gao et al. After $GdFeO_3$ nanoparticles with a size of 60 nm were synthesized by gel combustion method,

Fe_3O_4 was deposited on it and then coated with silica. The room-temperature magnetization of $GdFeO_3/Fe_3O_4/SiO_2$ nanocomposite was 48.7 emu/g, and results indicated that the developed nanocomposite could be helpful for GdNCT applications (Gao et al. 2013). In a different study, Pylypchuk et al. tried the absorption of Gd-DTPA (Gd-diethylenetriaminepentaacetic) into the chitosan-magnetic nanocomposite prepared by the coprecipitation method. According to the thermal analysis results, it was determined that there was 4.6% chitosan by mass in the chitosan-magnetic nanocomposite. Additionally, despite the low chitosan content, an absorption rate of 0.06 mmol Gd-DTPA per gram of composite was achieved, resulting that it may be a potential candidate for GdNCT application (Pylypchuk et al. 2016).

In order to increase the effectiveness of the treatment, various studies have also been carried out on the development of functional magnetic nanocarriers with both [10]B and [157]Gd isotopes (Icten et al. 2018; Korolkov et al. 2021). For instance, Korolkov et al. reported the new NCT agent formed by gadolinium and carborane modification on the superparamagnetic iron oxide nanoparticles. After the iron oxide nanoparticles were synthesized by the coprecipitation method, the same modifications such as polyelectrolyte poly(acrylic acid) (PAA)/poly(allylamine) (PALAm) formation and immobilization of gadolinium and carborane were carried out on the surface, respectively. Boron and gadolinium amounts in the modified sample were calculated as 0.077 and 0.632 mg/g, and the coercivity and room-temperature magnetization values were determined as 8.1 and 43.1 emu/g, respectively. In addition, the toxic effects of the final sample on different cancer cell lines such as BxPC-3, PC-3 MCF-7, HepG2, and L929 and normal cells such as human skin fibroblasts were investigated with different concentrations. Gadolinium-immobilized nanoparticles were found to have more toxic effects than non-gadolinium-immobilized nanoparticles. Moreover, it was observed that the cell line, which was most sensitive to toxicity, was BxPC-3, and the most likely mechanism for the toxic effect could be oxidative stress. The result obtained from this study is that carborane and gadolinium-modified Fe_3O_4 nanoparticles could be appropriate candidates for NCT (Korolkov et al. 2021). In another study by Icten et al., citric acid or folic acid-modified $GdBO_3$-Fe_3O_4 nanocomposites were prepared, and cell experiments were carried out against different cell lines such as Mia-Pa-Ca-2, HeLa, and A549. The [10]B and [157]Gd contents in the prepared nanocomposite were approximately 10^{14} atom/μg samples, making these nanocomposites suitable candidates for BNCT, GdNCT, and even GdBNCT (Icten et al. 2018).

4.3.2 IMAGING APPLICATIONS

MRI is a noninvasive imaging technique widely used in the medical field and offers advantages such as superior anatomical detail and soft-tissue image, high spatial resolution, and usage of nonionizing radiation. These advantages of this technique not only allow for the identification of diseases and lesions that are difficult to define but also provide opportunities to evaluate some treatment outcomes. However, it is necessary to use contrast agents that increase the image contrast in order to determine the differences between tissues in MRI. Gadolinium (III) chelate structures are primarily used as contrast agents in clinical applications, but there are concerns over toxicity problems of these structures (Bañobre-López et al. 2017; Peixoto et al. 2020). In this respect, magnetic nanoparticles, especially superparamagnetic iron oxide nanoparticles, which are utilized as T2 contrast agents, are considered an alternative to gadolinium (III) chelate structures owing to their biocompatibility, easy surface modification, and high magnetic moment (Bañobre-López et al. 2017). Superparamagnetic iron oxide nanoparticles also contribute to tumor imaging, metastatic breast cancer imaging, imaging of lymph nodes, detection of stem cells, and Alzheimer's disease (Yildiz 2016).

Demin et al. prepared N-(phosphonomethyl)-iminodiacetic acid (PMIDA) modified Fe_3O_4 nanoparticles (MNPs-PMIDA) for MRI application. Compared to PEG and bovine serum albumin (BSA), the best stabilizer, PMIDA, was chosen to stabilize the iron oxide nanoparticles. As it is known, phosphonic derivatives can interact strongly with the surfaces of iron oxide nanoparticles by forming P-O-Fe bonds. While the phosphate groups in the PMIDA structure bind to the surfaces of the iron oxide nanoparticles, the carboxyl groups provide colloidal stability. Optimum conditions for modifying Fe_3O_4 nanoparticles were determined as a reaction temperature of 40°C, a reaction time of 3.5 hours, and the equimolar amount of PMIDA. The MNPs-PMIDA nanoparticles exhibited good colloidal stability and magnetic features for biological applications. The relaxivity values of r_2 and r_1 for MNPs-PMIDA nanoparticles were calculated as 341 and 102 mmol^{-1} s^{-1}, respectively, which were higher than some commercial samples such as Resovist (r_2 151.0 and r_1 25.4 mmol^{-1} s^{-1}), Feridex (r_2 98.3 and r_1 23.9 mmol^{-1} s^{-1}) (Demin et al. 2018). In another example study, Sheng et al. developed a multi-functional magnetic nanoplatform that allows fluorescence imaging and MRI with magnetic hyperthermia. The magnetic $(Mn_xFe_{1-x})Fe_2O_4$ nanoparticles were prepared with the thermal composition method and then coated with a SiO_2 shell incorporated with fluorescein isothiocyanate. Finally, the magnetic nanoparticles were conjugated with methoxy poly(ethylene glycol) (mPEG) to provide high colloidal stability. $(Mn_xFe_{1-x})Fe_2O_4$ nanoparticles with dimensions of 10 and 14 nm had saturation magnetization of 66 and 80 emu/g, respectively, which were higher than magnetite nanoparticles. After all the surface modifications, the saturation magnetization values decreased due to nonmagnetic coatings, as expected. In this study, the PEGylated $(Mn_xFe_{1-x})Fe_2O_4@SiO_2$ and bare $(Mn_xFe_{1-x})Fe_2O_4@SiO_2$ nanoparticles were tested for MRI and hyperthermia applications. $(Mn_xFe_{1-x})Fe_2O_4@SiO_2$ nanoparticles possessed the r_2 relaxivity value of 175 mM [Fe+Mn]$^{-1}$ s^{-1}, while PEGylated $(Mn_xFe_{1-x})Fe_2O_4@SiO_2$ showed that of 150.3 mM [Fe+Mn]$^{-1}$ s^{-1}, which were close to the value of commercial MRI agents. Similarly, the SAR values of bare $(Mn_xFe_{1-x})Fe_2O_4@SiO_2$ and PEGylated $(Mn_xFe_{1-x})Fe_2O_4@SiO_2$ nanoparticles were calculated as 1635 and 1346 Wg^{-1} (at 41.98 KA m^{-1} alternating magnetic field), respectively. PEGylated $(Mn_xFe_{1-x})Fe_2O_4@SiO_2$ nanoparticles did not cause significant cytotoxicity on NIH/3T3 cell lines in the 2 to 200 µg mL^{-1} concentration range. Moreover, the fluorescence signal was observed from NIH/3T3 cell labeled with PEGylated $(Mn_xFe_{1-x})Fe_2O_4@SiO_2$ nanoparticles, which indicated that these nanoparticles pass through the cell membrane. As a result, the obtained results revealed that the developed magnetic nanoplatform could be suitable for multi-purpose biomedical applications (Sheng et al. 2018).

There are also magnetic nanoparticles in clinical trials for MRI applications, which are still being studied. For example, Ferumoxtran-10 is a contrast agent formed by coating superparamagnetic iron oxide nanoparticles (10–20 nm) with dextran T-10. This contrast agent has attracted attention, especially in Europe, and its clinical studies are continuing to improve the images of lymph mode in some cancer types, such as bladder and prostate cancer. Unfortunately, it has not yet been approved by the Food and Drug Administration (FDA) (Huang et al. 2022; Leenders 2003). Another type of magnetic nanoparticle with ongoing clinical trials in the MRI is the siloxane-coated superparamagnetic iron oxide nanoparticle (AMI-121). AMI-121 consists of a bulk structure with a total size of 300 nm, composed of iron oxide nanoparticles with dimensions of approximately 10 nm. The siloxane coating on the iron oxide acts as a protective agent in case of uptake by the gastrointestinal tract (Huang et al. 2022; Min et al. 2015).

The magnetic particle imaging (MPI) technique has received increasing attention in recent years owing to no background signal, high tissue penetration depth, and no need for ionizing radiation (Bulte 2019). Compared to the signal for MRI, the signal of MPI directly depends on the magnetic moment of magnetic nanoparticles, so it is more sensitive. Superparamagnetic iron oxide nanoparticles are utilized as tracer material in the MPI because of their superior magnetic properties and widespread use in clinical applications. The basis of MPI technique is based on the interaction of superparamagnetic nanoparticles with an externally applied magnetic field. When a strong magnetic field is applied to superparamagnetic nanoparticles, they are magnetized and reach the saturation point. There is no linear relationship between the strength of the magnetic field and the magnetization of superparamagnetic nanoparticles, but the mathematical model defines this relationship. The slope of the curve obtained from the relationship describes a narrow magnetic field intensity range, the oscillation of which can cause a sudden and large magnetization shift of superparamagnetic nanoparticles. The MPI signal consists of the shift of magnetization of nanoparticles when exposed to an oscillating magnetic field (Bakenecker et al. 2018; Meola et al. 2019). MPI has been applied to various medical applications such as cancer imaging, cardiovascular imaging, neuroimaging, and cell tracking (Billings et al. 2021). Irfan et al. prepared the citric acid and polyacrylic acid-coated $NiFe_2O_4$ nanoparticles ($NiFe_2O_4$@CA and $NiFe_2O_4$@PAA, respectively) as a potential tracer agent for MPI applications. Hydrodynamic sizes of $NiFe_2O_4$@CA and $NiFe_2O_4$@PAA nanoparticles were calculated as 93 and 70 nm with dynamic light scattering (DLS). The MPI performance of synthesized nanoparticles was examined via magnetic particle spectroscopy (MPS) at 9.9 kHz with a 15 mT excitation field. $NiFe_2O_4$@PAA sample exhibited a relaxation time of 3.10 µs and resolution of 7.75 mT, which were superior to commercial samples such as Vivotrax and Perimag. The developed nanoparticles could have the potential as MPI tracer agents (Irfan et al. 2021).

A different study for MPI tracer was carried out by Horwat et al. In this study, poloxamer Pluronics F127 (PF127) was modified to provide water-dispersibility for magnetic particles prepared via thermal composition. After further modification (PF127DA: acrylate groups modified PF127), the surfaces of the nanoparticles were also functionalized poly(glycidol)s (PG-SH) to improve the performance of MPI. The MPI performance of PF127 and PF127DAPG modified magnetic nanoparticles was examined via MPS. As shown in Figure 4.5a and b, it was observed that the decay of the spectra of higher harmonics decreased with increasing temperature, which could be due to the change in the viscosity of the medium depending on the temperature. Although both measurements included the same amount of magnetic nanoparticles, PF127DAPG modified magnetic nanoparticles were markedly different, with better

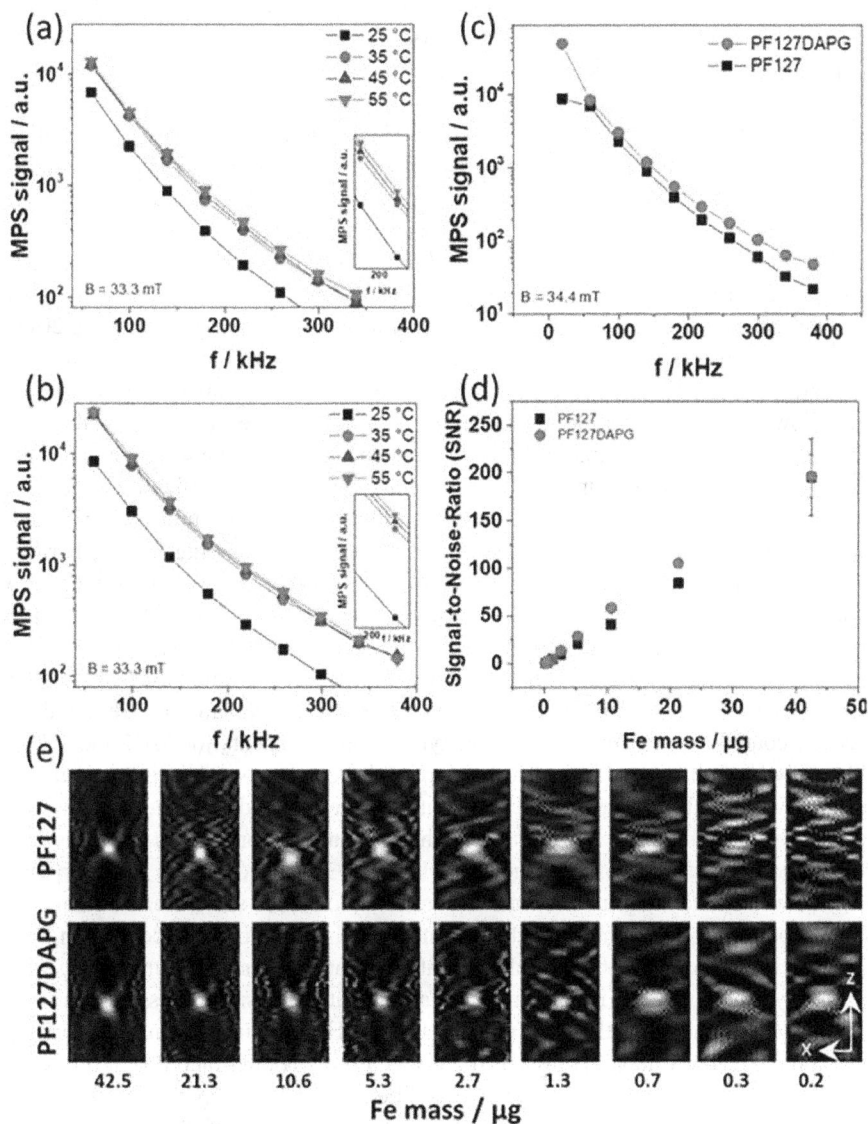

FIGURE 4.5 Temperature-dependent MPS performances of PF127 (a) and PF127DAPG (b) modified magnetic nanoparticles, higher response of samples in MPS measurement at 25°C (c), signal-to-noise ratio (SNR) (d), and MPI images (e) of samples. (Horvat et al. 2020), Wiley-VCH.

performance (Figure 4.5c), and exhibited an excellent signal-to-noise ratio (SNR) (Figure 4.5d). In addition, higher SNR in higher spatial resolution was obtained for PF127DAPG-modified magnetic nanoparticles using 1.3 µg of iron in MPI images. The same ratio was obtained for the PF127 modified sample using 5.3 µg of iron (Figure 4.5e). Magnetic nanoparticles modified with only PF127 showed cytotoxicity because of their high amphiphilic nature, causing the solubilization of the lipids from the cell membrane, whereas when functionalized with biocompatible PG, cytotoxicity decreased. The results obtained from this study indicated that these modified magnetic nanoparticles could be used in the future for diagnostics (Horvat et al. 2020).

The MPI technique can also be used with other imaging techniques. For instance, Lemaster et al. developed 1,1′-dioctadecyl-3,3,3′, 3′-tetramethylindotricarbocyanine iodide (DiR) labeled and a poly(lactic-co-glycolic acid) (PLGA)-based iron oxide nanobubbles for ultrasound imaging, photoacoustic imaging, and MPI. MPI density of functional nanoparticles was 20 times higher in stem cell imaging compared to control experiments. Similarly, ultrasound and photoacoustic imaging displayed a 3.8- and 10.2-fold increase in imaging intensity after injection (Lemaster et al. 2018).

Another imaging technique commonly used in the medical field is computed tomography (CT), which reflects anatomy, tissues, and functions in high resolution and allows the reconstruction of images. The basic principle of CT is based on the differences in the attenuation of the X-rays sent depending on the weight and density of the tissues. However, some low-density and soft tissues are difficult to visualize, and contrast agents possessing high atomic numbers are used to improve the tissues' contrast. The most common CT agents are compounds including iodine, barium, gold, bismuth, ytterbium, and gadolinium (Molkenova et al. 2022). The studies show that magnetic nanoparticles are mainly used for combining CT with other imaging techniques such as MRI and positron emission tomography (PET) or treatment methods such as drug delivery and hyperthermia rather than alone CT application (Chouhan et al. 2021; Thomas et al. 2013). For example, Naha et al. presented the dextran-coated bismuth-iron oxide nanohybrid contrast agents for MRI and CT. In this study, dextran-coated iron oxide nanoparticles containing changing amounts of bismuth were synthesized via the coprecipitation method and characterized by various analytical techniques. It was observed that the magnetic properties and transverse relaxivities (T_2) decreased the amount of bismuth in the iron oxide nanoparticles, which could be the integration of bismuth into the iron oxide structure. CT images of agents containing different amounts of bismuth at the same iron concentration (9.37 mg ml^{-1}) were taken at 140 kV, and X-ray attenuation was also measured using various X-ray tube voltages. Iron oxide formulations without bismuth possessed a weak X-ray attenuation; this value was only around 50 HU (Hounsfield unit) depending on applying X-ray voltage. However, there was a proportional increase in the X-ray attenuation with the increase in the amount of bismuth. The strongest attenuation was obtained in the Bi-30 sample using an 80 X-ray tube voltage of 80 kV. As a result, dextran-coated bismuth-iron oxide nanoparticles could be used as T2-weighted MR and CT agents due to their properties (Naha et al. 2014).

Yang et al. developed a single multi-functional nanocomposite for multimodal imaging and therapy applications. WS$_2$ nanosheets were modified with iron oxide nanoparticles (WS_2-OI) coated with mesoporous SiO_2 and PEG; then DOX as an anti-cancer drug was loaded into the composite structure to prepare the multi-functional nanocomposite (WS_2-IO@MS-PEG/DOX). In this nanocomposite, WS_2 nanosheets have NIR and X-ray absorbance properties so that they can be used in photothermal therapy and CT imaging applications. Additionally, iron oxide nanoparticles are T2 contrast agents in MRI, and mesoporous SiO_2 provides the space for drug transport. HU values of WS_2-IO@MS-PEG/DOX solutions increased with increasing concentrations for CT images, and the slope obtained from the HU values versus concentrations was calculated as 31.9 Lg^{-1}. Moreover, In vivo CT imaging was performed using 4T1 tumor-bearing Balb/c mice, and a noticeable contrast difference was seen similar to other imaging techniques. WS_2-IO@MS-PEG/DOX nanocomposite showed a synergistic effect by combining imaging and treatment methods on a single platform when other detailed results were examined (Yang et al. 2015).

4.3.3　SEPARATION AND DETECTION APPLICATIONS

The most significant challenges in medicine are the extraordinarily rapid and accurate separation and determination of nucleic acids, proteins, and cells in biological samples. In particular, various materials have been developed that can separate and detect biomolecules and cells accurately and precisely. These materials make detecting diseases early or providing valuable information about biological properties possible. Techniques using magnetic nanomaterials offer many advantages compared to other techniques (Haun et al. 2010). Magnetic isolation is of great interest in biomedical studies because it allows selective and sensitive capture of target molecules. Compared to traditional chromatographic methods, magnetic separation, purification, or detection method provides faster and more efficient separation of target molecules by the external magnetic field and affinity interaction (Yildiz 2016).

Nucleic acids are one of life's fundamental substances and have essential biological functions. They store, copy and transmit genetic information, so separation and detection of nucleic acids play a significant role in diagnosing and treating various diseases. Typically, the separation of nucleic acid from biosamples requires complex and multi-step procedures (Garner 2000). For example, the polymerase chain reaction (PCR) method is generally used for nucleic acid detection, which is time-consuming and expensive. In this respect, it is thought that it would be appropriate to use magnetic nanoparticles for the fast and economic separation and detection of nucleic acids (Tang et al. 2020). For instance, Zandieh et al. designed the nucleic acids (spherical "SNA" and linear "SH-DNA" functionalized Fe_3O_4@PDA nanoparticles for use in DNA extraction and detection. DNA extraction with "SNA" functionalized Fe_3O_4@PDA (Fe_3O_4@PDA@SNA) was found to be five times faster than with "SH-DNA" functionalized Fe_3O_4@PDA (Fe_3O_4@PDA@SH-DNA) because of the more appropriate conformation of DNA strands in SNA. About 80% of the isolated DNA with Fe_3O_4@PDA@SNA probe could be recovered at weak ionic strength. The same probe was employed as DNA sensing after incubation with 12-mer FAM-cDNA (reporter DNA) to form a nanoflare probe. The limit of detection (LOD) values of Fe_3O_4@PDA@SNA and Fe_3O_4@PDA@SH-DNA probes were determined to be 4.4 and 10.7 nM, respectively, and the probe signal of Fe_3O_4@PDA@SNA was stronger at the same target concentration. The authors consider that SNA functionalization may be an essential method in bioanalytical chemistry because of the positive results obtained in the study (Zandieh and Liu 2021). Tiwari et al. synthesized Fe_3O_4@silica@chitosan nanoparticles for magnetic separation of DNA. Various experimental parameters such as pH, temperature, and time were optimized for the adsorption of DNA on functional magnetic nanoparticles. The maximum DNA absorption was 50 μg at pH 5, and DNA binding efficiency was 88% at the same pH. The maximum elution efficiency of DNA was 98% at pH 8.5, and the amount of desorbed DNA remained constant above pH 8.5. Additionally, it was determined that the DNA isolation was 95% in the 10-minute time and 55% even after using five cycles. The results indicated that Fe_3O_4@silica@ chitosan nanoparticles had superior advantages such as high adsorption & elution capacity and selectivity compared to other methods (Tiwari et al. 2015).

Wang et al. prepared the silica-coated magnetic nanoparticles (Fe_3O_4@SiO_2) to simultaneously extract DNA and RNA from hepatocellular carcinoma (Hep G2). Average DNA and RNA isolation yields were calculated as 9.7 and 14.7 μg per 1 mL Hep G2 cell, respectively. This method allows for the automatic isolation of nucleic acids (Wang et al. 2017). In the literature, there are several studies using functional magnetic nanoparticles for the separation and detection of nucleic acids (Fei et al. 2022; Tavallaie et al. 2018; Yajima et al. 2022).

The isolation, separation, and detection of proteins using functional nanoparticles also attract great attention in biomedical fields (Eivazzadeh-Keihan et al. 2021; Safarik and Safarikova 2004). For example, Protein A has an affinity for binding to immunoglobulins, fibrinogen, and C-reactive protein, so it can be helpful in detecting target proteins in disease diagnosis. Thanh et al. published a study on the immobilization of protein A to superparamagnetic iron oxide nanoparticles for biomedical applications. Superparamagnetic iron oxide nanoparticles synthesized by the coprecipitation method with a size of 10 and 30 nm were modified with SiO_2, NH_2, and CHO groups using TEOS, APTES, and glutaraldehyde (GA), respectively. The final functional sample (Fe_3O_4/SiO_2/NH_2/CHO), including 10 nm iron oxide nanoparticles, possessed the protein A binding efficiency of 82%. This high binding capacity introduces that Fe_3O_4/SiO_2/NH_2/CHO nanoparticles may be used to diagnose some fibrinogen-based diseases (Thanh et al. 2019).

For vascular endothelial growth factor 165 ($VEGF_{165}$) protein detection, Zhang et al. modified the magnetic glassy carbon electrode (MGCE) with Fe_3O_4/Fe_2O_3@ Au nanoparticles as an electrochemical aptasensor (MGCE/Fe_3O_4/Fe_2O_3@Au). In the study, Fe_3O_4/Fe_2O_3@Au nanoparticles were utilized to increase the electron transfer required for detecting trace amounts of a biological analyte. Moreover, the surface of MGCE/Fe_3O_4/Fe_2O_3@Au was functionalized with a thiol-modified DNA aptamer to detect the $VEGF_{165}$ protein and the nonspecific sites on the final product (MGCE/Fe_3O_4/Fe_2O_3@Au-DNA) were blocked with the treatment of BSA. The electrochemical aptasensor showed a good linear relationship between 0.01 and 10 pg mL^{-1} with $R^2 = 0.9973$, and the LOD and the limit of quantitation (LOQ) were calculated to be 0.01 and 0.03 pg mL^{-1}. The electrochemical aptasensor had excellent selectivity, stability, and reproducibility, so it is considered an efficient, accurate, and cost-effective sensor for the early detection of some diseases based on the $VEGF_{165}$ protein (Zhang et al. 2022).

As another example of protein separation, Pham et al. first synthesized gold-containing iron oxide nanoparticles with a size of 15 to 40 nm for magnetic separation of protein immunoglobulin G (IgG). In this study, the magnetic separation process was carried out using a 3000 G magnet, and protein IgG was attached to the gold surface with a 35% yield by electrostatic interaction. Since protein IgG in different animals can bind effectively to protein A, magnetic nanoparticles presenting IgG can act as affinity probes to target bacteria (Pham et al. 2008). Ma et al. prepared Ti^{4+} modified, and PDA-coated iron oxide nanoparticles (Ti^{4+}-PDA@Fe_3O_4) to identify and detect phosphorylated proteins. Reversible protein phosphorylation is critical in many biological processes such as cell division and cellular metabolism. It is known that abnormal phosphorylation of proteins is associated with many diseases, and their identification and determination are necessary for disease diagnosis. Ti^{4+} ions on the surface of Ti^{4+}-PDA@Fe_3O_4 nanoparticles can bind to phosphate groups in phosphorylated proteins. Therefore, Ti^{4+}-PDA@Fe_3O_4 nanoparticles showed highly effective and selective adsorption for phosphorylated proteins (1273.9 mg g^{-1} for β-Cas) depending on the immobilized metal ion affinity chromatography (IMAC) (Ma and Jia 2016).

In another study related to protein detection, poly(4-styrenesulfonic acid-co-maleic acid) (PSS-MA) coated gold magnetic (GoldMag) nanoparticles forming Fe_3O_4, and Au nanoparticles were prepared by Zhang et al. for the detection of D-dimer, which are degradation product of fibrin. Detecting D-dimer in the blood is essential to identify thromboembolic events and myocardial infarction. Before detection, the surface of PSS-MA modified GoldMag was conjugated with an anti-D-dimer antibody. The detection range of D-dimer was estimated between 0.3 and 6 μg mL^{-1} (Zhang et al. 2017).

As with biomolecules, separating and detecting biological cells is critical in various fields, including diagnosis, therapy, and cell biology (Ma et al. 2019). Currently, methods based on different properties such as size, density, electric, magnetic, and adhesive have been developed to separate and identify cells. However, cells possessing similar sizes and shapes cannot be effectively separated considering these features (Plouffe and Murthy 2014; Plouffe et al. 2014). On the other hand, functional magnetic nanoparticles can be utilized in cell isolation after modification with specific molecules. For instance, Saei et al. demonstrated that the human epithelial growth factor receptor 2 (HER2+) breast cancer cells were effectively separated using an antibody (Ab)-modified iron oxide nanoparticles under a magnetic field. The anti-HER2 Ab was attached to the surface of nanoparticles with 1-ethyl-3-(3-dimethylaminopropyl) carbodiimide/N-hydroxysuccinimide crosslinkers. Targeting cell separation efficiencies of Ab-modified iron oxide nanoparticles were calculated to be 94.5% ± 0.8% for BT474 cells, which are HER2+, and 70.6% ± 0.4% for a mixture of BT474 and MCF7 cells (HER2-). Ab-modified iron oxide nanoparticles may be promising for the early and rapid detection of cancer cells (Saei et al. 2020). In another study, anti-CD3 monoclonal antibody (mAb) conjugated Fe_3O_4@Au, presented by Cui et al. was utilized to separate $CD3^+$ T cells from the whole splenocytes. Anti-CD3 mAb conjugated Fe_3O_4@Au separated $CD3^+$ T cells from the whole splenocytes with high efficiency (98.4%) (Cui et al. 2011). Lu et al. prepared PEI-coated Fe_3O_4 nanoparticles to separate and enrich lung cancer cells from sputum samples.

The analysis demonstrated that the percentage of positive cells in sputum samples increased from 6.3% to 38.5% after treatment with the PEI-coated Fe$_3$O$_4$ (Lu et al. 2014). In another study performed by Vajhadin et al., after HER2-positive cells were isolated with high efficiency using CoFe$_2$O$_4$@Ag-HB5 aptamer, they were detected with an MXene-based cytosensor based on the metallic conductivity (with 75 minutes total analysis time). Electrochemical detection of cells is mainly based on the interaction between SK-BR-3 cells captured with CoFe$_2$O$_4$@Ag-HB5 and modified MXene nanolayers. This hybrid system can be used not only for detecting cancer cells but also for drug delivery (Vajhadin et al. 2022). Digehsaraei et al. (2021) developed a fluorescence immunosensor consisting of graphene quantum nanoparticles (GQNPs) and magnetic nanoparticles (MNPs) for or detection of HER2-positive breast cancer cells (Figure 4.6a). Firstly, silica-coated MNPs were functionalized with an amino group via APTES, and then, Herceptin was bonded to the surface of nanoparticles to provide the separation of cancer cells. Similarly, GQNPs were conjugated with

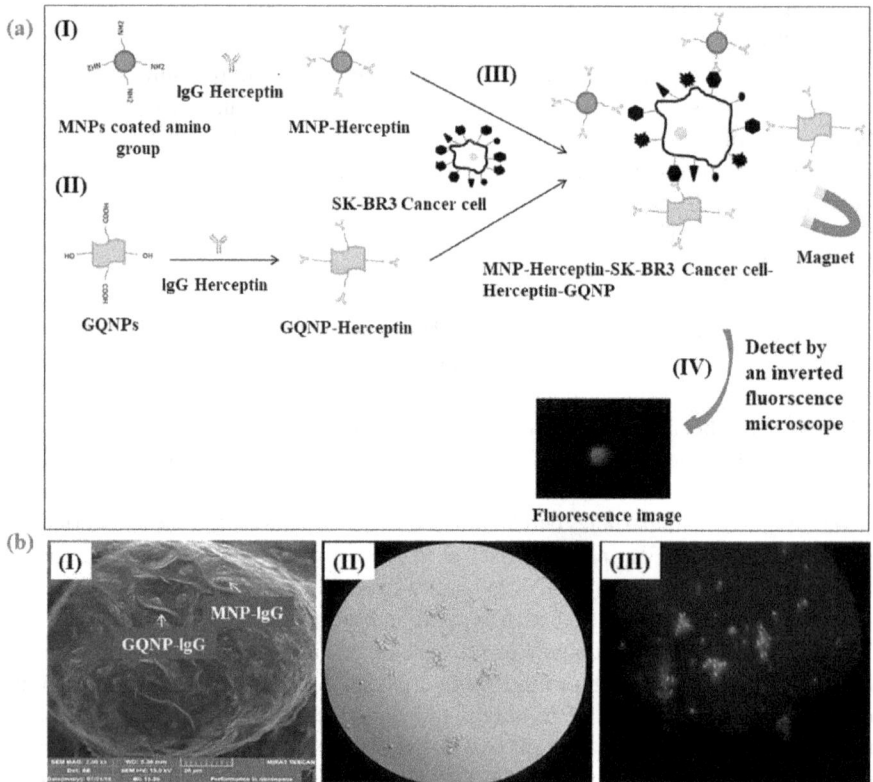

FIGURE 4.6 (a) Schematic descriptions of the conjugation of Herceptin to MNPs (I) and GQNPs (II), incubation of MNPs-Herceptin and GQNPs-Herceptin structures with breast cancer cells, and separation from the medium (III) and detection of breast cancer cells by an inverted fluorescence microscope (IV). (b) SEM (I), visible (II), and fluorescence (III) images of MNPs-Herceptin-Breast cancer cell-Herceptin-GQNPs structure. Adapted with permission from (Digehsaraei et al. 2021), Copyright (2021), Elsevier.

Herceptin to form a sandwich structure, which creates a signal for the detection of cancer cells. After incubation of MNPs-Herceptin and GQNPs-Herceptin structures with breast cancer cells, the sandwich structure forming MNPs-Herceptin-Breast cancer cell-Herceptin-GQNPs was separated from the medium via an external magnet and detected by an inverted fluorescence microscope. It was confirmed that Herceptin was successfully attached to the surfaces of MNPs and GQNPs in the different imaging methods obtained as shown in Figure 4.6b. The results revealed that the HER2 receptor could be detected selectively and with a detection limit of 1 cell mL^{-1}.

REFERENCES

Alkhayal, Anoud, Arshia Fathima, Ali H. Alhasan and Edreese H. Alsharaeh. 2021. "PEG coated Fe$_3$O$_4$/RGO nano-cube-like structures for cancer therapy via magnetic hyperthermia." *Nanomaterials* 11 (9): 2398.

Assa, Farnaz, Hoda Jafarizadeh-Malmiri, Hossein Ajamein, Hamideh Vaghari, Navideh Anarjan, Omid Ahmadi and Aydin Berenjian. 2017. "Chitosan magnetic nanoparticles for drug delivery systems." *Critical Reviews in Biotechnology* 37 (4): 492–509.

Avedian, Ninet, Farzaaneh Zaaeri, Mohammad Porgham Daryasari, Hamid Akbari Javar and Mehdi Khoobi. 2018. "pH-sensitive biocompatible mesoporous magnetic nanoparticles labeled with folic acid as an efficient carrier for controlled anticancer drug delivery." *Journal of Drug Delivery Science and Technology* 44: 323–332.

Bakenecker, Anna C., Mandy Ahlborg, Christina Debbeler, Christian Kaethner, Thorsten M. Buzug and Kerstin Lüdtke-Buzug. 2018. "Magnetic particle imaging in vascular medicine." *Innovative Surgical Sciences* 3 (3): 179–192.

Bañobre-López, Manuel, Cristina Bran, Carlos Rodriguez-Abreu, Juan Gallo, Manuel Vazquez and Jose Rivas. 2017. "A colloidally stable water dispersion of Ni nanowires as an efficient T 2-MRI contrast agent." *Journal of Materials Chemistry B* 5 (18): 3338–3347.

Billings, Caroline, Mitchell Langley, Gavin Warrington, Farzin Mashali and Jacqueline Anne Johnson. 2021. "Magnetic particle imaging: current and future applications, magnetic nanoparticle synthesis methods and safety measures." *International Journal of Molecular Sciences* 22 (14): 7651.

Blanco, Elvin, Haifa Shen and Mauro Ferrari. 2015. "Principles of nanoparticle design for overcoming biological barriers to drug delivery." *Nature Biotechnology* 33 (9): 941–951.

Bloemen, Maarten, Ward Brullot, Tai Thien Luong, Nick Geukens, Ann Gils and Thierry Verbiest. 2012. "Improved functionalization of oleic acid-coated iron oxide nanoparticles for biomedical applications." *Journal of Nanoparticle Research* 14 (9): 1–10.

Bulte, Jeff WM. 2019. "Superparamagnetic iron oxides as MPI tracers: a primer and review of early applications." *Advanced Drug Delivery Reviews* 138: 293–301.

Cardoso, Vanessa Fernandes, António Francesko, Clarisse Ribeiro, Manuel Bañobre-López, Pedro Martins and Senentxu Lanceros-Mendez. 2018. "Advances in magnetic nanoparticles for biomedical applications." *Advanced Healthcare Materials* 7 (5): 1700845.

Chiang, Po-Chieh, Dung-Shing Hung, Jeng-Wen Wang, Chih-Sung Ho and Yeong-Der Yao. 2007. "Engineering water-dispersible FePt nanoparticles for biomedical applications." *IEEE Transactions on Magnetics* 43 (6): 2445–2447.

Chouhan, Raghuraj Singh, Milena Horvat, Jahangeer Ahmed, Norah Alhokbany, Saad M. Alshehri and Sonu Gandhi. 2021. "Magnetic nanoparticles—a multifunctional potential agent for diagnosis and therapy." *Cancers* 13 (9): 2213.

Cui, Yi-Ran, Chao Hong, Ying-Lin Zhou, Yue Li, Xiao-Ming Gao and Xin-Xiang Zhang. 2011. "Synthesis of orientedly bioconjugated core/shell Fe$_3$O$_4$@Au magnetic nanoparticles for cell separation." *Talanta* 85 (3): 1246–1252.

Daboin, Viviana, Sarah Briceño, Jorge Suárez, Lila Carrizales-Silva, Olgi Alcalá, Pedro Silva and Gema Gonzalez. 2019. "Magnetic SiO_2-$Mn_{1-x}Co_xFe_2O_3$ nanocomposites decorated with Au@Fe_3O_4nanoparticles for hyperthermia." *Journal of Magnetism and Magnetic Materials* 479: 91–98.

De Stasio, Gelsomina, Deepika Rajesh, Patrizia Casalbore, Matthew J. Daniels, Robert J. Erhardt, Bradley H. Frazer, Lisa M. Wiese, Katherine L. Richter, Brandon R. Sonderegger and Benjamin Gilbert. 2005. "Are gadolinium contrast agents suitable for gadolinium neutron capture therapy?" *Neurological Research* 27 (4): 387–398.

Demin, Alexander M., Alexandra G. Pershina, Artem S. Minin, Alexander V. Mekhaev, Vladimir V. Ivanov, Sofiya P. Lezhava, Alexandra A. Zakharova, Iliya V. Byzov, Mikhail A. Uimin and Victor P. Krasnov. 2018. "PMIDA-modified Fe_3O_4 magnetic nanoparticles: synthesis and application for liver MRI." *Langmuir* 34 (11): 3449–3458.

Dhavale, Rp, Pp Waifalkar, Apoorva Sharma, Subasa C. Sahoo, P Kollu, Ad Chougale, Drt Zahn, G Salvan, Ps Patil and Pb Patil. 2018. "Monolayer grafting of aminosilane on magnetic nanoparticles: an efficient approach for targeted drug delivery system." *Journal of Colloid and Interface Science* 529: 415–425.

Digehsaraei, Sepideh Yektaniroumand, Mojtaba Salouti, Bahram Amini, Sanaz Mahmazi, Mohsen Kalantari, Alireza Kazemizadeh and Jamshid Mehrvand. 2021. "Developing a fluorescence immunosensor for detection of HER2-positive breast cancer based on graphene and magnetic nanoparticles." *Microchemical Journal* 167: 106300.

Duan, Xiaopin and Yaping Li. 2013. "Physicochemical characteristics of nanoparticles affect circulation, biodistribution, cellular internalization, and trafficking." *Small* 9 (9–10): 1521–1532.

Dukenbayev, Kanat, Ilya V. Korolkov, Daria I. Tishkevich, Artem L. Kozlovskiy, Sergey V. Trukhanov, Yevgeniy G. Gorin, Elena E. Shumskaya, Egor Y. Kaniukov, Denis A. Vinnik and Maxim V. Zdorovets. 2019. "Fe_3O_4 nanoparticles for complex targeted delivery and boron neutron capture therapy." *Nanomaterials* 9 (4): 494.

Efremova, Maria V., Yulia A. Nalench, Eirini Myrovali, Anastasiia S. Garanina, Ivan S. Grebennikov, Polina K. Gifer, Maxim A. Abakumov, Marina Spasova, Makis Angelakeris and Alexander G. Savchenko. 2018. "Size-selected Fe_3O_4–Au hybrid nanoparticles for improved magnetism-based theranostics." *Beilstein Journal of Nanotechnology* 9 (1): 2684–2699.

Eguía-Eguía, Sandra I., Lorenzo Gildo-Ortiz, Mario Pérez-González, Sergio A. Tomas, Jesús A. Arenas-Alatorre and Jaime Santoyo-Salazar. 2021. "Magnetic domains orientation in (Fe_3O_4/γ-Fe_2O_3) nanoparticles coated by Gadolinium-diethylenetriaminepentaacetic acid (Gd^{3+}-DTPA)." *Nano Express* 2 (2): 020019.

Ehteshamzadeh, Taraneh, Saeed Kakaei, Mehdi Ghaffari and Ali Reza Khanchi. 2021. "Doxorubicin Embedded Polyvinylpyrrolidone-Coated Fe_3O_4 nanoparticles for targeted drug delivery system." *Journal of Superconductivity and Novel Magnetism* 34 (12): 3345–3360.

Eivazzadeh-Keihan, Reza, Hossein Bahreinizad, Zeinab Amiri, Hooman Aghamirza Moghim Aliabadi, Milad Salimi-Bani, Athar Nakisa, Farahnaz Davoodi, Behnam Tahmasebi, Farnoush Ahmadpour and Fateme Radinekiyan. 2021. "Functionalized magnetic nanoparticles for the separation and purification of proteins and peptides." *TRAC Trends in Analytical Chemistry* 141: 116291.

Fatima, Hira, Tawatchai Charinpanitkul and Kyo-Seon Kim. 2021. "Fundamentals to apply magnetic nanoparticles for hyperthermia therapy." *Nanomaterials* 11 (5): 1203.

Fei, Zhongjie, Chu Cheng, Rongbin Wei, Guolei Tan and Pengfeng Xiao. 2022. "Reversible superhydrophobicity unyielding magnetic beads of flipping-triggered (SYMBOL) regulate the binding and unbinding of nucleic acids for ultra-sensitive detection." *Chemical Engineering Journal* 431: 133953.

Fraceto, Leonardo Fernandes and Daniele Ribeiro De Araujo. 2014. *Microspheres: Technologies, Applications and Role in Drug Delivery Systems*. Nova Science Pub Inc., p.1–273.

Gao, Shanmin, Xin Liu, Tao Xu, Xuehua Ma, Zheyu Shen, Aiguo Wu, Yinghuai Zhu and Narayan S. Hosmane. 2013. "Synthesis and characterization of $Fe^{10}BO_3/Fe_3O_4/SiO_2$ and $GdFeO_3/Fe_3O_4/SiO_2$: nanocomposites of biofunctional materials." *ChemistryOpen* 2 (3): 88–92.

Gao, Yu, Yi Liu and Chenjie Xu. 2014. Magnetic nanoparticles for biomedical applications: from diagnosis to treatment to regeneration. Engineering in translational medicine, London: Springer, p.567–583.

Garner, Ian. 2000. Isolation of high-molecular-weight DNA from animal cells. *The nucleic acid protocols handbook*, Springer, p. 3–7.

Guo, Zhiyang, Yi Sun, Lirui Zhang, Qian Ding, Wei Chen, Hao Yu, Qingyun Liu and Min Fu. 2022. "Surface imprinted core–shell nanorod for selective extraction of glycoprotein." *Journal of Colloid and Interface Science* 615: 597–605.

Gupta, Jagriti, Pa Hassan and Kc Barick. 2021. "Core-shell $Fe_3O_4@ZnO$ nanoparticles for magnetic hyperthermia and bio-imaging applications." *AIP Advances* 11 (2): 025207.

Hammad, Mohaned, Valentin Nica and Rolf Hempelmann. 2016. "Synthesis and characterization of Bi-magnetic core/shell nanoparticles for hyperthermia applications." *IEEE Transactions on Magnetics* 53 (4): 1–6.

Haun, Jered B., Tae-Jong Yoon, Hakho Lee and Ralph Weissleder. 2010. "Magnetic nanoparticle biosensors." *Wiley Interdisciplinary Reviews: Nanomedicine and Nanobiotechnology* 2 (3): 291–304.

Hergt, Rudolf and Silvio Dutz. 2007. "Magnetic particle hyperthermia—biophysical limitations of a visionary tumour therapy." *Journal of Magnetism and Magnetic Materials* 311 (1): 187–92.

Horvat, Sonja, Patrick Vogel, Thomas Kampf, Andreas Brandl, Aws Alshamsan, Hisham A. Alhadlaq, Maqusood Ahamed, Krystyna Albrecht, Volker C. Behr and Andreas Beilhack. 2020. "Crosslinked coating improves the signal-to-noise ratio of iron oxide nanoparticles in magnetic particle imaging (MPI)." *ChemNanoMat* 6 (5): 755–58.

Hosmane, Narayan S. 2011. *Boron and gadolinium neutron capture therapy for cancer treatment*. Singapore: World Scientific Pub. Co.

Hu, Yong, Jia Yang, Ping Wei, Jingchao Li, Ling Ding, Guixiang Zhang, Xiangyang Shi and Mingwu Shen. 2015. "Facile synthesis of hyaluronic acid-modified Fe_3O_4/Au composite nanoparticles for targeted dual mode MR/CT imaging of tumors." *Journal of Materials Chemistry B* 3 (47): 9098–9108.

Huang, Ruijie, Xingyu Zhou, Guiyuan Chen, Lanhong Su, Zhaoji Liu, Peijie Zhou, Jianping Weng and Yuanzeng Min. 2022. "Advances of functional nanomaterials for magnetic resonance imaging and biomedical engineering applications." *Wiley Interdisciplinary Reviews: Nanomedicine and Nanobiotechnology* 14 (4): e1800.

Huang, S., Sy Wang, A. Gupta, Da Borca-Tasciuc and Sj Salon. 2012. "On the measurement technique for specific absorption rate of nanoparticles in an alternating electromagnetic field." *Measurement Science and Technology* 23 (3): 035701.

Icten, Okan. 2021a. "The design of gold decorated iron borates (Fe_3BO_6 and $FeBO_3$) for photothermal therapy and boron carriers." *European Journal of Inorganic Chemistry* 2021 (21): 1985–1992.

Icten, Okan. 2021b. Functional nanocomposites: promising candidates for cancer diagnosis and treatment. In *Synthetic Inorganic Chemistry*, 279–340: Elsevier.

Icten, Okan. 2021c. "Preparation of Gadolinium-based metal-organic frameworks and the modification with Boron-10 isotope: a potential dual agent for MRI and neutron capture therapy applications." *ChemistrySelect* 6 (8): 1900–1910.

Icten, Okan, Beril Erdem Tuncdemir and Hatice Mergen. 2022. "Design and development of gold-loaded and boron-attached multicore manganese ferrite nanoparticles as a potential agent in biomedical applications." *ACS Omega.*

Icten, Okan, Dursun Ali Kose, Stephan J. Matissek, Jason A. Misurelli, Sherine F. Elsawa, Narayan S. Hosmane and Birgul Zumreoglu-Karan. 2018. "Gadolinium borate and iron oxide bioconjugates: nanocomposites of next generation with multifunctional applications." *Materials Science and Engineering: C* 92: 317–28.

Irfan, M., N Dogan, Ayhan Bingolbali and F. Aliew. 2021. "Synthesis and characterization of NiFe$_2$O$_4$ magnetic nanoparticles with different coating materials for magnetic particle imaging (MPI)." *Journal of Magnetism and Magnetic Materials* 537: 168150.

Keshtkar, Mohammad, Daryoush Shahbazi-Gahrouei and Alireza Mahmoudabadi. 2020. "Synthesis and application of Fe$_3$O$_4$@Au composite nanoparticles as magnetic resonance/computed tomography dual-modality contrast agent." *Journal of Medical Signals and Sensors* 10 (3): 201.

Kim, Dokyoon, Kwangsoo Shin, Soon Gu Kwon and Taeghwan Hyeon. 2018. "Synthesis and biomedical applications of multifunctional nanoparticles." *Advanced Materials* 30 (49): 1802309.

Korolkov, Ilya V., Alexandr V. Zibert, Lana I. Lissovskaya, K Ludzik, M Anisovich, Artem L. Kozlovskiy, Ae Shumskaya, M Vasilyeva, Dmitriy I. Shlimas and Monika Jażdżewska. 2021. "Boron and gadolinium loaded Fe$_3$O$_4$ nanocarriers for potential application in neutron capture therapy." *International Journal of Molecular Sciences* 22 (16): 8687.

Korolkov, Ilya V., K. Ludzik, Al Kozlovskiy, Ms Fadeev, Ae Shumskaya, Ye G. Gorin, M. Jazdzewska, M. Anisovich, Vs Rusakov and Mv Zdorovets. 2020. "Immobilization of carboranes on Fe$_3$O$_4$-polymer nanocomposites for potential application in boron neutron cancer therapy." *Colloids and Surfaces A: Physicochemical and Engineering Aspects* 601: 125035.

Leenders, William. 2003. "Ferumoxtran-10 advanced magnetics." *Idrugs: The Investigational Drugs Journal* 6 (10): 987–993.

Lemaster, Jeanne E., Fang Chen, Taeho Kim, Ali Hariri and Jesse V. Jokerst. 2018. "Development of a trimodal contrast agent for acoustic and magnetic particle imaging of stem cells." *ACS Applied Nano Materials* 1 (3): 1321–1331.

Li, D., Sj Li, Y. Zhang, Jj Jiang, Wj Gong, Jh Wang and Zd Zhang. 2015. "Monodisperse water-soluble-Fe$_2$O$_3$/polyvinylpyrrolidone nanoparticles for a magnetic resonance imaging contrast agent." *Materials Research Innovations* 19 (sup 3): S58–S62.

Li, Xiaoming, Jianrong Wei, Katerina E. Aifantis, Yubo Fan, Qingling Feng, Fu-Zhai Cui and Fumio Watari. 2016. "Current investigations into magnetic nanoparticles for biomedical applications." *Journal of Biomedical Materials Research Part A* 104 (5): 1285–1296.

Linh, Ph, Nx Phuc, Lv Hong, Ll Uyen, Nv Chien, Ph Nam, Nt Quy, Htm Nhung, Pt Phong and In-Ja Lee. 2018. "Dextran coated magnetite high susceptibility nanoparticles for hyperthermia applications." *Journal of Magnetism and Magnetic Materials* 460: 128–136.

Liu, Gang, Jinhao Gao, Hua Ai and Xiaoyuan Chen. 2013. "Applications and potential toxicity of magnetic iron oxide nanoparticles." *Small* 9 (9–10): 1533–1545.

Liu, Yating, Yan Li and Yun Wei. 2014. "Highly selective isolation and purification of heme proteins in biological samples using multifunctional magnetic nanospheres." *Journal of Separation Science* 37 (24): 3745–3752.

Lu, An-Hui, EL Salabas and Ferdi Schüth. 2007. "Magnetic nanoparticles: synthesis, protection, functionalization, and application." *Angewandte Chemie International Edition* 46 (8): 1222–1244.

Lu, Wei, Min Ling, Min Jia, Ping Huang, Chengkui Li and Biao Yan. 2014. "Facile synthesis and characterization of polyethylenimine-coated Fe$_3$O$_4$ superparamagnetic nanoparticles for cancer cell separation." *Molecular Medicine Reports* 9 (3): 1080–1084.

Ma, Xiangdong and Li Jia. 2016. "Polydopamine assisted preparation of Ti^{4+}-decorated magnetic particles for selective and rapid adsorption of phosphorylated proteins." *Journal of Chemical Technology and Biotechnology* 91 (4): 892–900.

Ma, Yuanyuan, Tianxiang Chen, Muhammad Zubair Iqbal, Fang Yang, Norbert Hampp, Aiguo Wu and Liqiang Luo. 2019. "Applications of magnetic materials separation in biological nanomedicine." *Electrophoresis* 40 (16–17): 2011–2028.

Majidi, Sima, Fatemeh Zeinali Sehrig, Samad Mussa Farkhani, Mehdi Soleymani Goloujeh and Abolfazl Akbarzadeh. 2016. "Current methods for synthesis of magnetic nanoparticles." *Artificial Cells, Nanomedicine, and Biotechnology* 44 (2): 722–734.

Markides, Han, M. Rotherham and Aj El Haj. 2012. "Biocompatibility and toxicity of magnetic nanoparticles in regenerative medicine." *Journal of nanomaterials* 2012: 11.

Materón, Elsa M., Celina M. Miyazaki, Olivia Carr, Nirav Joshi, Paulo Hs Picciani, Cleocir J. Dalmaschio, Frank Davis and Flavio M. Shimizu. 2021. "Magnetic nanoparticles in biomedical applications: a review." *Applied Surface Science Advances* 6: 100163.

Meola, Antonio, Jianghong Rao, Navjot Chaudhary, Guosheng Song, Xianchuang Zheng and Steven D. Chang. 2019. "Magnetic particle imaging in neurosurgery." *World Neurosurgery* 125: 261–270.

Min, Yuanzeng, Joseph M. Caster, Michael J. Eblan and Andrew Z. Wang. 2015. "Clinical translation of nanomedicine." *Chemical Reviews* 115 (19): 11147–11190.

Molaei, Mohammad Jafar and Esmaeil Salimi. 2022. "Magneto-fluorescent superparamagnetic $Fe_3O_4@SiO_2@$alginate/carbon quantum dots nanohybrid for drug delivery." *Materials Chemistry and Physics*: 126361.

Molkenova, Anara, Timur Sh Atabaev, Suck Won Hong, Chuanbin Mao, Dong-Wook Han and Ki Su Kim. 2022. "Designing inorganic nanoparticles into computed tomography and magnetic resonance (CT/MR) imaging-guidable photomedicines." *Materials Today Nano*: 100187.

Monteserín, Maria, Silvia Larumbe, Alejandro V. Martínez, Saioa Burgui and L. Francisco Martín. 2021. "Recent advances in the development of magnetic nanoparticles for biomedical applications." *Journal of Nanoscience and Nanotechnology* 21 (5): 2705–2741.

Naha, Pratap C., Ajlan Al Zaki, Elizabeth Hecht, Michael Chorny, Peter Chhour, Eric Blankemeyer, Douglas M. Yates, Walter Rt Witschey, Harold I. Litt and Andrew Tsourkas. 2014. "Dextran coated bismuth–iron oxide nanohybrid contrast agents for computed tomography and magnetic resonance imaging." *Journal of Materials Chemistry B* 2 (46): 8239–8248.

Nguyen, Hoa Du, Thi Ngoc Linh Nguyen, Thien Vuong Nguyen, Phan Thi Hong Tuyet, Thi Hai Hoa Nguyen, Quoc Thang Nguyen, Thu Ha Hoang, Tran Chien Dang, Bui Le Minh and Le Trong Lu. 2020. "Biological durability, cytotoxicity and MRI image contrast effects of chitosan modified magnetic nanoparticles." *Journal of Nanoscience and Nanotechnology* 20 (9): 5338–5348.

Peixoto, L., R Magalhães, D Navas, S Moraes, C Redondo, R Morales, Jp Araújo and Ct Sousa. 2020. "Magnetic nanostructures for emerging biomedical applications." *Applied Physics Reviews* 7 (1): 011310.

Perlstein, Benny, Tammy Lublin-Tennenbaum, Inbal Marom and Shlomo Margel. 2010. "Synthesis and characterization of functionalized magnetic maghemite nanoparticles with fluorescent probe capabilities for biological applications." *Journal of Biomedical Materials Research Part B: Applied Biomaterials* 92 (2): 353–360.

Pham, Thao Thi Hien, Cuong Cao and Sang Jun Sim. 2008. "Application of citrate-stabilized gold-coated ferric oxide composite nanoparticles for biological separations." *Journal of Magnetism and Magnetic Materials* 320 (15): 2049–2055.

Plouffe, Brian D and Shashi K. Murthy. 2014. "Perspective on microfluidic cell separation: a solved problem?" *Analytical chemistry* 86 (23): 11481–11488.

Plouffe, Brian D., Shashi K. Murthy and Laura H. Lewis. 2014. "Fundamentals and application of magnetic particles in cell isolation and enrichment: a review." *Reports on Progress in Physics* 78 (1): 016601.

Prasad, Ai, Ak Parchur, Rr Juluri, N Jadhav, Bn Pandey, Rs Ningthoujam and Rk Vatsa. 2013. "Bi-functional properties of Fe_3O_4@YPO_4:Eu hybrid nanoparticles: hyperthermia application." *Dalton Transactions* 42 (14): 4885–4896.

Pylypchuk, Ie V., D Kołodyńska, M Kozioł and Pp Gorbyk. 2016. "Gd-DTPA adsorption on chitosan/magnetite nanocomposites." *Nanoscale research letters* 11 (1): 1–10.

Reddy, Alavala Matta, Byung Kook Kwak, Hyung Jin Shim, Chiyoung Ahn, Sun Hang Cho, Byung Jin Kim, Sang Young Jeong, Sung-Joo Hwang and Soon Hong Yuk. 2009. "Functional characterization of mesenchymal stem cells labeled with a novel PVP-coated superparamagnetic iron oxide." *Contrast Media & Molecular Imaging* 4 (3): 118–126.

Ruiz, Amalia, Paulo César Morais, Ricardo Bentes De Azevedo, Zulmira Gm Lacava, Angeles Villanueva and María Del Puerto Morales. 2014. "Magnetic nanoparticles coated with dimercaptosuccinic acid: development, characterization, and application in biomedicine." *Journal of Nanoparticle Research* 16 (11): 1–20.

Saei, Arezoo, Shima Asfia, Hasan Kouchakzadeh and Moones Rahmandoust. 2020. "Antibody-modified magnetic nanoparticles as specific high-efficient cell-separation agents." *Journal of Biomedical Materials Research Part B: Applied Biomaterials* 108 (6): 2633–2642.

Safarik, Ivo and Mirka Safarikova. 2004. "Magnetic techniques for the isolation and purification of proteins and peptides." *Biomagnetic Research and Technology* 2 (1): 1–17.

Sanad, Mohamed F., Bianca P. Meneses-Brassea, Dawn S. Blazer, Shirin Pourmiri, George C. Hadjipanayis and Ahmed A. El-Gendy. 2021. "Superparamagnetic Fe/Au nanoparticles and their feasibility for magnetic hyperthermia." *Applied Sciences* 11 (14): 6637.

Santana, Sara Df, Vijaykumar L. Dhadge and Ana Ca Roque. 2012. "Dextran-coated magnetic supports modified with a biomimetic ligand for IgG purification." *ACS Applied Materials & Interfaces* 4 (11): 5907–5914.

Saputra, Kormil, Arif Hidayat, Chusnana Insjaf Yogihati, Sigit Tri Wicaksono, Nurul Hidayat, Samsul Hidayat and Siriwat Soontaranon. 2019. Magneto-thermal effect in MnO. 25Fe2. 75O4-PEG nanoparticles and their potential as hyperthermia therapy. In *IOP Conference Series: Materials Science and Engineering*, 012008: IOP Publishing.

Shabatina, Tatyana I., Olga I. Vernaya, Vladimir P. Shabatin and Mikhail Ya Melnikov. 2020. "Magnetic nanoparticles for biomedical purposes: modern trends and prospects." *Magnetochemistry* 6 (3): 30.

Sheng, Yang, Shuai Li, Zongquan Duan, Rong Zhang and Junmin Xue. 2018. "Fluorescent magnetic nanoparticles as minimally-invasive multi-functional theranostic platform for fluorescence imaging, MRI and magnetic hyperthermia." *Materials Chemistry and Physics* 204: 388–396.

Song, Xiaoli, Xiadan Luo, Qingqing Zhang, Aiping Zhu, Lijun Ji and Caifeng Yan. 2015. "Preparation and characterization of biofunctionalized chitosan/Fe_3O_4 magnetic nanoparticles for application in liver magnetic resonance imaging." *Journal of Magnetism and Magnetic Materials* 388: 116–122.

Soni, Abhishek Kumar, Rashmi Joshi, Bheeshma Pratap Singh, N Naveen Kumar and Raghumani Singh Ningthoujam. 2019. "Near-infrared-and magnetic-field-responsive $NaYF_4$: Er^{3+}/Yb^{3+}@ SiO_2@AuNP@Fe_3O_4 nanocomposites for hyperthermia applications induced by fluorescence resonance energy transfer and surface plasmon absorption." *ACS Applied Nano Materials* 2 (11): 7350–7361.

Talegaonkar, Sushama and Arundhati Bhattacharyya. 2019. "Potential of lipid nanoparticles (SLNs and NLCs) in enhancing oral bioavailability of drugs with poor intestinal permeability." *AAPS PharmSciTech* 20 (3): 1–15.

Tang, Congli, Ziyu He, Hongmei Liu, Yuyue Xu, Hao Huang, Gaojian Yang, Ziqi Xiao, Song Li, Hongna Liu and Yan Deng. 2020. "Application of magnetic nanoparticles in nucleic acid detection." *Journal of Nanobiotechnology* 18 (1): 1–19.

Tavallaie, Roya, Joshua Mccarroll, Marion Le Grand, Nicholas Ariotti, Wolfgang Schuhmann, Eric Bakker, Richard David Tilley, David Brynn Hibbert, Maria Kavallaris and John Justin Gooding. 2018. "Nucleic acid hybridization on an electrically reconfigurable network of gold-coated magnetic nanoparticles enables microRNA detection in blood." *Nature Nanotechnology* 13 (11): 1066–1071.

Thanh, Bui Trung, Nguyen Van Sau, Heongkyu Ju, Mohammed Jk Bashir, Hieng Kiat Jun, Thang Bach Phan, Quang Minh Ngo, Ngoc Quyen Tran, Tran Hoang Hai and Pham Hung Van. 2019. "Immobilization of protein A on monodisperse magnetic nanoparticles for biomedical applications." *Journal of Nanomaterials* 2019.

Thomas, Reju, In-Kyu Park and Yong Yeon Jeong. 2013. "Magnetic iron oxide nanoparticles for multimodal imaging and therapy of cancer." *International Journal of Molecular Sciences* 14 (8): 15910–15930.

Tiwari, Arpita P., Rajshri K. Satvekar, Sonali S. Rohiwal, Vidya A. Karande, Abhinav V. Raut, Priti G. Patil, Prajakta B. Shete, Sj Ghosh and Sh Pawar. 2015. "Magneto-separation of genomic deoxyribose nucleic acid using pH responsive Fe_3O_4@silica@chitosan nanoparticles in biological samples." *Rsc Advances* 5 (11): 8463–8470.

Torresan, Veronica, Andrea Guadagnini, Denis Badocco, Paolo Pastore, Guillermo Arturo Muñoz Medina, Marcela B. Fernàndez Van Raap, Ian Postuma, Silva Bortolussi, Marina Bekić and Miodrag Čolić. 2021. "Biocompatible iron–boron nanoparticles designed for neutron capture therapy guided by magnetic resonance imaging." *Advanced Healthcare Materials* 10 (6): 2001632.

Tran, Phuong, Sang-Eun Lee, Dong-Hyun Kim, Yong-Chul Pyo and Jeong-Sook Park. 2020. "Recent advances of nanotechnology for the delivery of anticancer drugs for breast cancer treatment." *Journal of Pharmaceutical Investigation* 50 (3): 261–270.

Vajhadin, Fereshteh, Mohammad Mazloum-Ardakani, Maryamsadat Shahidi, Seyed Mohammad Moshtaghioun, Fateme Haghiralsadat, Azar Ebadi and Abbas Amini. 2022. "MXene-based cytosensor for the detection of HER2-positive cancer cells using $CoFe_2O_4$@Ag magnetic nanohybrids conjugated to the HB5 aptamer." *Biosensors and Bioelectronics* 195: 113626.

Wang, Bangjin, Aihong Duan, Shengming Xie, Junhui Zhang, Liming Yuan and Qiue Cao. 2021. "The molecular imprinting of magnetic nanoparticles with boric acid affinity for the selective recognition and isolation of glycoproteins." *RSC Advances* 11 (41): 25524–25529.

Wang, Guangshuo, Dexing Zhao, Yingying Ma, Zhixiao Zhang, Hongwei Che, Jingbo Mu, Xiaoliang Zhang and Zheng Zhang. 2018. "Synthesis and characterization of polymer-coated manganese ferrite nanoparticles as controlled drug delivery." *Applied Surface Science* 428: 258–263.

Wang, Jiuhai, Zeeshan Ali, Jin Si, Nianyue Wang, Nongyue He and Zhiyang Li. 2017. "Simultaneous extraction of DNA and RNA from hepatocellular carcinoma (Hep G2) based on silica-coated magnetic nanoparticles." *Journal of Nanoscience and Nanotechnology* 17 (1): 802–806.

Wong, Ray M., Dustin A. Gilbert, Kai Liu and Angelique Y. Louie. 2012. "Rapid size-controlled synthesis of dextran-coated, 64Cu-doped iron oxide nanoparticles." *ACS Nano* 6 (4): 3461–3467.

Wu, Wei, Zhaohui Wu, Taekyung Yu, Changzhong Jiang and Woo-Sik Kim. 2015. "Recent progress on magnetic iron oxide nanoparticles: synthesis, surface functional strategies and biomedical applications." *Science and Technology of Advanced Materials* 16 (2): 023501.

Yajima, Shuto, Ayako Koto, Maho Koda, Hiroaki Sakamoto, Eiichiro Takamura and Shin-Ichiro Suye. 2022. "Photo-cross-linked probe-modified magnetic particles for the selective and reliable recovery of nucleic acids." *ACS Omega* 7 (15): 12701–12706.

Yan, Yulan, Chunmei Yang, Guidong Dai, Yu Zhang, Guojian Tu, Yuwei Li, Lu Yang and Jian Shu. 2021. "Folic acid-conjugated $CuFeSe_2$ nanoparticles for targeted T2-weighted magnetic resonance imaging and computed tomography of tumors in vivo." *International Journal of Nanomedicine* 16: 6429.

Yang, Guangbao, Hua Gong, Teng Liu, Xiaoqi Sun, Liang Cheng and Zhuang Liu. 2015. "Two-dimensional magnetic $WS_2@Fe_3O_4$ nanocomposite with mesoporous silica coating for drug delivery and imaging-guided therapy of cancer." *Biomaterials* 60: 62–71.

Yildiz, Ibrahim. 2016. "Applications of magnetic nanoparticles in biomedical separation and purification." *Nanotechnology Reviews* 5 (3): 331–340.

Zandieh, Mohamad and Juewen Liu. 2021. "Spherical nucleic acid mediated functionalization of polydopamine-coated nanoparticles for selective DNA extraction and detection." *Bioconjugate Chemistry* 32 (4): 801–809.

Zhang, Xiaomei, Qinlu Zhang, Ting Ma, Qian Liu, Songdi Wu, Kai Hua, Chao Zhang, Mingwei Chen and Yali Cui. 2017. "Enhanced stability of gold magnetic nanoparticles with Poly (4-styrenesulfonic acid-co-maleic acid): tailored optical properties for protein detection." *Nanoscale Research Letters* 12 (1): 1–9.

Zhang, Yanling, Min Liu, Shuai Pan, Lulu Yu, Shaoshuai Zhang and Ruijiang Liu. 2022. "A magnetically induced self-assembled and label-free electrochemical aptasensor based on magnetic Fe3O4/Fe2O3@ Au nanoparticles for VEGF165 protein detection." *Applied Surface Science* 580: 152362.

Zhao, Guanghui, Jianzhi Wang, Xiaomen Peng, Yanfeng Li, Xuemei Yuan and Yingxia Ma. 2014. "Facile solvothermal synthesis of mesostructured Fe_3O_4/chitosan nanoparticles as delivery vehicles for ph-responsive drug delivery and magnetic resonance imaging contrast agents." *Chemistry–An Asian Journal* 9 (2): 546–553.

Zibert, Alexandr V., Lana I. Lissovskaya, Ilya V. Korolkov and Maxim V. Zdorovets. 2022. $Gd_xFe_{3-x}O_4$ nanoparticles with silane shell as potential theranostic agent for cancer treatment. In *Journal of Physics: Conference Series*, 012006: IOP Publishing.

5 Induction of Physicochemical Effects from Functional Magnetic Nanoparticles in Biological Media and Their Potential for Alternative Medical Therapies

*Manoj Kumar Srinivasan, Nivedha Jayaseelan, Saravanan Alamelu, and Kamalesh Balakumar Venkatesan**
Annamalai University

Kalist Shagirtha
St. Josephs College of Arts and Science

CONTENTS

* Manoj Kumar Srinivasan and Nivedha Jayaseelan contributed equally.

DOI: 10.1201/9781003335580-5

5.1   INTRODUCTION

Nanotechnology is currently most suitable in biomedical applications, due to recent developments in our understanding of materials and the addition of new materials that may be safely included for medical diagnosis and therapies (Patra et al. 2018). Owing to their superior physical, chemical, and, biological properties, nanoparticles provide a multitude of advantages for numerous bioapplications (Yetisgin et al. 2020). Synthesis and functionalization processes influence the compositions (polymers, peptides, metals, and lipids) and morphologies (flowers, rods, pyramids, and spheres) of nanoparticles (Jeevanandam et al. 2018).

Over the past 10 years, magnetic nanoparticles (MNPs) have been recognized as potentially useful tools in the field of nanomedicine due to their superior physicochemical qualities. These qualities include ease of surface modification, superparamagnetism, excellent colloidal stability, good relaxation performance, and outstanding biocompatibility in *in vivo* environment (Darroudi et al. 2021).

MNPs are notable for their versatility in a variety of biomedical applications, including hyperthermia, medication administration, cell isolation magnetic resonance imaging (MRI), and magnetic resonance spectroscopy (MRS) (Wu et al. 2019). Several synthetic approaches, including microemulsion, coprecipitation, high-temperature decomposition, and solvothermal procedures, have been developed and employed to manufacture shape- and size-controlled MNPs. High-quality MNPs, on the other hand, are commonly produced utilizing organic phases and completed with hydrophobic surface ligands, limiting their colloidal stability in physiological fluids and limiting their biological applications (Kyeong et al. 2021). To address these challenges with MNP *in vivo* application, the surface of MNPs is routinely altered by capping individual layers to boost their stability and biocompatibility, inhibit nonspecific protein absorption, and lengthen the time that they circulate in the circulation. To achieve precise surface functionalization (e.g., quantum dot. silica, gold), a variety of design components, including organic ligands (such as biomacromolecules, synthetic polymers, and small molecules), organic–inorganic hybrid NPs (such as metal-organic frameworks [MOFs]), organic–inorganic NPs, and inorganic NPs, can be used (Ali et al. 2021). Organic ligands, for example, are extremely beneficial in a variety of biological and therapeutic applications. MNPs may have the appropriate size, controllable structure, surface charge, and programmable responses to tumor microenvironments (such as pH, redox, and ROS) due to their properties (Kankala et al. 2022). MNPs are sometimes contained in polyethylene glycol (PEG) and its derivatives (RES) for enhanced stability and biocompatibility, as well as to avoid rapid clearance by the reticuloendothelial system (Cole et al. 2011). Charge-switchable

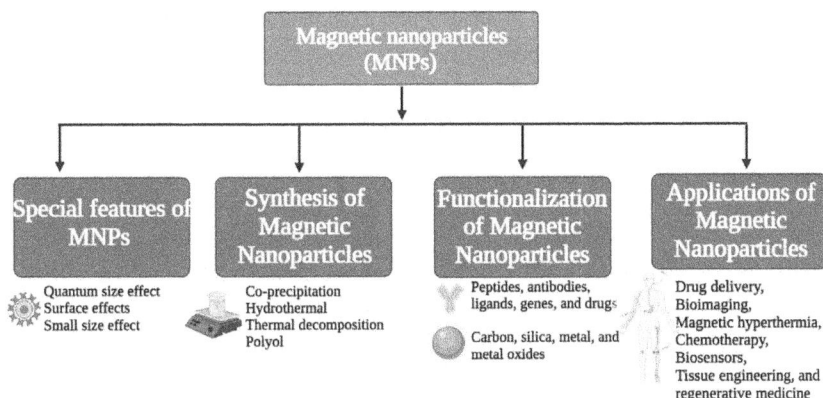

FIGURE 5.1 Schematic illustration of magnetic nanoparticle's (MNP) special features, synthesis, functionalization, and biomedical applications.

polymers can be used with MNPs to activate their surface charge switch, converting it from a negative to a positive charge, hence extending circulation duration and boosting cellular absorption (Ooi et al. 2020). To achieve tumor targeting, folic acid (FA), antibodies, Arg-Gly-Asp (RGD) peptides, hyaluronic acid (HA), and other organic target ligands can be decorated onto the surfaces of MNPs (Rizwanullah et al. 2021). These ligands have the ability to connect with the receptors produced by tumor cells at the target sites. Several tumor microenvironment-responsive polymers have also been developed and widely employed to enhance MNP self-assembly into smart nanocarriers for drug administration, controlled release, and increased MR contrast in diseased areas (Rizwanullah et al. 2021). Various strategies, such as ligand substitution, self-assembly, and in-situ synthesis, have been described to change the surface of MNPs. These approaches enable the integration of the functions and components of organic ligands with MNPs, and they offer a promising method of changing the properties of MNPs for use in biomedical applications (Baki et al. 2021).

This chapter presents special features, synthesis, surface engineering (functionalization), and biomedical applications (bioimaging, drug delivery, chemotherapy, magnetic hyperthermia, biosensors, and regenerative medicine) of MNPs (Figure 5.1).

5.2 SPECIAL FEATURES OF MNPs

MNP characteristics are distinct from general magnetic materials and magnetic properties, which are only associated with the physical length for the nanoscale in the 1–100 nm orders of magnitude. Alternatively, magnetic body size and physical length characteristics show anomalous magnetic and electrical properties. These characteristics vary quite a bit from one MNP class to the next. The chemical stability of magnetic nanoparticles, large specific surface areas of magnetic nanoparticles, high loading capacities, low intraparticle concentrations, and superparamagnetism are all key aspects of magnetic nanoparticles. When the size is less than the requisite critical dimensions, a phenomenon known as superparamagnetism can take place (Kolhatkar et al. 2013). It is important to note that the critical diameter might change depending on the substance. For instance, the critical diameter of Fe_3O_4 is 35 nm (Lim et al. 2011). Compared to the saturation magnetization values for paramagnetic materials, superparamagnetic nanoparticles are orders of magnitude higher (Faraudo et al. 2010). In addition, they have a high level of resistance to an external field (Furlani 2010). Additionally, the orientations of the magnetic moment vectors begin to relax. They have very little attraction to one another when no magnetic field is present, which lessens the likelihood that particles will gather together (Furlani 2010; Sahoo et al. 2005).

5.2.1 QUANTUM SIZE EFFECT

Material level spacing and atomic number N are inversely related. The number of atoms within the particles is constrained when the particle size is tiny. Close to the fermi level, the electrical energy levels of nanometals transition from continuous to discrete. The most and least occupied molecular orbitals are discontinuous in nanoscale semiconductor particle energy gap expansion. When the energy gap separation is greater than the magnetic, thermal, photonic, electrostatic, and other energy characteristics of the material, the nanoparticle and macroscopic physical properties will be separate. Conductive metal, for example, can become an insulator in the presence of ultrafine particles. The macroscopic quantum size effect affects the electron's magnetic moment and particle size, affecting the amount of anomalous heat generated.

5.2.2 SURFACE EFFECTS

Due to nanomaterials composition and small particle size, the atomic number of a particle's surface corresponds to that of a nanoparticle's surface. The geometric surface of particle number grows as particle size decreases. Due to the particle's huge surface area and surface energy, the particle's surface atoms are highly active, interact swiftly with the surrounding gas, and readily absorb the gas. Unit mass paired with an increase in particle surface area and an increase in surface atomic number results in a significant deficiency in atomic coordination number. This phenomenon is known as a nanoscale material surface effect. People can use nanoparticles in a variety of ways to make better use of materials and find new applications for them (Batlle et al. 2002). When the ratio of surface atoms to bulk atoms is high, the surface spins contribute significantly to the nanoparticles' magnetic characteristics. Surface effects have a significant impact on the performance of magnetic nanoparticles.

5.2.3 SMALL SIZE EFFECT

The original crystal periodic boundary conditions are destroyed because the magnetic nanoparticles are so small that they can be affected by the magnetic exchange of wavelength, length, conduction electrons DE Broglie wavelength, the width of the magnetic domain wall, more physical characteristic length, or superconducting state coherence length. Some examples of the new effects are the change from magnetic order to magnetic disorder and the change in the magnetic coercive force.

5.3 SYNTHESIS OF MAGNETIC NANOPARTICLES

For the multistep synthesis of MNPs, meticulous attention to detail is required to produce the desired results (Majidi et al. 2016). Developing MNPs can be done in a variety of different methods. Building MNPs can be approached from either the "top-down" or the "bottom-up" perspective (Mehta 2017). The "top-down" method involves using high-energy ball milling to reduce the size of magnetic samples to the requisite nanoscale dimensions. This method has the advantage of enabling the manufacture of a large number of particles using only a single batch of ingredients. A disadvantage is that it does not control particle size and form, which is necessary for many biological applications (Öztürk et al. 2022). Starting with either a ferrous (Fe^{2+}) or ferric (Fe^{3+}) ion salt would be the first step in the "bottom-up" strategy. The salt is then subjected to its unique chemical process to nucleate and expedite seeded growth to develop particles to their appropriate hydrodynamic sizes (Mosayebi et al. 2017). This is done to create particles with the necessary hydrodynamic sizes. In the research that has been done, numerous "bottom-up" solutions have been mentioned. The procedures of hydrothermal synthesis, coprecipitation, polyol synthesis, and thermal breakdown (Table 5.1) are the ones that have been mentioned in the scientific literature the most frequently (Anik et al. 2021). Other techniques include the production of micelles, microemulsions, solgel, solvothermal processes, sonochemical processes, flow injection techniques, physical vapor deposition, and microwave-assisted processes. Other techniques include flow injection techniques, flow injection techniques, flow injection techniques, and flow injection techniques. Some processes that can be used include chemical vapor deposition, laser pyrolysis, electrodeposition, carbon ARC, and combustion (Akbarzadeh et al. 2012a and 2012b).

TABLE 5.1

Advantages and Disadvantages of the Most Popular Methods Used to Synthesize MNPs (Alromi et al. 2021)

Synthesis Method	Advantages	Disadvantages
Polyol	Cost-effective industrial application Biocompatibility	A complex synthesis of small particles Unstable oxidation
Hydrothermal	Excellent control of size, shape, and dispersion Magnetic controllability	Prolonged synthesis duration Adsorption of capping agents
Thermal decomposition	Excellent size distribution Great reproducibility	Soluble in organic solvents Toxicity
Coprecipitation	Easily scale up the production. Fast reaction	Poor reproducibility Surface oxidation

5.4 FUNCTIONALIZATION OF MAGNETIC NANOPARTICLES

MNPs must be functionalized. It permits biomedical MNP use (Table 5.2). Without functionalization, MNPs would be unstable in acidic conditions and leach, reducing their lifespan and reusability (Mollarasouli et al. 2021). Functionalizing MNPs improves their solubility, biocompatibility, surface catalytic activity, agglomeration, physiochemical, and mechanical properties (Zhu et al. 2018). MNP functionalization prevents oxidation. Iron, nickel, and cobalt, as well as their alloys, are sensitive to oxygen (Schladt et al. 2011). Oxidation can change maghemite (Fe_2O_3) and magnetite's magnetic properties (Laurent et al. 2008). As illustrated in Figure 5.2, the stability of colloids may be confirmed so that oxidation, agglomeration, and sedimentation can be avoided by modifying the surface of nanoparticles with antibodies, peptides, genes, ligands, and drugs to develop hybrid multifunctional nanomaterials.

In-situ and post-synthesis functionalization is possible. In contrast to the in-situ procedure, post-synthesis functionalizes MNPs following synthesis, ligand addition, ligand exchange, and encapsulation (Bohara et al. 2016). Due to the range of coating agents, encapsulation is most often used. Organic materials like polymers and surfactants can be encapsulated, as can inorganic materials like carbon, silica, metal, and metal oxides. Functionalizing polymers is a common biomedical procedure, especially in nanomedicine. Silica is a common inorganic encapsulating covering. Silica encapsulating techniques include the Stöber method, aerosol pyrolysis, microemulsion, and sodium silicate. Functionalized MNPs have been used for a number of biological applications, including drug delivery (Zhu et al. 2018; Uribe Madrid et al. 2015). Carbon, metals, and metal oxides have limited biological applications (Zhu et al. 2018).

5.5 APPLICATIONS OF MAGNETIC NANOPARTICLES

5.5.1 MNPs IN DRUG FORMULATION FOR DRUG DELIVERY

Therapeutic components are either loosely bound to MNPs or encased in composites of MNPs and polymers in magnetic drug delivery systems. Targeted medication delivery is significant because, as illustrated in Figure 5.3, it operates in a wide range of environmental conditions, including ultrasound, light, temperature, and magnetic fields to deliver pharmaceuticals directly to the site of the disease in an external magnetic field associated with nano-sized iron oxide (Fe_3O_4) particles (Chen et al. 2010). They are then placed inside a porous silica shell that has been stabilized with polyethylene glycol (Fe_3O_4-DOX/pSiO$_2$-PEG). The drug molecule is protected within the composite carrier by a thin layer of porous silica gel that functions as a barrier to escape the active ingredient, dioxane (DOX). The polymer Eudragit coats medications for intestinal absorption since its pH is sensitive. A possible application is transporting medicines to different sites (Linares et al. 2019).

In a petri dish containing ethanol, superparamagnetic nanoparticles coated with Eudragit and capable of retaining DOX were evaluated (Khizar et al. 2020). Drugs were released slowly from the colloidal particles but gradually increased in rate. Injecting IONPs coated in amino-silane into a rat intravenous cancer model caused

TABLE 5.2

Table Summarizing the Recent Applications of Functionalized MNPs in Various Biomedical Applications (Dash et al. 2022)

Magnetic Nanoparticle	Functionalization	Application	Applicable Model Systems	
			In vitro	*In vivo*
Magnablate	Iron nanoparticles	Thermal ablation		Prostate cancer
Ferriheme/ ferumoxytol	Iron oxide-carboxymethyl dextran	Vascular imaging		Migraine
		Monitor response to bevacizumab therapy		Glioma
		Detect recent myocardial infarction		Myocardial infarction
		Localize lymph node metastases		Pancreatic cancer
	VEGF-165 peptide coupling	Cardiovascular imaging/VEGF Delivery	HUVEC	
Ferumoxide	Iron oxide-dextran	Labeling of inflamed cells		MRI imaging of patient forearm
FIONs	PEG-phospholipid	Pancreatic islet graft imaging		Rat liver
(Gal-PEI-SPIO)	Galactose (Gal) and polyethylenimine (PEI)-modified MNP	siRNA duplexes targeting c-Met	Hepa1–6 cells	Hepatic tumor model in C57BL/6 mice
(ZnFe$_2$O$_4$-mSi) core nanoparticle	Magnetic zinc-doped iron oxide with mesoporous silica	let-7a micro-RNA + doxorubicin	Hela cells	Xenografted nude mice

(Continued)

TABLE 5.2 (Continued)

Table Summarizing the Recent Applications of Functionalized MNPs in Various Biomedical Applications (Dash et al. 2022)

Magnetic Nanoparticle	Functionalization	Application	Applicable Model Systems	
			In vitro	*In vivo*
SPION	T40 dextran coated, epichlorohydrine stabilized	MRI of lymph node, liver, intestine		Pig model
	Gemcitabine (Gem)- loaded PLGA-PEG functionalized	Targeted delivery		
	Daunomycin-loaded		MCF-7	
	PSMA targeted docetaxel-loaded		HeLa	
			PC-3	
PEI nanoparticle	Fe$_3$O$_4$-PEG-LAC-chitosan functionalized	Survivin-siRNA targeted delivery	K562, MCF-7cells	
Ferumoxide	Poly-L-lysine coated	Magnetic targeting in stroke	HB1.F3	Rat
IONPs	Polyacrylic acid-co-maleic acid (PAM) coated	tPA delivery	HUVEC	Rats, human blood
	Amine PEG coated with BSA/ATF protein surface conjugation	Urokinase delivery	PANC02	Mice
	Streptavidin-coated and HepB Ab conjugated	Electrochemical detection of HepB	Sandwich ELISA- based	
	Streptavidin-coated and biotinylated HIV-DNA probe conjugate	Viral detection	Electrical impedance based	

(Continued)

TABLE 5.2 (Continued)
Table Summarizing the Recent Applications of Functionalized MNPs in Various Biomedical Applications (Dash et al. 2022)

Magnetic Nanoparticle	Functionalization	Application	Applicable Model Systems	
			In vitro	*In vivo*
Fe_3O_4-virus-magnetic-MIPs (virus-MMIPs)	Green self-polymerization strategy using dopamine imprinting	Hep A virus detection	CHO cells	Human serum sample
Fe_3O_4 NPs	PEI coated	Genome editing by CRISPR/Cas9	HEK-293 cells	
	Amino (−NH$_2$) modified and poly amino coated to generate poly-NH$_2$-MNPs	SARS-CoV-2 RNA capture and spike protein detection	Spectroscopy and qPCR-based	Nasopharyngeal swab samples
Superparamagnetic particles	Tosyl group and Influenza protein coated	Viral detection by immunomagnetic assay		Saliva sample
Graphene oxide MNP	Carboxyl group	miR-122	Fluorescein-labeled HRP-CRET	
Zinc ferrite (ZnFe)	Silica coated and amino (−NH2) modified with carboxylic polymers	SARS-CoV-2 RNA capture	Automated *in vitro* RNA extraction	
BNF-80	Coated with protein A and SARS-CoV-2 antibody	SARS-CoV-2 (virus) detection	Spectroscopy-based	

Magnetic nanoparticles
coated with biocompatible
polymer (Dextran,
Polyethylene glycol,
Polyethylene oxide,
Polooxamines)
OR
Inorganic coatings
(SiO₂)

Biocompatible coating
can be functionalized
by carboxylic groups,
biotins, amino groups
avidin
They act as attachment
points for further
modification by drug or
antibodies

MNP

Drug molecules

MNP
Magnetic nanoparticles

Bio targeting agent
Antibodies, peptides,
genes, ligands

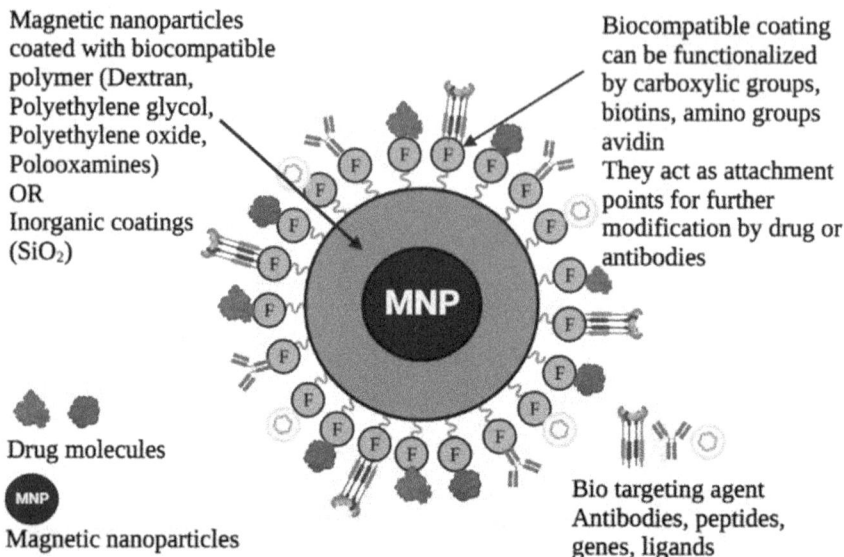

FIGURE 5.2 Surface stabilization of MNPs by peptides, antibodies, ligands, genes, and drugs for the fabrication of multifunctional MNPs for theranostic applications.

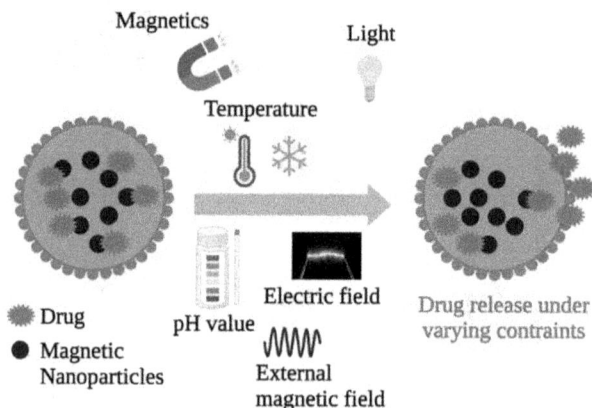

Magnetics   Light

Temperature

Drug release under
varying contraints

Drug
Magnetic
Nanoparticles

pH value

Electric field

External
magnetic field

FIGURE 5.3 Drug release from a nanocomposite in the presence of an external magnetic field, depicted under a variety of experimental conditions (magnetics, electricity, light, temperature, and pH).

the tumor's temperature to rise to 43°C (Jordan et al. 2006). In an AMF, Zhao et al. observed that administering an IONP solution to a mouse with tumors in its head and neck resulted in no damage to healthy cells (Zhao et al. 2012). With the help of the biocompatible polymer PEG, the mixture was able to escape the reticuloendothelial system, allowing for extended observation. The DOX was soaked up by the MNS and incorporated into the hydrogel. When RF fields are applied to MNS, the medicine is released twice as quickly since the drug is activated by heat. More than 80% of the cells were killed in the RF field, compared to 40% when there was no RF field when using a hydrogel/MNS composite loaded with DOX to deliver the medication locally to HeLa cell lines in a test tube. The antimicrobial cipro-floxacin and metal nanoparticles (MNPs) were rapidly liberated from PCL micro-spheres when they were magnetically stimulated. Using a single emulsion procedure, researchers developed a nanocarrier of magnetite nanoparticles loaded with DOX and encased in a biocompatible PLGA shell. Intravenous medications with a con-trolled release. Fe_3O_4 nanoparticles and DOX combined to create microbubbles with broad diagnostic potential (Nandwana et al. 2015; Sirivisoot et al. 2015; Jia et al. 2012). Anticancer theranostic agents were created by encapsulating paclitaxel and SPIONs in PLGA nanoparticles. A total of magnetite nanoparticles contain-ing DOX were created using a double-emulsion method and PLGA/PEG (Schleich et al. 2013). Nanoparticles made of polymers and drugs are employed for sustained release. 5-fluorouracil (5-FU)-encapsulated MNPs were modified with biocompati-ble PCLPEGPCL for *in vivo* and *in vitro* research (Asadi et al. 2018). With the help of a modified double-emulsion technique, oleic acid-coated superparamagnetic magne-tite nanoparticles are encapsulated in PLGA nanospheres of the same size (Mosafer et al. 2017). Magnetite nanoparticles (MNPs) were coated with PEGylated curcumin to create MNP@PEG-Cur composites. Nanoparticles release medicine at varying rates depending on the pH of the surrounding environment, with a greater proportion being released in an acidic environment (pH = 5.4). The hemolysis and cell survival tests performed at a pH of 7.4 demonstrated that MNP@PEG-Cur was biocompat-ible (Ayubi et al. 2019). Methotrexate was conjugated to iron oxide MNPs with an arginine cap to provide a novel drug delivery system. *In vitro* studies have shown that HFF-2 MCF-7, and 4T1 cells resist nanoparticles (Attari et al. 2019). Magnetic Fe_3O_4 nanoparticles were held together with chitosan and imatinib to create a medication delivery system. The pH-sensitive Fe_3O_4@CS produced imatinib, which inhibited the growth of MCF-7 breast cancer cells. It's more efficient than standard chemotherapy (Karimi Ghezeli et al. 2019). IONPs coated in silica delivered dopamine to malignant cells. In a pH 7.4 solution, Fe_3O_4 nanoparticles loaded with DA and SiO_2 released significantly more medication than uncoated Fe_3O_4 nanoparticles. Because of MNPs, access to the anticancer medication camptothecin was expanded (Dey et al. 2019). Drugs are combined with MNPs to alter drug release in living organisms, which is useful for cancer research. In biotechnology and biology, MNPs are utilized as car-riers (Patil et al. 2020).

Researchers have developed cancer drug delivery systems employing probes from a wide variety of disciplines. Some examples of effective and relatively safe nanopar-ticles include raspberry-shaped gold-coated magnetite nanoparticles and poly(-lactide-co-glycolide) (PLGA) nanocomposite particles. Boosting the magneto- and photothermal responses of the final probe by including gold and Fe_3O_4 nanoparticles was achieved. In a phosphate-buffered saline solution at 37°C, the nanocomposite particles released 5-FU in a manner that suggested they could be employed in photo-thermal therapy for DU145 prostate cancer cells (Keyvan Rad et al. 2019). However, whether or not photoresponsive magnetic composites are biodegradable and whether or not they represent significant safety issues is still a matter of discussion (Linsley and Wu 2017). Several naturally occurring chromophores, including oxy- and deoxy-hemoglobin, lipids, and water can absorb ultraviolet light. This suggests that the maximum penetration depth of UV radiation into tissues is less than 10mm. It's possible that the sun's rays, when exposed for extended periods, could cause harm. Thus, mucosal surfaces like the skin, eyes, and other mucosal organs should be avoided while using light-sensitive nanotherapeutics (Sahle et al. 2018). The new silica-coated iron oxide and polyaniline nanocomposites (Si-MNPs/PANI) have bio-medical uses and the ability to respond to dual stimulation. The nanocomposites Si-MNPs/PANI reacted to both magnetic and electrical fields. Their cytocompatibility and hemocompatibility further support these composites' potential for usage in dose-controlled drug delivery *in vitro*. Because of the risk of harming healthy tissue during deep tissue penetration, electrical stimulation has not been widely employed in medicine despite being simple and inexpensive to implement (attenuation of a stimulus) (Lalegül-Ülker et al. 2021).

5.5.2 MNPs for Chemotherapy

Chemotherapy is cytostatic and cytotoxic but cannot target cells. It's used to treat cancer and kill tumor cells. MNPs can be focused on after internalization. These therapeutic chemicals were released from nanoparticles into the cell's cytoplasm. The therapeutic molecule is made by combining chemotherapeutic pharmaceutical drugs with MNPs, which are delivered intravenously beneath an external magnetic field (Patra et al. 2018). First-stage chemotherapy involves releasing a chemical to damage cancer cell structure. Magnetic fields promote MNP biodistribution for safe drug delivery to the microenvironment. The use of medicines at low enough concentrations is made feasible by MNPs, which decreases the negative effects of drugs. These medicinal chemicals improve intracellular penetration, drug circulation duration, and hydrophobic solubility over standard drugs. Controlled drug delivery systems reduce nonspecific absorption, are more efficient, offer greater targeting capabilities, and regulate drug release (parenteral, transdermal, oral, and pulmonary) (Kadian 2018). The generated nanocarrier based on IONPs worked as an actual and efficient transport carrier for the medication DOX. It could destroy cancer cells remotely by creating heat in an AMF. The LHRH peptide-functionalized PEG-coated nanocarrier enhances the medicines' targeting ability and stability in physiological media (Taratula et al. 2013). The IONP-based technology is reliable for treating ovarian cancer using a combinatorial strategy (Figure 5.4).

MCF-7 was supplied using a DOX-loaded magnetic alginate-chitosan microsphere. SPIONs were used to promote the DOX release. MCF-7 cells were killed *in vitro* by magnetic hyperthermia or chemotherapy. *In vivo* studies with anticancer medicines and treatment indicated the tumor vanished after 12days (Xue et al. 2018). To kill cancer cells, GO-coated Fe_3O_4 nanospheres containing DOX were PEI-functionalized. *In vivo* and *in vitro* studies proposed a drug release method (DOXGO/IONP/PEI) that eliminated tumor cells due to chemotherapeutic effects. This system didn't destroy healthy tissues (Zhu et al. 2015). DOX is functionalized with -Fe_2O_3 nanoparticles and encapsulated in PLGA microspheres to prevent tumor

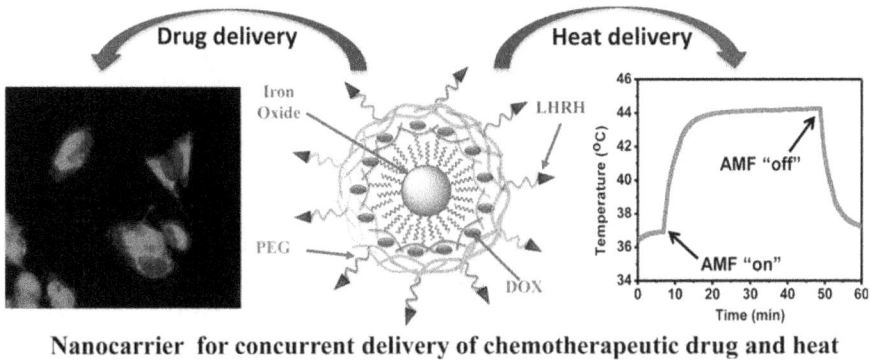

Nanocarrier for concurrent delivery of chemotherapeutic drug and heat

FIGURE 5.4 Nanocarrier for combination chemotherapy and hyperthermia caused by therapeutic drug release and heat, respectively. Reprinted from ref Taratula et al. 2013 Copyright 2022 Elsevier.

growth. MNPs inside the shell caused heat and drug release permeability. Heat and chemotherapy improve breast cancer cell (4T1) apoptosis *in vitro* (Fang et al. 2015). MNPs and carmustine were added to the oleosome's lipid core to treat cancer. Magnetic oleosomes release anticancer medicine in a magnetic field (Cho et al. 2018). Localized pancreatic cancer requires thermotherapy and chemotherapy (PANC-1 and BxPC3). Combining MNPs with gemcitabine and nucAnt (N6L) regulated the release of chemotherapeutics in a magnetic field since heat slows cell growth (Sanhaji et al. 2019). Encasing a chemotherapeutic medication in a magnetic core and coating it with a "Heat Shock Protein Inhibitor" against lung cancer stem cells (CSCs) created a silica-coated MNP multifunctional system. These nanoparticles were utilized topically for AMF-based chemotherapy and thermotherapy. After 30 minutes, chemotherapeutic medication and heat therapy killed 98% of lung CSCs *in vitro*. In *in vivo* trials with lung CSC xenograft-bearing mice, this combination therapy reduced tumor development and spread with few side effects (Liu et al. 2020). A DOX-loaded SPION-based hydrogel displayed regulated drug release and long-lasting efficacy. This novel hydrogel composition may prevent cancer relapses (Gao et al. 2019).

Fe_3O_4-cored nanoparticles in an AMF can release DOX from an alginate shell, releasing heat that can be used with chemotherapy. Folate was linked to nanoparticles, allowing the nanocomposite to enter and persist in cancer cells. *In vivo* proof of reciprocal therapy for mouse lung cancer required a folate component (Ha et al. 2019).

5.5.3 BIOIMAGING

Important data about the therapeutic cells' principle, durability, and activity can be gleaned through molecular imaging after transplantation (Chang et al. 2016; Ye et al. 2017). Imaging methods can be divided into three categories: radionuclide radiation, magnetic resonance imaging (MRI), and light (optical). To this end, we typically employ contrast chemicals to label cells and signal subsequent separation from the host tissue. However, optical imaging methods are limited by both the wavelength of the applied light and the depth to which it may penetrate the human body. Radionuclide imaging techniques are the most sensitive in medical practice. However, there are still problems, such as the inability to follow cells over time and the lack of anatomical knowledge, which put people's health at risk. Researchers de Vries et al. (2005) found that MRI might efficiently monitor the progress of therapeutic cells in patients without requiring intrusive treatments. The use of MRI to track cells for therapeutic purposes is not without its limitations. So far, we have a fragmented grasp of how these transplanted creatures are distributed. The biggest issue is that after stem cells have been engrafted, we have no idea what their purpose is supposed to be (Gao et al. 2013). Using MRI and other sensors, scientists want to develop a versatile platform that can reveal spatial and functional details (Xu et al. 2008).

5.5.4 MAGNETIC HYPERTHERMIA

Hyperthermia destroys tumor cells by elevating a tumor's temperature to 41°C to 45°C (Somvanshi et al. 2020). When metastatic disease cells have spread throughout the body, whole-body hyperthermia is employed (Patade et al. 2020). Warming a tumor is one approach to produce limited hyperthermia. Radiofrequency, ultrasound, infrared radiation, microwaves, magnetic thermo-seeds, etc. produce hyperthermia. Traditional hyperthermia therapy causes tissue damage or lacks heat penetration. Ferromagnetic or superparamagnetic particles heat tumor tissue in MNPs-based hyperthermia therapy. MNPs have some advantages over hyperthermia. Because MNPs utilized for hyperthermia are a few nanometers in size, they can easily reach tumors after intravenous infusion with the help of a magnetic field, boosting hyperthermia's explicitness. The MNPs distinctive architecture includes explicit malignant growth restriction operators targeting tumor tissues. The far-shifting attractive field used to heat MNPs kills tumor cells while inflicting minimal damage to healthy tissue. Figure 5.5 shows how MNPs are injected near tumors and heated by an AC magnetic field to kill cancer cells.

5.5.5 DISEASE THERAPY

Weak target site capabilities and impaired drug dispersion across bio-barriers are common problems with conventional drug administration, leading to reduced pharmacological activity and an increased risk of side effects (Kang et al. 2016; Xie et al. 2016; Chen et al. 2016a; Zhou et al. 2014). Drug packaging and transport of nanodevices offer potential solutions to these problems. Recent research has focused on vector-mediated drug delivery for its potential to increase treatment efficacy and stability by affecting pharmacokinetics and pharmacodynamics (Yang et al. 2014, 2015, 2016; Xie et al. 2013). Nanocarriers can transport drugs across the body and into sick cells more effectively (Yu et al. 2014; Wang et al. 2014; Chen et al. 2016b; Song et al. 2014). Increasing the drug's exposure and lengthening its half-life would boost its therapeutic efficacy. One of the active drug-targeting approaches is using stimuli-sensitive carriers, which allows for the targeted delivery of pharmaceuticals within the human body. By manipulating the physical features of colloidal systems in response to an external stimulus, we can increase the concentration of the medicine

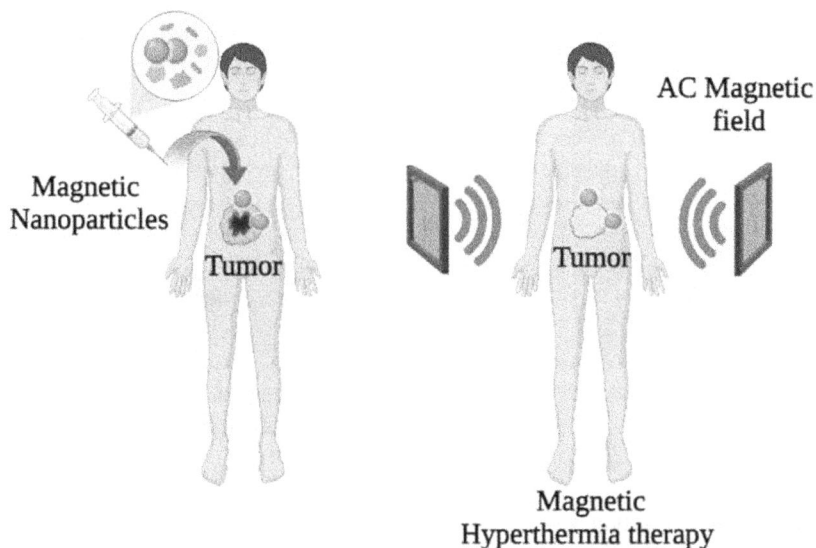

FIGURE 5.5   Schematics of MNPs-based hyperthermia therapy for cancer treatment.

at the site where it is to be administered. Because of their unique magnetic reactivity in this setting, MNPs have the potential as stimuli-sensitive drug carriers. Thus, by directing such nanodevices to the intended location, medications can be kept there for an extended period before being released, greatly minimizing the adverse effects of drug use due to nonspecific distribution (Ibrahim et al. 1983; Kohler et al. 2005). Larger particles have stronger sensitivity to magnetic guiding. They are thus more typically utilized as MNPs as magnetic targets, but smaller particles have good tissue permeability and are hence the most widely used MNPs as heat emitters (Arias 2008). Using magnetic nanoparticles to administer targeted magnetic hyperthermia (TMH) is a promising new approach to treating tumors *in vivo*. Due to its substantial therapeutic benefits, high level of safety, and lack of adverse effects, the therapy may gain a great deal of attention in future clinical research. However, improving the therapeutic efficacy of MNPs-based passive or active targeted hyperthermia treatment via intravenous distribution *in vivo* is a challenge.

5.5.6   TISSUE ENGINEERING AND REGENERATIVE MEDICINE

A possible method of reconstructing damaged tissues using tissue engineering. Over the past decade, many efforts have been made to utilize SPION's unique properties in regenerative medicine applications, including bone, nerve, cartilage, and tendon regeneration (Liu et al. 2020; Gonçalves et al. 2016). Magnetic nanoparticles have been shown to alter cellular biological processes by numerous research organizations. An intense magnetic field could stimulate the cell cycle and cell proliferation. Magnetomechanical contact with cells also allows for morphology, proliferation, and differentiation modifications. Tissue engineers have used magnetically actuated cell manipulation (MAM) in four main ways: (ia) magnetic targeting to guide cells to the intended location; (ii) magnetic seeding of the cells into the scaffold; (iii) magnetic scaffold; and (iv) magnetic labeling of cells to generate flattened structures (Zhang et al. 2020). Magnetic scaffolds' distinctive properties have attracted much interest in their potential uses. It has been shown that the magnetic field in magnetized scaffolds affects several important facets of tissue engineering, including vascularization, the induction of appropriate cellular mechano- and electro-transduction processes, the enhancement of cellular adhesion, proliferation, and differentiation, and the formation of a significant bioactive interface. For bone healing, IONPs were used to create PLGA/PCL scaffolds (Tao et al. 2020). Using MNPs can enhance the vital bioactive cell-scaffold interaction in tissue engineering. SPION serves as an overlying layer to alter the scaffold's exterior. The study found that IONPs enhanced the osteogenic differentiation of adipose-derived stem cells and enhanced the mechanical characteristics of scaffolds (Chen et al. 2018). Preosteoblasts' biological reaction to magnetic materials' physical stimulation has been studied.

A combination of piezoelectric polymers with magneto-responsive NPs of $CoFe_2O_4$ creates the required scaffolds. Magnetoactive materials with a trabecular bone-like structure provide the mechano- and electro-transduction required to stimulate preosteoblast adhesion and proliferation in bone tissue engineering (Figure 5.6) (Fernandes et al. 2019). Three-dimensional (3D) magnetic nanocomposite scaffolds of poly(-caprolactone) supported by iron-doped hydroxyapatite (FeHA) NPs, magnetized polyethylene glycol (PEG)-based hydrogels, and chitosan scaffolds loaded with SPION plasmid gene microspheres have all been used in studies showing enhanced bone regeneration and vascularization. The glycosylated SPIONs-agarose hydrogel scaffolds for tissue engineering have been created in various research using MNPs (Li et al. 2018; Aliramaji et al. 2017; Zhao et al. 2019; Bin et al. 2020; Lai et al. 2018; Díaz et al. 2016). Results for nerve regeneration using magnetic fields and flexible scaffolds are encouraging (Karimi et al. 2021). The introduction of IONPs to nerve

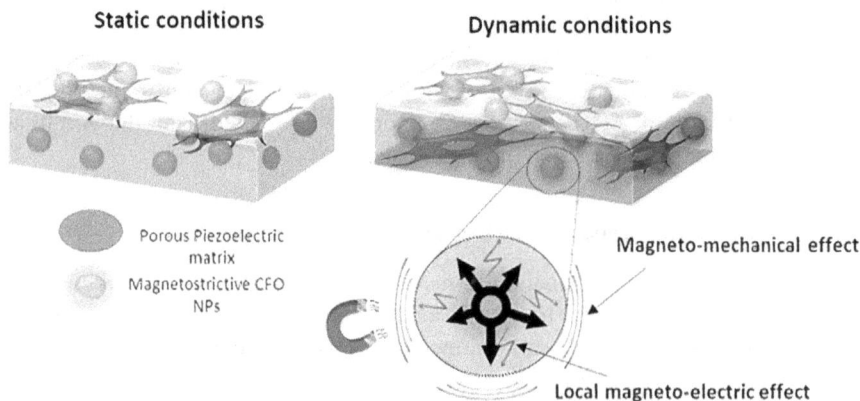

FIGURE 5.6 Schematic illustration of magnetomechanical and local magnetoelectrical features of three-dimensional PVDF/CFO scaffolds upon the application of magnetic stimuli. Reprinted from Fernandes et al. 2019. Copyright 2022 ACS.

scaffolds led to dramatic changes in neurite outgrowth, axonal extension, and signaling pathways that promote neuronal differentiation due to direct interactions between SPIONs and proteins (Karimi et al. 2021; Chen et al. 2020). The usage of a composite NGC scaffold containing PCL, melatonin, and SPIONs was found to promote axonal development and inhibit inflammation (Chen et al. 2020). In addition, a nanocomposite scaffold made of chitosan-glycerophosphate polymers was shown to stimulate nerve regeneration and improve Schwann cell survival in a study conducted by Liu et al. (2017). Tissue engineers have found many uses for magnetically sensitive smart hydrogels, including wound healing and bone, neuron, muscle, and heart repair. The results showed that tissue regeneration might be sped up by combining IONPs with biomaterials such as chitosan, collagen, polyacrylamide, poly(lactide-coglycolide) dextran, agarose, and poly(polyacrylamide) (Liu et al. 2020).

5.5.7 BIOSENSORS

The sensor and MNPs authoritative to the sensor make up the functionalized unit of magneto-resistive sensors used in biosensing applications (Tu 2013). The magnetic fields created by the MNPs that modify the sensor's magnetic fields cause variations in electrical flow or opposition within the sensor, which is the basic basis for this class of magneto-resistive sensors (Hasanzadeh et al. 2015). When compared to electrochemical, radioactive, or colorimetric methods, sensors based on MNPs have shown considerable biomarker recognition and capability focus points, as well as high affectability, long-term security, and the potential for equipment downsizing. The approach has so found use in a number of fields, including clinical analysis, the food industry, and environmental protection. Direct marking and incorrect naming are examples of binding methods used by MNPs to stay put on the sensor surface. These MNPs disrupt magnetic fields, and magnetic sensors pick up on this. The sandwich immunoassay is the gold standard for indirect marking. Antibodies specific to the target protein are attached to the outside of the device.

5.6 CONCLUSION

Functionalized MNPs have been studied in biomedicine for years. Rapid progress in synthesizing MNPs with organic ligands on their surfaces has been made. Organic ligand-mediated MNPs improve stability, solubility, biocompatibility, cancer targeting, and cellular absorption.

Magnetic nanoparticles are employed in medicine, imaging, and medication delivery in biomedicine. Recent breakthroughs in synthesizing and functionalizing monodispersed, tunable, and adjustable magnetic particle size and shape have helped the scientific community address and resolve various therapeutic magnetic particle issues. Magnetic nanoparticles are good for targeted drug delivery due to their high surface-to-volume ratio and magnetic characteristics. Magnetic nanocomposites, including magnetic dendrimers, magnetic hydrogel, magnetic liposomes, etc., helped scientists overcome challenges like premature burst release of loaded drugs and enhanced targeting efficacy. Magnetic hyperthermia-aided apoptosis and necrosis can fight numerous disorders, including cancer. Combining localized magnetic hyperthermia with drug administration may treat the disease. Magnetic nanoparticle-based theranostics can diagnose and treat diseases. CT, MRI, PET, and MPI have great potential for accurate illness diagnosis and therapy. Low magnetization, intrinsic magnetism's tendency to combine, and magnetic nanoparticle toxicity demand quick attention. The future of magnetic particles in biomedicine lies in knowing their therapeutic efficacy and toxicity in complicated biological systems.

REFERENCES

Akbarzadeh A, Mikaeili H, Zarghami N, Mohammad R, Barkhordari A, Davaran S. Preparation and in vitro evaluation of doxorubicin-loaded Fe_3O_4 magnetic nanoparticles modified with biocompatible copolymers. *Int J Nanomedicine*. 2012a;7:511–526. doi:10.2147/IJN.S24326.

Akbarzadeh A, Samiei M, Davaran S. Magnetic nanoparticles: preparation, physical properties, and applications in biomedicine. *Nanoscale Res Lett*. 2012b;7(1):144. doi:10.1186/1556-276X-7-144.

Ali A, Shah T, Ullah R, Zhou P, Guo M, Ovais M, Tan Z, Rui Y. Review on recent progress in magnetic nanoparticles: Synthesis, characterization, and diverse applications. *Front Chem*. 2021;9:629054. doi:10.3389/fchem.2021.629054.

Aliramaji S, Zamanian A, Mozafari M. Super-paramagnetic responsive silk fibroin/chitosan/magnetite scaffolds with tunable pore structures for bone tissue engineering applications. *Mater Sci Eng C Mater Biol Appl*. 2017;70(Pt 1):736–744. doi:10.1016/j.msec.2016.09.039.

Alromi DA, Madani SY, Seifalian A. Emerging application of magnetic nanoparticles for diagnosis and treatment of cancer. *Polymers (Basel)*. 2021;13(23):4146. doi:10.3390/polym13234146.

Anik MI, Hossain MK, Hossain I, Mahfuz AM, Rahman MT, Ahmed I. Recent progress of magnetic nanoparticles in biomedical applications: A review. *Nano Select*. 2021;2(6):1146–1186. doi:10.1002/nano.202000162.

Arias JL. Novel strategies to improve the anticancer action of 5-fluorouracil by using drug delivery systems. *Molecules*. 2008;13(10):2340–2369. doi:10.3390/molecules13102340.

Asadi N, Annabi N, Mostafavi E, et al. Synthesis, characterization and in vitro evaluation of magnetic nanoparticles modified with PCL-PEG-PCL for controlled delivery of 5FU [retracted in: Artif Cells Nanomed Biotechnol. 2022 Dec;50(1):108]. *Artif Cells Nanomed Biotechnol*. 2018;46(sup 1):938–945. doi:10.1080/21691401.2018.1439839.

Attari E, Nosrati H, Danafar H, Kheiri Manjili H. Methotrexate anticancer drug delivery to breast cancer cell lines by iron oxide magnetic based nanocarrier. *J Biomed Mater Res A*. 2019;107(11):2492–2500. doi:10.1002/jbm.a.36755.

Ayubi M, Karimi M, Abdpour S, et al. Magnetic nanoparticles decorated with PEGylated curcumin as dual targeted drug delivery: Synthesis, toxicity and biocompatibility study. *Mater Sci Eng C Mater Biol Appl*. 2019;104:109810. doi:10.1016/j.msec.2019.109810.

Baki A, Wiekhorst F, Bleul R. Advances in magnetic nanoparticles engineering for biomedical applications: A review. *Bioengineering (Basel)*. 2021;8(10):134. doi:10.3390/bioengineering8100134.

Batlle X, Labarta A. Finite-size effects in fine particles: magnetic and transport properties. *J Phys D Appl Phys*. 2002;35(6):201. doi:10.1088/0022-3727/35/6/201.

Bin S, Wang A, Guo W, Yu L, Feng P. Micro magnetic field produced by Fe3O4 nanoparticles in bone scaffold for enhancing cellular activity. *Polymers (Basel)*. 2020;12(9):2045. doi:10.3390/polym12092045.

Bohara RA, Thorat ND, Pawar SH. Role of functionalization: strategies to explore potential nano-bio applications of magnetic nanoparticles. *RSC Advances*. 2016;6(50):43989–4012. doi:10.1039/C6RA02129H.

Chang L, Hu J, Chen F, et al. Nanoscale bio-platforms for living cell interrogation: current status and future perspectives. *Nanoscale*. 2016;8(6):3181–3206. doi:10.1039/c5nr06694h.

Chen FH, Zhang LM, Chen QT, Zhang Y, Zhang ZJ. Synthesis of a novel magnetic drug delivery system composed of doxorubicin-conjugated Fe3O4 nanoparticle cores and a PEG-functionalized porous silica shell. *Chem Commun (Camb)*. 2010;46(45):8633–8635. doi:10.1039/c0cc02577a

Chen H, Sun J, Wang Z, et al. Magnetic cell-scaffold interface constructed by superparamagnetic IONP enhanced osteogenesis of adipose-derived stem cells. *ACS Appl Mater Interfaces*. 2018;10(51):44279–44289. doi:10.1021/acsami.8b17427.

Chen X, Ge X, Qian Y, Tang H, Song J, Qu X, Yue B, Yuan WE. Electrospinning multilayered scaffolds loaded with melatonin and Fe3O4 magnetic nanoparticles for peripheral nerve regeneration. *Adv Funct Mat*. 2020;30(38):2004537. doi:10.1002/adfm.202004537.

Chen Z, Chen Z, Zhang A, Hu J, Wang X, Yang Z. Electrospun nanofibers for cancer diagnosis and therapy. *Biomater Sci*. 2016a;4(6):922–932. doi:10.1039/c6bm00070c.

Chen Z, Zhang A, Yang Z, Wang X, Chang L, Chen Z, James Lee L. Application of DODMA and derivatives in cationic nanocarriers for gene delivery. *Curr Organ Chem*. 2016b;20(17):1813–1819. doi:10.2174/1385272820666160202004348.

Cho HY, Lee T, Yoon J, et al. Magnetic oleosome as a functional lipophilic drug carrier for cancer therapy. *ACS Appl Mater Interfaces*. 2018;10(11):9301–9309. doi:10.1021/acsami.7b19255.

Cole AJ, David AE, Wang J, Galbán CJ, Hill HL, Yang VC. Polyethylene glycol modified, cross-linked starch-coated iron oxide nanoparticles for enhanced magnetic tumor targeting. *Biomaterials*. 2011;32(8):2183–2193. doi:10.1016/j.biomaterials.2010.11.040.

Darroudi M, Gholami M, Rezayi M, Khazaei M. An overview and bibliometric analysis on the colorectal cancer therapy by magnetic functionalized nanoparticles for the responsive and targeted drug delivery. *J Nanobiotechnol*. 2021;19(1):399. doi:10.1186/s12951-021-01150-6.

Dash S, Das T, Patel P, Panda PK, Suar M, Verma SK. Emerging trends in the nanomedicine applications of functionalized magnetic nanoparticles as novel therapies for acute and chronic diseases. *J Nanobiotechnol*. 2022;20(1):393. doi:10.1186/s12951-022-01595-3.

de Vries IJ, Lesterhuis WJ, Barentsz JO, et al. Magnetic resonance tracking of dendritic cells in melanoma patients for monitoring of cellular therapy. *Nat Biotechnol*. 2005;23(11):1407–1413. doi:10.1038/nbt1154.

Dey C, Das A, Goswami MM. Dopamine loaded SiO2 coated Fe3O4 magnetic nanoparticles: a new anticancer agent in pH-dependent drug delivery. *ChemistrySelect*. 2019;4(41):12190–12196. doi:10.1002/slct.201902909.

Díaz E, Valle MB, Barandiarán JM. Magnetic composite scaffolds of polycaprolactone/nFeHA, for bone-tissue engineering. *Int J Poly Mat Poly Biomat*. 2016;65(12):593–600. doi:1080/00914037.2016.1149848.

Fang K, Song L, Gu Z, Yang F, Zhang Y, Gu N. Magnetic field activated drug release system based on magnetic PLGA microspheres for chemo-thermal therapy. *Colloids Surf B Biointerfaces*. 2015;136:712–720. doi:10.1016/j.colsurfb.2015.10.014.

Faraudo J, Camacho J. Cooperative magnetophoresis of superparamagnetic colloids: theoretical aspects. *Colloid Polymer Sci*. 2010;288(2):207–215. doi:10.1007/s00396-011-2454-4.

Fernandes MM, Correia DM, Ribeiro C, Castro N, Correia V, Lanceros-Mendez S. Bioinspired three-dimensional magnetoactive scaffolds for bone tissue engineering. *ACS Appl Mater Interfaces*. 2019;11(48):45265–45275. doi:10.1021/acsami.9b14001.

Furlani EP. Magnetic biotransport: analysis and applications. *Materials*. 2010;3(4):2412–2446. doi:10.3390/ma3042412.

Gao F, Xie W, Miao Y, et al. Magnetic hydrogel with optimally adaptive functions for breast cancer recurrence prevention. *Adv Healthc Mater*. 2019;8(14): e1900203. doi:10.1002/adhm.201900203.

Gao Y, Cui Y, Chan JK, Xu C. Stem cell tracking with optically active nanoparticles. *Am J Nucl Med Mol Imaging.* 2013;3(3):232–246.

Gonçalves AI, Rodrigues MT, Carvalho PP, et al. Exploring the potential of starch/ polycaprolactone aligned magnetic responsive scaffolds for tendon regeneration. *Adv Healthc Mater.* 2016;5(2):213–222. doi:10.1002/adhm.201500623.

Ha PT, Le TT, Bui TQ, Pham HN, Ho AS, Nguyen LT. Doxorubicin release by magnetic inductive heating and in vivo hyperthermia-chemotherapy combined cancer treatment of multifunctional magnetic nanoparticles. *New J Chem.* 2019;43(14):5404–5413. doi:10.1039/C9NJ00111E.

Hasanzadeh M, Shadjou N, de la Guardia M. Iron and iron-oxide magnetic nanoparticles as signal-amplification elements in electrochemical biosensing. *TrAC Trends Anal Chem.* 2015;72:1–9. doi:10.1016/j.trac.2015.03.016.

Ibrahim A, Couvreur P, Roland M, Speiser P. New magnetic drug carrier. *J Pharm Pharmacol.* 1983;35(1):59–61. doi:10.1111/j.2042-7158.1983.tb04269.x.

Jeevanandam J, Barhoum A, Chan YS, Dufresne A, Danquah MK. Review on nanoparticles and nanostructured materials: history, sources, toxicity and regulations. *Beilstein J Nanotechnol.* 2018;9:1050–1074. doi:10.3762/bjnano.9.98.

Jia Y, Yuan M, Yuan H, et al. Co-encapsulation of magnetic Fe3O4 nanoparticles and doxorubicin into biodegradable PLGA nanocarriers for intratumoral drug delivery. *Int J Nanomedicine.* 2012;7:1697–1708. doi:10.2147/IJN.S28629.

Jordan A, Scholz R, Maier-Hauff K, et al. The effect of thermotherapy using magnetic nanoparticles on rat malignant glioma. *J Neurooncol.* 2006;78(1):7–14. doi:10.1007/s11060-005-9059-z.

Kadian R. Nanoparticles: A promising drug delivery approach. *Asian J Pharm Clin Res.* 2018;11(1):30–35. doi:10.22159/ajpcr. 2018.v11i1.22035.

Kang C, Sun Y, Zhu J, et al. Delivery of nanoparticles for treatment of brain tumor. *Curr Drug Metab.* 2016;17(8):745–754. doi:10.2174/1389200217666160728152939.

Kankala RK, Han YH, Xia HY, Wang SB, Chen AZ. Nanoarchitectured prototypes of mesoporous silica nanoparticles for innovative biomedical applications. *J Nanobiotechnol.* 2022;20(1):126. doi:10.1186/s12951-022-01315-x.

Karimi Ghezeli Z, Hekmati M, Veisi H. Synthesis of imatinib-loaded chitosan-modified magnetic nanoparticles as an anti-cancer agent for pH responsive targeted drug delivery. *Appl Organometallic Chem.* 2019;33(4):e4833. doi:10.1002/aoc.4833.

Karimi S, Bagher Z, Najmoddin N, Simorgh S, Pezeshki-Modaress M. Alginate-magnetic short nanofibers 3D composite hydrogel enhances the encapsulated human olfactory mucosa stem cells bioactivity for potential nerve regeneration application. *Int J Biol Macromol.* 2021;167:796–806. doi:10.1016/j.ijbiomac.2020.11.199.

Keyvan Rad J, Alinejad Z, Khoei S, Mahdavian AR. Controlled release and photothermal behavior of multipurpose nanocomposite particles containing encapsulated gold-decorated magnetite and 5-FU in Poly(lactide-*co*-glycolide). *ACS Biomater Sci Eng.* 2019;5(9):4425–4434. doi:10.1021/acsbiomaterials.9b00790.

Khizar S, Ahmad NM, Ahmed N, Manzoor S, Elaissari A. Encapsulation of doxorubicin in magnetic-polymer hybrid colloidal particles of Eudragit E100 and their hyperthermia and drug release studies. *Polymers Adv Technol.* 2020;31(8):1732–1743. doi:10.1002/pat.4900.

Kohler N, Sun C, Wang J, Zhang M. Methotrexate-modified superparamagnetic nanoparticles and their intracellular uptake into human cancer cells. *Langmuir.* 2005;21(19):8858–8864. doi:10.1021/la0503451.

Kolhatkar AG, Jamison AC, Litvinov D, Willson RC, Lee TR. Tuning the magnetic properties of nanoparticles. *Int J Mol Sci.* 2013;14(8):15977–16009. doi:10.3390/ijms140815977.

Kyeong S, Kim J, Chang H. Magnetic nanoparticles. *Adv Exp Med Biol.* 2021;1309:191–215. doi:10.1007/978-981-33-6158-4_8.

Lai WY, Feng SW, Chan YH, Chang WJ, Wang HT, Huang HM. In vivo investigation into effectiveness of Fe$_3$O$_4$/PLLA nanofibers for bone tissue engineering applications. *Polymers (Basel)*. 2018;10(7):804. doi:10.3390/polym10070804.

Lalegül-Ülker Ö, Elçin YM. Magnetic and electrically conductive silica-coated iron oxide/polyaniline nanocomposites for biomedical applications. *Mater Sci Eng C Mater Biol Appl*. 2021;119:111600. doi:10.1016/j.msec.2020.111600.

Laurent S, Forge D, Port M, Roch A, Robic C, Vander Elst L, Muller RN. Magnetic iron oxide nanoparticles: synthesis, stabilization, vectorization, physicochemical characterizations, and biological applications. *Chemical Reviews*. 2008;108(6):2064–2110. doi:10.1021/cr068445e.

Li C, Armstrong JP, Pence IJ, et al. Glycosylated superparamagnetic nanoparticle gradients for osteochondral tissue engineering. *Biomaterials*. 2018;176:24–33. doi:10.1016/j.biomaterials.2018.05.029.

Lim J, Lanni C, Evarts ER, Lanni F, Tilton RD, Majetich SA. Magnetophoresis of nanoparticles. *ACS Nano*. 2011;5(1):217–226. doi:10.1021/nn102383s.

Linares V, Yarce CJ, Echeverri JD, Galeano E, Salamanca CH. Relationship between degree of polymeric ionisation and hydrolytic degradation of Eudragit® E polymers under extreme acid conditions. *Polymers (Basel)*. 2019;11(6):1010. doi:10.3390/polym11061010.

Linsley CS, Wu BM. Recent advances in light-responsive on-demand drug-delivery systems. *Ther Deliv*. 2017;8(2):89–107. doi:10.4155/tde-2016-0060.

Liu D, Hong Y, Li Y, et al. Targeted destruction of cancer stem cells using multifunctional magnetic nanoparticles that enable combined hyperthermia and chemotherapy. *Theranostics*. 2020;10(3):1181–1196. doi:10.7150/thno.38989.

Liu Z, Liu J, Cui X, Wang X, Zhang L, Tang P. Recent advances on magnetic sensitive hydrogels in tissue engineering. *Front Chem*. 2020;8:124. doi:10.3389/fchem.2020.00124.

Liu Z, Zhu S, Liu L, et al. A magnetically responsive nanocomposite scaffold combined with Schwann cells promotes sciatic nerve regeneration upon exposure to magnetic field. *Int J Nanomedicine*. 2017;12:7815–7832. doi:10.2147/IJN.S144715.

Majidi S, Zeinali Sehrig F, Farkhani SM, Soleymani Goloujeh M, Akbarzadeh A. Current methods for synthesis of magnetic nanoparticles. *Artif Cells Nanomed Biotechnol*. 2016;44(2):722–734. doi:10.3109/21691401.2014.982802.

Mehta RV. Synthesis of magnetic nanoparticles and their dispersions with special reference to applications in biomedicine and biotechnology. *Mater Sci Eng C Mater Biol Appl*. 2017;79:901–916. doi:10.1016/j.msec.2017.05.135.

Mollarasouli F, Zor E, Ozcelikay G, Ozkan SA. Magnetic nanoparticles in developing electrochemical sensors for pharmaceutical and biomedical applications. *Talanta*. 2021;226:122108. doi:10.1016/j.talanta.2021.122108.

Mosafer J, Abnous K, Tafaghodi M, Jafarzadeh H, Ramezani M. Preparation and characterization of uniform-sized PLGA nanospheres encapsulated with oleic acid-coated magnetic-Fe3O4 nanoparticles for simultaneous diagnostic and therapeutic applications. *Colloids Surf A Physicochem Eng Asp*. 2017;514:146–154. doi:10.1016/j.colsurfa.2016.11.056.

Mosayebi J, Kiyasatfar M, Laurent S. Synthesis, functionalization, and design of magnetic nanoparticles for theranostic applications. *Adv Healthc Mater*. 2017;6(23). doi:10.1002/adhm.201700306.

Nandwana V, De M, Chu S, et al. Theranostic magnetic nanostructures (MNS) for cancer. *Cancer Treat Res*. 2015;166:51–83. doi:10.1007/978-3-319-16555-4_3.

Ooi YJ, Wen Y, Zhu J, Song X, Li J. Surface charge switchable polymer/DNA nanoparticles responsive to tumor extracellular pH for tumor-triggered enhanced gene delivery. *Biomacromolecules*. 2020;21(3):1136–1148. doi:10.1021/acs.biomac.9b01521.

Öztürk Er E, Dalgıç Bozyiğit G, Büyükpınar Ç, Bakırdere S. Magnetic nanoparticles based solid phase extraction methods for the determination of trace elements. *Crit Rev Anal Chem*. 2022;52(2):231–249. doi:10.1080/10408347.2020.1797465.

Patade SR, Andhare DD, Somvanshi SB, Jadhav SA, Khedkar MV, Jadhav KM. Self-heating evaluation of superparamagnetic MnFe2O4 nanoparticles for magnetic fluid hyperthermia application towards cancer treatment. *Ceramics Int.* 2020;46(16):25576-25583. doi:10.1016/j.ceramint.2020.07.029.

Patil AR, Nimbalkar MS, Patil PS, Chougale AD, Patil PB. Controlled release of poorly water soluble anticancerous drug camptothecin from magnetic nanoparticles. *Materials Today Proc.* 2020;23:437–443. doi:10.1016/j.matpr.2020.02.064.

Patra JK, Das G, Fraceto LF, et al. Nano based drug delivery systems: recent developments and future prospects. *J Nanobiotechnol.* 2018;16(1):71. doi:10.1186/s12951-018-0392-8.

Rizwanullah M, Ahmad MZ, Ghoneim MM, Alshehri S, Imam SS, Md S, Alhakamy NA, Jain K, Ahmad J. Receptor-mediated targeted delivery of surface-modified nanomedicine in breast cancer: recent update and challenges. *Pharmaceutics.* 2021;13(12):2039. doi:10.3390/pharmaceutics13122039.

Sahle FF, Gulfam M, Lowe TL. Design strategies for physical-stimuli-responsive programmable nanotherapeutics. *Drug Discov Today.* 2018;23(5):992–1006. doi:10.1016/j.drudis.2018.04.003.

Sahoo Y, Goodarzi A, Swihart MT, et al. Aqueous ferrofluid of magnetite nanoparticles: Fluorescence labeling and magnetophoretic control. *J Phys Chem B.* 2005;109(9):3879–3885. doi:10.1021/jp045402y.

Sanhaji M, Göring J, Couleaud P, et al. The phenotype of target pancreatic cancer cells influences cell death by magnetic hyperthermia with nanoparticles carrying gemcitabine and the pseudo-peptide NucAnt. *Nanomedicine.* 2019;20:101983. doi:10.1016/j.nano.2018.12.019.

Schladt TD, Schneider K, Schild H, Tremel W. Synthesis and bio-functionalization of magnetic nanoparticles for medical diagnosis and treatment. *Dalton Transactions.* 2011;40(24):6315–6343. doi:10.1039/C0DT00689K.

Schleich N, Sibret P, Danhier P, et al. Dual anticancer drug/superparamagnetic iron oxide-loaded PLGA-based nanoparticles for cancer therapy and magnetic resonance imaging. *Int J Pharm.* 2013;447(1–2):94–101. doi:10.1016/j.ijpharm.2013.02.042.

Sirivisoot S, Harrison BS. Magnetically stimulated ciprofloxacin release from polymeric microspheres entrapping iron oxide nanoparticles. *Int J Nanomedicine.* 2015;10:4447–4458. doi:10.2147/IJN.S82830.

Somvanshi SB, Patade SR, Andhare DD, Jadhav SA, Khedkar MV, Kharat PB, Khirade PP, Jadhav KM. Hyperthermic evaluation of oleic acid coated nano-spinel magnesium ferrite: enhancement via hydrophobic-to-hydrophilic surface transformation. *J Alloys Compd.* 2020;835:155422. doi:10.1016/j.jallcom.2020.155422.

Song J, Xie J, Li C, et al. Near infrared spectroscopic (NIRS) analysis of drug-loading rate and particle size of risperidone microspheres by improved chemometric model. *Int J Pharm.* 2014;472(1–2):296–303. doi:10.1016/j.ijpharm.2014.06.033.

Tao C, Lina X, Changxuan W, et al. Orthogonal test design for the optimization of superparamagnetic chitosan plasmid gelatin microspheres that promote vascularization of artificial bone. *J Biomed Mater Res B Appl Biomater.* 2020;108(4):1439–1449. doi:10.1002/jbm.b.34491.

Taratula O, Dani RK, Schumann C, et al. Multifunctional nanomedicine platform for concurrent delivery of chemotherapeutic drugs and mild hyperthermia to ovarian cancer cells. *Int J Pharm.* 2013;458(1):169–180. doi:10.1016/j.ijpharm.2013.09.032.

Tu L. Detection of magnetic nanoparticles for bio-sensing applications. Retrieved from the University of Minnesota Digital Conservancy. 2013, https://hdl.handle.net/11299/175355.

Uribe Madrid SI, Pal U, Kang YS, Kim J, Kwon H, Kim J. Fabrication of Fe3O4@ mSiO2 core-shell composite nanoparticles for drug delivery applications. *Nanoscale Res Lett.* 2015;10(1):1–8. doi:10.1186/s11671-015-0920-5.

Wang X, Huang X, Yang Z, et al. Targeted delivery of tumor suppressor microRNA-1 by transferrin-conjugated lipopolyplex nanoparticles to patient-derived glioblastoma stem cells. *Curr Pharm Biotechnol*. 2014;15(9):839–846. doi:10.2174/1389201015666141031105234.

Wu K, Su D, Liu J, Saha R, Wang JP. Magnetic nanoparticles in nanomedicine: a review of recent advances. *Nanotechnology*. 2019;30(50):502003. doi:10.1088/1361-6528/ab4241.

Xie J, Teng L, Yang Z, et al. A polyethylenimine-linoleic acid conjugate for antisense oligonucleotide delivery. *Biomed Res Int*. 2013;2013:710502. doi:10.1155/2013/710502.

Xie J, Yang Z, Zhou C, Zhu J, Lee RJ, Teng L. Nanotechnology for the delivery of phytochemicals in cancer therapy. *Biotechnol Adv*. 2016;34(4):343–353. doi:10.1016/j.biotechadv.2016.04.002.

Xu C, Xie J, Ho D, Wang C, Kohler N, Walsh EG, Morgan JR, Chin YE, Sun S. Au–Fe3O4 dumbbell nanoparticles as dual-functional probes. *Angew Chem Int Ed Engl*. 2008;47(1):173–176. doi:10.1002/anie.200704392.

Xue W, Liu XL, Ma H, et al. AMF responsive DOX-loaded magnetic microspheres: transmembrane drug release mechanism and multimodality postsurgical treatment of breast cancer. *J Mater Chem B*. 2018;6(15):2289–2303. doi:10.1039/c7tb03206d.

Yang X, Yang S, Chai H, et al. A novel isoquinoline derivative anticancer agent and its targeted delivery to tumor cells using transferrin-conjugated liposomes. *PLoS One*. 2015;10(8):e0136649. doi:10.1371/journal.pone.0136649.

Yang Z, Xie J, Zhu J, et al. Functional exosome-mimic for delivery of siRNA to cancer: in vitro and in vivo evaluation. *J Control Release*. 2016;243:160–171. doi:10.1016/j.jconrel.2016.10.008.

Yang Z, Yu B, Zhu J, et al. A microfluidic method to synthesize transferrin-lipid nanoparticles loaded with siRNA LOR-1284 for therapy of acute myeloid leukemia. *Nanoscale*. 2014;6(16):9742–9751. doi:10.1039/c4nr01510j.

Ye Y, Xing J, Zeng L, Yu Z, Chen T, Hosmane NS, Lu G, Wu A. Fluorescent/magnetic nanoprobes of high specificity for detection of triple negative breast cancer. *J Biomed Nanotech*. 2017;13(8):980–988. doi:10.1166/jbn.2017.2414.

Yetisgin AA, Cetinel S, Zuvin M, Kosar A, Kutlu O. Therapeutic nanoparticles and their targeted delivery applications. *Molecules*. 2020;25(9):2193. doi:10.3390/molecules25092193.

Yu B, Wang X, Zhou C, et al. Insight into mechanisms of cellular uptake of lipid nanoparticles and intracellular release of small RNAs. *Pharm Res*. 2014;31(10):2685–2695. doi:10.1007/s11095-014-1366-7.

Zhang C, Cai YZ, Lin XJ, Wang Y. Magnetically actuated manipulation and its applications for cartilage defects: characteristics and advanced therapeutic strategies. *Front Cell Dev Biol*. 2020;8:526. doi:10.3389/fcell.2020.00526.

Zhao Q, Wang L, Cheng R, et al. Magnetic nanoparticle-based hyperthermia for head & neck cancer in mouse models. *Theranostics*. 2012;2(1):113–121. doi:10.7150/thno.3854.

Zhao Y, Fan T, Chen J, et al. Magnetic bioinspired micro/nanostructured composite scaffold for bone regeneration. *Colloids Surf B Biointerfaces*. 2019;174:70–79. doi:10.1016/j.colsurfb.2018.11.003.

Zhou C, Yang Z, Teng L. Nanomedicine based on nucleic acids: pharmacokinetic and pharmacodynamic perspectives. *Curr Pharm Biotechnol*. 2014;15(9):829–838. doi:10.2174/1389201015666141020155620.

Zhu N, Ji H, Yu P, Niu J, Farooq MU, Akram MW, Udego IO, Li H, Niu X. Surface modification of magnetic iron oxide nanoparticles. *Nanomaterials*. 2018;8(10):810. doi:10.3390/nano8100810.

Zhu X, Zhang H, Huang H, Zhang Y, Hou L, Zhang Z. Functionalized graphene oxide-based thermosensitive hydrogel for magnetic hyperthermia therapy on tumors. *Nanotechnology*. 2015;26(36):365103. doi:10.1088/0957-4484/26/36/365103.

6 Magnetic Nanomaterials for Microwave Absorption for Health, Electronic Safety, and Military Applications

F. Ruiz Perez, and F. Caballero-Briones
Instituto Politécnico Nacional, Materiales y Tecnologías
para Energía, Salud y Medio Ambiente (GESMAT)

S.M. López-Estrada
Unidad de Investigación y Desarrollo Tecnológico
(UNINDETEC), Secretaría de Marina-Armada de México

CONTENTS

6.1 INTRODUCTION

The fast and continuous development of electronic technologies such as cell phones, laptops, GPS, radio, television, X-ray machines, appliances, and radars, with

DOI: 10.1201/9781003335580-6

applications in entertainment, communication, health industry, and defense has caused electromagnetic radiation to become an environmental pollution problem with serious threats to human health and effects that hinder or impede the operation or reduce the life of electronic devices (Niu et al. 2020). For example, the increase in the number of electronic devices causes a higher interaction of concurrent electromagnetic fields increasing the possibility of operation failures due to cross interference. (Yahyaei and Mohseni 2018). For example, devices such as Wi-Fi modems, radio, or television, use electronic components such as coils to transmit and receive a specific signal, however, the same components generate unwanted additional signals. When the electromagnetic field of the additional signal interacts with another device or system, electromagnetic interference (EMI) appears (Kim et al. 2014). The solution proposed to avoid this problem has been not to place the devices too close to each other, which doesn't represent a solution of the real problem.

In the military field, radar systems allow to detect targets such as aircrafts, warships, and helicopters, in a specific space volume. Radars are widely used in activities such as air traffic control, weather monitoring, and speed control. Several studies have shown the harmful effects on human health of the exposure to electromagnetic signals from radar systems such as the incidence of testicular cancer, leukemia, brain tumors, among others (Variani et al. 2019).

To solve the drawbacks of electromagnetic pollution, microwave-absorbing materials (MAMs) have attracted attention in multiple investigations. MAMs are the base of electromagnetic shielding which involves the reflection and/or absorption of EM radiation converting the energy of electromagnetic waves into heat. Materials such as aluminum, copper, and nickel were the first used to block electromagnetic waves due to their electrical and thermal conductivities (Chung 2020). However, its use involves some disadvantages such as corrosion problems, weight, and processing difficulties. In addition, they don't have magnetic properties, so their main blocking mechanism is the reflection of radiation, which causes the electromagnetic waves to continue propagating in the space. Therefore, metals are used mostly to insulate sensitive components or enclosed environments to avoid electromagnetic interference (Pandey, Tekumalla, and Gupta 2020).

Nanotechnology is the science of developing and studying materials with dimensions below 100 nm. By reducing the particle size, the properties of the materials are modified and improved with respect to their bulk properties, for example, ferromagnetic magnetite became superparamagnetic when particle size is below 20 nm (Upadhyay, Parekh, and Pandey 2016). Metal nanoparticles are an alternative to bulk metals since they manage to solve various inconveniences of the bulk metals such as weight, and also, many metal nanoparticles have magnetic properties, which made them suitable for MAMs in conjunction with other materials (Jamkhande et al. 2019; Green and Chen 2019). However, metal nanoparticles still are subject to corrosion.

On the other hand, metamaterials are topological structures, which can interact with electromagnetic radiation. Metamaterials are defined as artificial materials, *that is*, they do not exist in nature, and also provide particle characteristics as negative refraction, perfect absorption through the periodic arrangement of structures with dimensions less than the wavelength of the incident radiation (Yoo et al. 2015). Since its invention, metamaterials have been of great interest in different applications such as invincibility cloak, zero-index materials, and electromagnetic absorbers. However, its development is subject to have specific properties, such as they must be light, with thin thickness, and cover wide electromagnetic frequency ranges (Huang et al. 2022). In order to achieve those metamaterials, meet the expected characteristics, the development with magnetic nanomaterials has become a great effort of the scientific community. The combination of magnetic nanoparticles and incident waves allows to trap the incident energy in a small space and to suppress the reflection and transmission of radiation, thus managing to absorb the energy and preventing it from reaching other areas, devices, or systems. Initially, materials such as silver, copper, titanium, chrome, and nickel were widely used. However, the current 2D materials such as graphene and nanoparticles of transition metals have drawn attention due to their remarkable optical, electrical, and magnetic properties (Zhou et al. 2020). Nanostructured metamaterials have synergetic effects on electromagnetic attenuation as discussed in Section 6.5.

Therefore, the present chapter is organized as follows: Section 6.2 discusses the electromagnetic spectrum, with emphasis on the dose effects on human health and the need for MAMs and the properties they must fulfill. Section 6.3 presents the microwave absorption theory in order to provide the reader with the theoretical basis for MAMs operation. Section 6.4 presents the use of several magnetic nanosized composite materials for microwave absorption and Section 6.5 depicts the properties and uses of nanostructured metamaterials, with an emphasis on those fabricated from magnetic nanomaterials.

6.2   ELECTROMAGNETIC SPECTRUM

Electromagnetic (EM) radiation consists of oscillating electric and magnetic fields perpendicular to each other that propagates through space. The whole that encompasses the electromagnetic radiation range as a function of the frequency or wavelength, from microwaves to cosmic rays, is called the EM spectrum, as shown in Figure 6.1 (NASA 2010).

The development of wireless devices and systems, such as cell phones, radio systems, data systems, satellites, medical devices, appliances, and Internet of things (IoT), among others, has led to high exposure to EM radiation, with the risks that this entails (Wdowiak et al. 2017; Jauchem 2008). The relationship between the radiation absorbed by the human body and its effects is determined by the equivalent or effective dose. The established units for this parameter are the roentgen equivalent man (rem) and sievert (Sv, J/kg). 1 rem equals to 0.01 Sv. In addition, millirem o mrem (1/1,000 rem) represents the biological equivalent dose (Commision n.d.). The equivalent dose is calculated by the following equation:

$$H_T = \sum_R w_R D_{T,R} \qquad (6.1)$$

Where $D_{T,R}$ is the absorbed dose averaged over the body and w_R the radiation weighting factor.

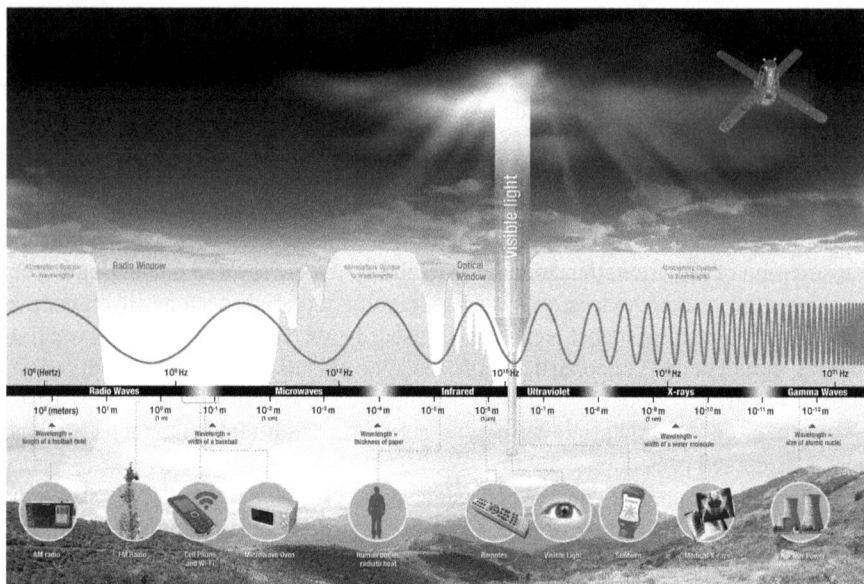

FIGURE 6.1 Electromagnetic spectrum. Source from (NASA 2010).

Multiple organizations set the radiation exposure limit for the human body which should not exceed 50 mSv (5 rem) per year, and the limits for different body parts are defined as follows: whole body (5 rem), eyes (15 rem), extremities (50 rem), and 9 months pregnant women (0.5 rem) (Canadian Nuclear Safety Commision n.d.; U.S.NRC n.d.). Therefore, there is a great interest in developing MAM's with specific properties such as lightness, flexibility, broadband attenuation, and heat dissipation, among others (Bi et al. 2019), to be implemented in wearable devices or textiles to ensure that the radiation dose remains below the permissible levels, for example, in exposed personnel in military, medical, scientific, or industrial facilities.

6.3 MICROWAVE ABSORPTION THEORY

Development of new materials with the capability to handle external EM waves is one of the proposals to mitigate the problems arising from EM pollution. These materials can absorb the incident EM radiation and convert the energy into Joule heat, reducing or eliminating the reflection of the waves. The reduction or elimination mechanisms of the reflected waves can be divided as follows (Ruiz-Perez et al. 2022):

1. Absorption: The ability to convert EM energy into heat is attributed to dielectric or magnetic losses which are related to the permeability and permittivity properties of the materials, *that is*, with the intrinsically magnetic or electric loss properties.

2. Multiple reflections: This mechanism allows EM waves to penetrate the material and then be reflected between the back and front surfaces within the material until dissipated; this process is also known as destructive interference.

According to the transmission line theory, the reflection loss (RL) characteristics allow to study the electromagnetic absorption properties of different materials through the relative complex permittivity ($\varepsilon_r = \varepsilon' - j\varepsilon''$) and relative complex permeability ($\mu_r = \mu' - j\mu''$) by the following equations:

$$Z_{in} = Z_0 \sqrt{\frac{\mu_r}{\varepsilon_r}} \tanh\left[j\left(\frac{2\pi fd}{c}\right)\sqrt{\mu_r \varepsilon_r} \right] \tag{6.2}$$

$$RL(dB) = 20 \, log\left|\frac{Z_{in} - Z_0}{Z_{in} + Z_0}\right| \tag{6.3}$$

where Z_{in} represents the impedance of the material medium, Z_0 is the impedance of the free space ($\approx 377 \, \Omega$), d is the thickness of the absorber material, f is the electromagnetic radiation frequency, and c is the speed of light. Then, the electromagnetic absorption response is related with the dielectric and magnetic loss as well as to the impedance matching.

6.3.1 DIELECTRIC LOSS

Dielectric materials are electrical insulators that under the action of an electric field can create electrical dipoles (entities with positive and negative charges separated by a distance), *that is*, can be polarized. Due to the electrical polarization of a material, positive charges will move in the same direction as the incident electric field, while negative charges will move in the opposite direction. Dielectric losses are related to the permittivity of the material: the real part (ε') represents the ability of the material to interact with the electrical component of the radiation, while the imaginary part (ε'') is related to the losses or attenuation. Dielectric losses are defined as the dissipation of energy due to the movement of charges by changing the direction of polarization. As shown in Figure 6.2, dielectric losses derive from different polarization processes including dipole/molecular, electronic, ionic, and atomic polarization.

When the incident electric field is removed, the dipoles return to their initial orientation due to the influence of thermal motion, and this process is named as polarization relaxation.

The ε'' parameter, according to the free electron theory, is related to the electrical conductivity of the material; from equation (6.4), a higher value of electrical conductivity carries a higher value of ε'' indicating higher attenuation capability.

$$\varepsilon'' = \frac{\sigma}{2\pi\varepsilon_0 f} \tag{6.4}$$

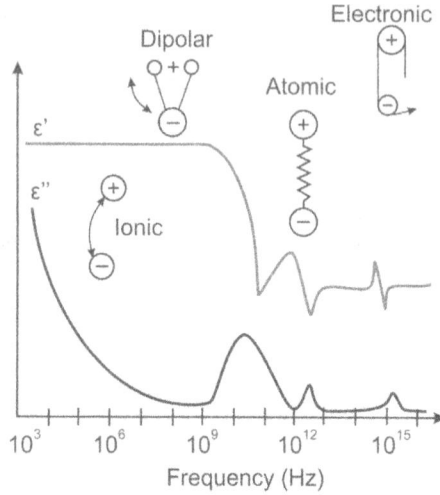

FIGURE 6.2 Polarization mechanism as function of frequency. Adapted from (El Khaled et al. 2016). This is an open access article distributed under the Creative Commons Attribution License which permits unrestricted use, distribution, and reproduction in any medium, provided the original work is properly cited.

Where σ (S/m) represents the electrical conductivity of the material, ε_0 is the dielectric constant in vacuum (8.854×10^{-12} F/m), and f the frequency (Hz) (Micheli et al. 2010).

However, it is necessary to consider that increasing the electrical conductivity will produce an impedance mismatch which will cause an attenuation response due to reflection of radiation and not to its absorption. On the other hand, the reduction of conductivity of the material may reduce the absorption response due to the decrement of the dielectric properties; the compromise point between absorption and conductivity is known as percolation threshold (Qin and Brosseau 2012).

6.3.2 MAGNETIC LOSS

Dielectric materials have remarkable attenuation responses at frequencies between 8 and 18 GHz even having low or no impedance matching value due to negligible magnetic permeability, *that is*, having no magnetic properties. In complex permeability as well as complex permittivity, the real part (μ') and the imaginary part (μ'') represent the storage and attenuation of the energy of a magnetic field capability respectively (Zeng et al. 2020).

Magnetic materials provide different loss mechanisms that complement the dielectric losses, *that is*, hysteresis loss, eddy current loss, and magnetic natural resonance. These mechanisms refer to different processes of energy dissipation when a magnetic material interacts with a time-varying external field (Bertotti and Fiorillo 2016). The magnetic properties of materials originate from the motion of electrons and the magnetic momentum of atoms. As with electric dipoles, magnetic dipoles are defined

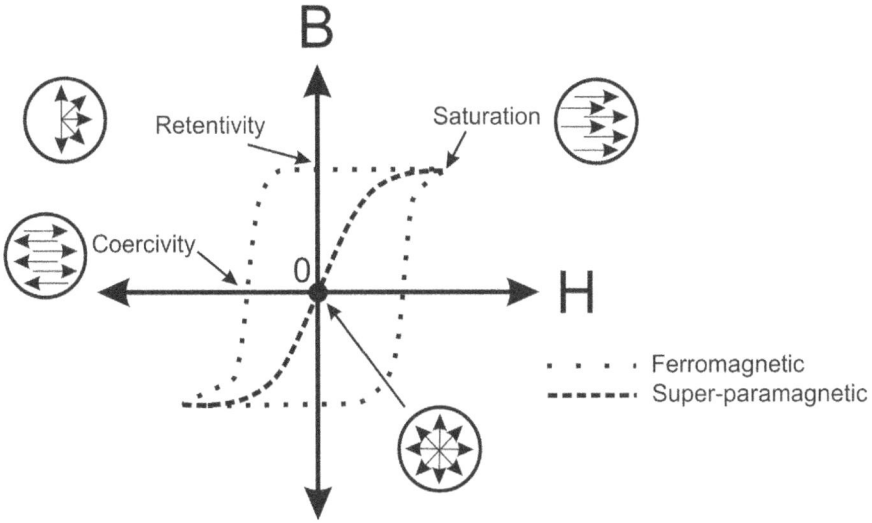

FIGURE 6.3 Hysteresis loop of magnetic materials. Adapted from (Ruiz-Perez et al. 2022) with permission from Elsevier.

as the North and South poles rather than positive and negative charges. Figure 6.3 shows the hysteresis curve of some types of magnetic materials: as can be seen, at the zero-field point the magnetic domains are randomly orientated. Subsequently, when the magnetic field increases, the domains begin to orient in the direction of the incident field to a point of saturation. When the magnetic field is removed, there may be two cases: (i) the domains keep a certain degree of orientation, *that is*, they continue being magnetic after removing the incident field, a property named field retentivity; in this case, the field required to remove the remnant magnetization, *that is*, to randomize the magnetic domains, is called coercivity field; (ii) the domains return to their initial random state just after removing the field, so very small or null coercivity field is required to remove the magnetization. The described cases represent different kinds of magnetic materials: (i) ferromagnetic or (ii) superparamagnetic materials, respectively.

The properties of superparamagnetic materials make them ideal candidates for implementation as radiation attenuators, since the energy of the incident wave will be reduced while orienting the magnetic domains; once eliminating the incident wave, the magnetic domains will return to the initial state spontaneously, to attenuate the energy of new incident electromagnetic fields.

6.3.3 INFLUENCE OF SIZE FACTOR IN MICROWAVE ABSORPTION

The term "nano" is used to define 10^{-9} (one-billionth of a meter). Nanomaterials are those with one of its physical dimensions less than 100 nanometers. In nanoscale, the bulk material properties change from classical physics to quantum mechanisms. Gold, for example, has been classified as a chemical inert material

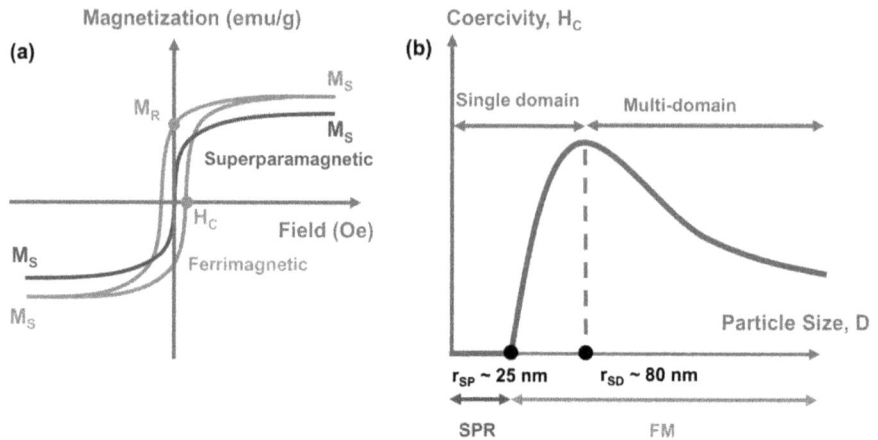

FIGURE 6.4   (a) Fe_3O_4 SPM and FM hysteresis loop. (b) Size and magnetic behavior relation. Reproduced from Nguyen et al. 2021. This is an open access article distributed under the Creative Commons Attribution License which permits unrestricted use, distribution, and reproduction in any medium, provided the original work is properly cited.

with remarkable electrical conductivity and a yellowish color. However, in the nanoscale, it has semiconductor properties and presents a reddish color (Sun et al. 2015). The dielectric and magnetic properties of materials are directly related to their size. By reducing the size of particles to nanoscale, the ability to interact with electromagnetic radiation is increased. In addition, due to the high specific surface area of nanoscale materials, attenuation caused by the relaxation processes is improved (Huo, Wang, and Yu 2009). Joule dissipation and dielectric loss are also improved in nanomaterials because of their large area/volume ratio and small sizes which allow an enhanced polarizability compared with the bulk state. Also, nanomaterials would improve the wave scattering and destructive interference. Multi-magnetic domain structures are present in bulky magnetic materials, but by reducing the material to nanoscale, the multi-magnetic domains are converted to single magnetic structures. Superparamagnetic response occurs in nanoparticles with diameters between 3 and 50 nm depending on the material (Soler and Paterno 2017). In particles with such sizes, there is no presence of coercive field in their hysteresis curves.

For example, Fe_3O_4 nanoparticles can display superparamagnetic (SPR) or ferromagnetic (FM) behavior, as shown in Figure 6.4a. The hysteresis curve of FM materials exhibits H_C and M_R values indicating the magnetic retention. On the other hand, the coercivity and remanent magnetization of SPM nanoparticles are equal to zero. The magnetic response depends on the particle size, as depicted in Figure 6.4b; the transition sizes from superparamagnetic and ferromagnetic response is around 25 nm, while the transition from single to multiple magnetic domain is around 80 nm, respectively (Nguyen et al. 2021). The control of particle size below the superparamagnetic transition then enables to enhance the attenuation response in MAMs.

6.4 MAGNETIC NANOSIZED COMPOSITE MATERIALS FOR MICROWAVE ABSORPTION

For the development of materials that attenuate electromagnetic radiation, the combination of dielectric and magnetic materials, which leads to a synergy between both types of loses, has been a widely studied route. Composite materials are the combination of two or more materials with different physical and chemical properties and the resulting material of this combination has different properties from the individuals. Therefore, the electromagnetic response of a composite material can be tailored by varying the fractions of the composite constituents. For example, nanomaterials have been dispersed in polymer-based matrixes for electromagnetic attenuation applications. Polymeric nanocomposites aim to cover the expected EMI-shielding material properties such as high surface area, lightness, corrosion resistance, and tunable electrical and magnetic properties. In the following sections, different examples of composites for MAMs are discussed.

6.4.1 CARBON MAGNETIC MATERIALS

Traditionally, metals such as copper, nickel, and aluminum are the most used materials in electromagnetic protection applications; however, they generate various disadvantages such as adding additional weight to the system and being susceptible to corrosion. Carbon is one of the most abundant materials on Earth and can be found naturally in two allotropes, graphite, and diamond. Other carbon allotropes are carbon nanotubes, graphene, fullerenes, and carbon fibers. As an alternative to metallic materials for EMI shielding, carbon-based materials have attracted attention due to their low density, hardness, high thermal conductivity, corrosion resistance, high and tunable electrical conductivity, high surface area, and potential EM wave absorption capacity (D.-C. Wang et al. 2021). However, the difference between permeability and permittivity values of carbon-based materials causes their electromagnetic attenuation response to be deficient. The combination of magnetic nanomaterials and carbon allotropes to form carbon magnetic composites has been implemented to improve the electromagnetic attenuation response of carbon-based materials (B. Wang et al. 2021). Among these materials, carbon nanotubes, carbon fibers, and graphene have been tested for the development of radiation-absorbing materials because of their high surface area and chemical modification possibilities (Y. Wang et al. 2020).

6.4.1.1 Carbon Fiber Composites

Carbon fiber (CF) has remarkable properties such as high tensile strength, high Young's modulus, low density, and high electrical conductivity. Ranjkesh and Nasouri studied microwave-absorbing materials based on Ni/C from the carbonization of $Ni(NO_3)_2$ $6H_2O$-impregnated cigarette buts (Ranjkesh and Nasouri 2022). The formation of nickel nanoparticles was performed on carbon fibers, and it was observed that the morphology of nanoparticles varied according to the synthesis parameters. As shown in Figure 6.5a, Ni nanoparticles range from 20 to 40 nm. The magnetic hysteresis curves presented in Figure 6.5b were S-like type, related to soft ferromagnetic behavior, *that is*, materials that can be magnetized and demagnetized

(a)

(b)   (c)

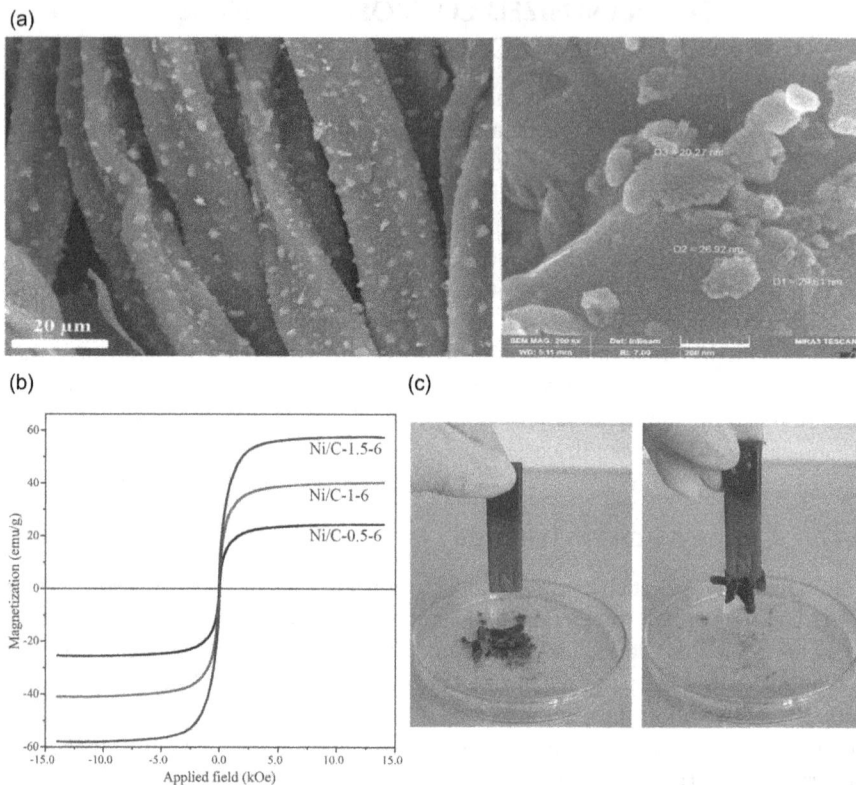

FIGURE 6.5 Ni/C composites. (a) FESEM micrographs. (b) Magnetic hysteresis loop. (c) Magnetic performance. Adapted from (Ranjkesh and Nasouri 2022), with permission from Elsevier.

easily, as depicted in Figure 6.5c. The magnetic properties increased with the nickel contents on the surface of the composite.

Pure carbon nanofibers have high dielectric properties and null magnetic properties, leading to weak impedance matching. By the incorporation of TiN, a semiconductor material with outstanding dielectric and magnetic properties into carbon nanofibers, the contribution of magnetic losses to EM attenuation is increased and generates a synergy with the dielectric losses (Yu et al. 2022). TiN nanoparticles generate defects in the structure of CF, increasing the charge carrier concentration and therefore the electric conductivity, demonstrating that the conductivity of the material can be modulated with the concentration of a dielectric material. On the other hand, the material has considerable magnetic properties attributed to the defects and vacancies caused by the TiN network. By increasing the number of defects and exceeding the equilibrium point, local magnetic moments are generated. For TiN/NCF, TiN-induced defects provide the magnetic properties that allow the impedance matching for electromagnetic attenuation. The microwave absorption of the composite was evaluated in the frequency range of 2–18 GHz at varying absorber coating thicknesses from 1 to 6 mm. A maximum attenuation value of −45.19 dB at

15.6 GHz was obtained with a coating thickness of 1.91 mm. However, attenuation values higher than −20 dB were obtained in the range of 3.89–18 GHz, covering part of the S-band and up to the C-Ku band, as shown in Figure 6.6a. This behavior was attributed to different loss mechanisms, including the morphology of TiN/NCNF which provides channels for the movement of the charges, as well as the TiN-induced defects which contribute to an increased carrier concentration that in turn, potentiated dielectric losses. Furthermore, the defects and bonds present in the composite contribute to dipolar relaxation. Finally, TiN defects and vacancies generate a ferromagnetic response that contributes to the processes of magnetic loss. The mentioned mechanisms are represented in Figure 6.6b.

Magnetic metal oxides with carbon-based materials have been widely proposed for electromagnetic attenuation. Among the different magnetic metal

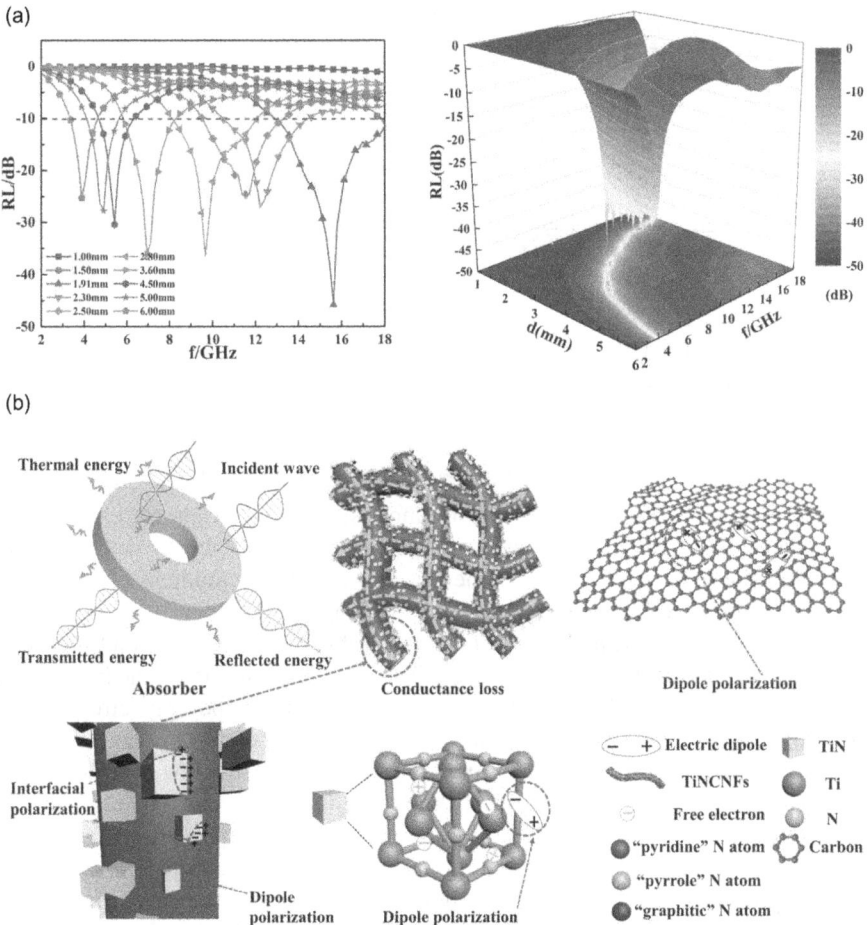

FIGURE 6.6 TiN/NCNF composite. (a) Frequency dependence and three-dimensional attenuation response. (b) Scheme of EM absorption mechanisms. Adapted from (Yu et al. 2022), with permission from Elsevier.

(a)

(b)   (c)

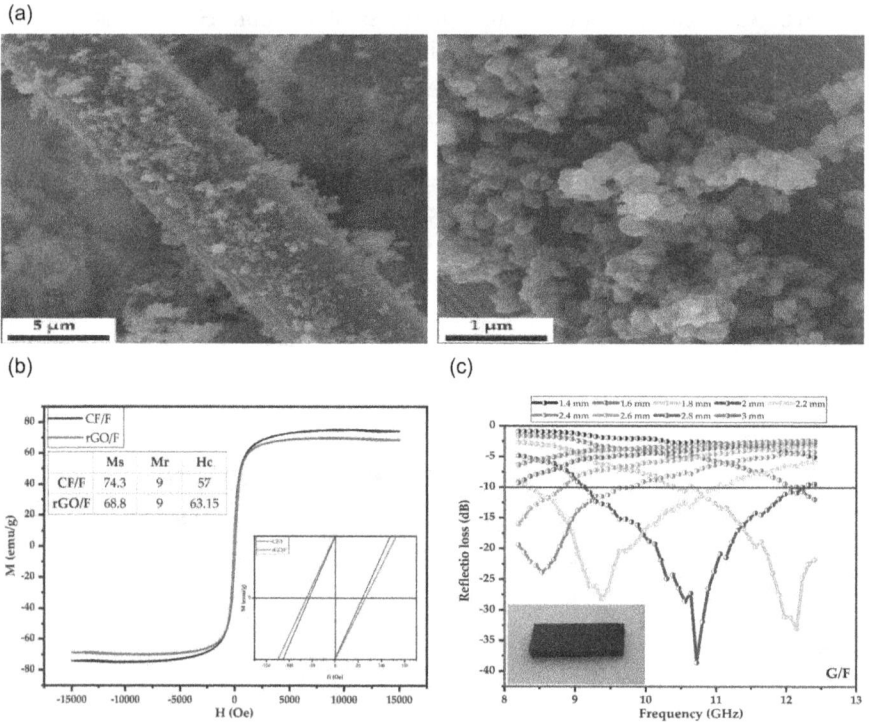

FIGURE 6.7   (a) FESEM images of CF/Fe$_3$O$_4$. (b) Magnetic hysteresis loop of CF/ Fe$_3$O$_4$ and rGO/Fe$_3$O$_4$. Reflection loss curves of double layer composites. Adapted from (Gang, Niaz Akhtar, and Boudaghi 2021), with permission from Elsevier.

oxides, magnetite (Fe$_3$O$_4$) is among the most studied materials for electromagnetic absorption due to its magnetic properties described above, and low cost. However, the lack of dielectric properties limits the development of absorbent materials using Fe$_3$O$_4$ nanoparticles alone. Gang et al. report the design and synthesis by a solvothermal route of a bilayer absorber based on Fe$_3$O$_4$@carbon nanofiber as an absorbing layer and Fe$_3$O$_4$/rGO composite as a matching layer. In Figure 6.7a, Fe$_3$O$_4$ NPs onto the CF are shown. Correspondingly, circular, and hexagonal particles with sizes around 75–155 nm agglomerated by Van der Waals forces, are observed in Figure 6.7b. In the hysteresis loop, the low coercivity field indicated the soft magnetic nature of the CF/Fe$_3$O$_4$ composite. The electromagnetic attenuation response of composites with thicknesses ranging from 1.4 to 3 mm was evaluated between 8 and 13 GHz. The composite with a thickness of 1.4 mm presented the highest attenuation, as shown in Figure 6.7c, the highest reflection loss corresponds to −52.5 dB at 10.8 GHz. In addition, by varying the thickness of the material, the response was modulated to other frequencies. The excellent response of electromagnetic attenuation was attributed to the synergy of the electrical and magnetic properties of the materials, allowing rotation of dipoles, multiple interfacial polarizations, and matching impedance (Gang, Niaz Akhtar, and Boudaghi 2021).

6.4.1.2 Magnetic Graphene

Graphene is another carbon allotrope that has received the attention of researchers for the development of radiation-absorbing material due to its remarkable properties such as the excellent electrical and thermal conductivity, high surface area, and the thickness of a carbon atom, making it an ultra-thin material (Zhu et al. 2010). Like CNFs, despite its excellent intrinsic properties, the use of graphene for electromagnetic attenuation makes impossible its implementation due to its high electrical conductivity and null magnetic properties, which causes an impedance mismatch. The combination of graphene with magnetic nanomaterials has covered the disadvantage of impedance mismatch allowing graphene to be used as an electromagnetic absorber material. The incorporation of nickel-based nanoparticles has proven its effectiveness for the development of microwave-absorbing materials. Hou et al. synthesized xNi/yNiO/rGO (NNG) composites by hydrothermal method and calcination reduction process (Hou et al. 2022). The magnetic characterization of xNi/yNi nanoparticles (NN), as shown in Figure 6.8a, demonstrates a high magnetic saturation of NN particles. Introducing rGO without magnetic properties results in a decrease in the magnetic response. The NNG samples show a soft ferromagnetic response. The study depicted that the incorporation of NN magnetic nanoparticles allows to adjust the impedance matching successfully. As a result, the composite displays electromagnetic attenuation at different frequencies with different thicknesses. As shown in Figure 6.8b, the material has the optimal response of $-46.5\,dB$ EN 3.57 GHz with thickness of 3.6 mm. Figure 6.8c illustrates the proposed attenuation mechanisms, the dielectric losses are improved by the polarization produced by the load on the surface of NNG composite, and the dipolar polarization by functional groups and defects in the graphene sheets. The NN nanoparticles dispersed into the graphene sheets generate a synergistic effect of electromagnetic dissipation.

Qu et al. prepared a graphene-based composite microwave-absorbing material under mild reaction conditions using dopamine (DA) to form polydopamine (PDA) through a mild self-polymerization reaction. The FeCoNiO$_x$-PDA-rGO composite was achieved by chemical reduction method and deposition of metal nanoparticles on PDA surface (Qu et al. 2021). In Figure 6.9a, TEM images show FeCoNiOx nanoparticles about 100 nm wrapped by a thin PDA-rGO layer, which provides corrosion resistance to the composite allowing to keep the magnetic properties and improve the impedance matching in different environments. Figure 6.9b shows the hysteresis loop diagram of the different samples. Due to the presence of magnetic nanoparticles, the composite has magnetic and dielectric losses, which leads to the dissipation of electromagnetic waves. In general, magnetic losses are relatively weak, so magnetic materials play a major role in impedance coupling due to the high conductivity of the graphene substrate. The composites show different absorbing performance which can be adjusted by modifying the amount of magnetic material into the composite. According to Figure 6.9c, the attenuation response between 5 and 18 GHz can be adjusted by modifying the thickness of the composite coating.

Conventionally, the response of materials developed for electromagnetic attenuation depends on the amount of the magnetic material in the matrix. As an unconventional proposal, polymer nanomaterials have been used with high efficiency,

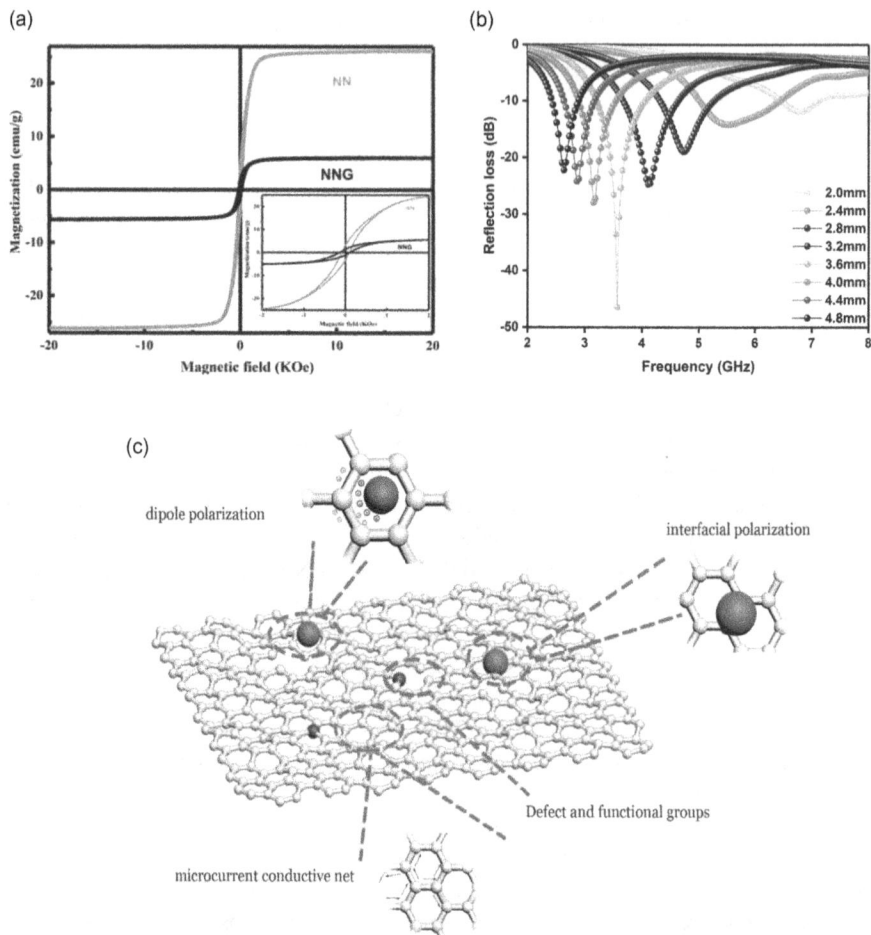

FIGURE 6.8 (a) Hysteresis loops graphics of NN and NNG. (b) Attenuation of electromagnetic reflection NNG sample. (c) Scheme of the proposal absorption mechanism in the NNG. Adapted from Hou et al. 2022, with permission from Elsevier.

flexibility, and corrosion resistance (Gupta et al. 2014). A graphene, magnetite (Fe_3O_4) nanoparticles and polyvinyl alcohol (PVA), PVA/Gr(x)/Fe_3O_4(0.1-x) hybrid, was studied by Khodiri et al. (2020). PVA is an interesting polymer due to its remarkable properties such as flexibility, hydrophilicity, and low cost. Fe_3O_4 nanoparticles provide the good magnetic properties. In addition, its low conductivity allows EM radiation to penetrate the material and be attenuated. The SEM image in Figure 6.10a shows spherical magnetite particles with size around 12 nm, which in combination with the graphene morphology improve the electromagnetic attenuation. As mentioned, superparamagnetic behavior is an important characteristic for the development of MAMs. To study the influence of magnetite nanoparticles, the hysteresis curves of the materials were obtained. Figure 6.10b shows the hysteresis loops of PVA/Gr(x)/Fe_3O_4 (with different graphene concentrations (0.08 and 0.02 wt%)). The curves have the S-like

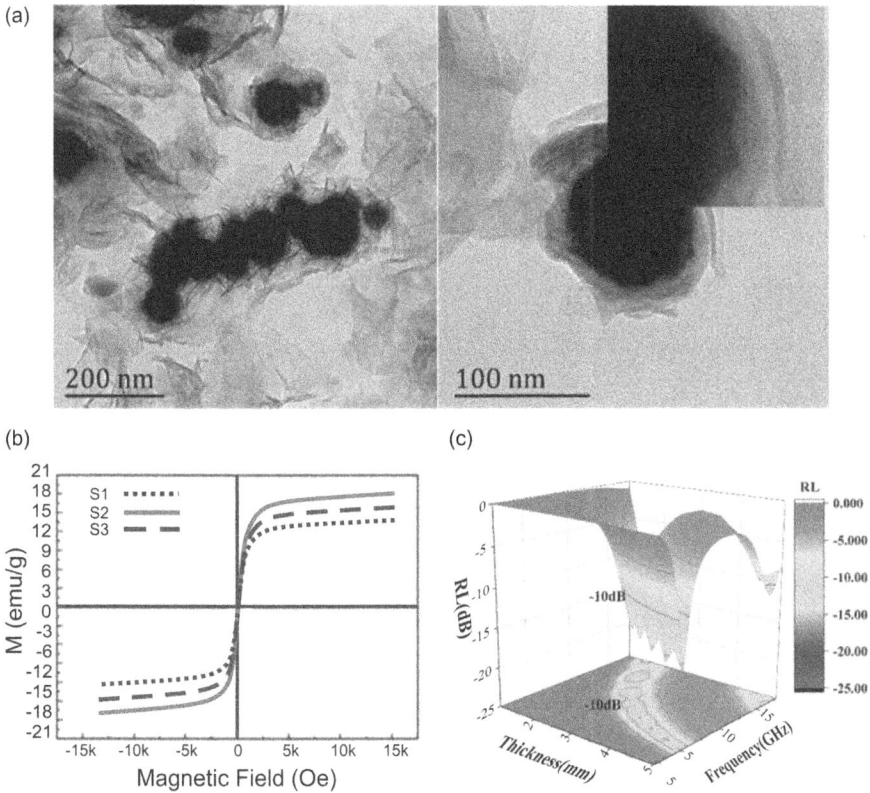

FIGURE 6.9 FeCoNiO$_X$-PDA-rGO composite. (a) TEM images. (b) Hysteresis loops. (c) Reflection loss plot Fe: Co: Ni = 1: 2: 1. Adapted from (Qu et al. 2021), with permission from Elsevier. Figure 7 before.

shape characteristic of SPM materials. The reduction of saturation of magnetization values can be attributed to the mayor graphene content which has more OH groups, which affect the magnetic response. On the other hand, increasing the concentration of Fe$_3$O$_4$ nanoparticles results in higher M$_S$ value. According with Figure 6.10c, the shielding effectiveness (SE$_T$) of the nanocomposite in the band X-band (8–12 GHz) is increased by increasing the concentration of graphene in the polymer network, which improves the interaction between particles. The electromagnetic attenuation response, as shown in Figure 6.10d, is attributed to different phenomena, *that is*, part of the incident radiation is reflected while another part penetrates the structure and is absorbed due to the interaction with the materials and the multiple internal reflections.

6.5 NANOSTRUCTURED METAMATERIALS

Despite the remarkable response of magnetic nanomaterials in MAMs, they have as main disadvantage the short range of frequency attenuation of the electromagnetic radiation, which limits its implementation in practical applications. To overcome this

(a)

(b)

(c)

(d)

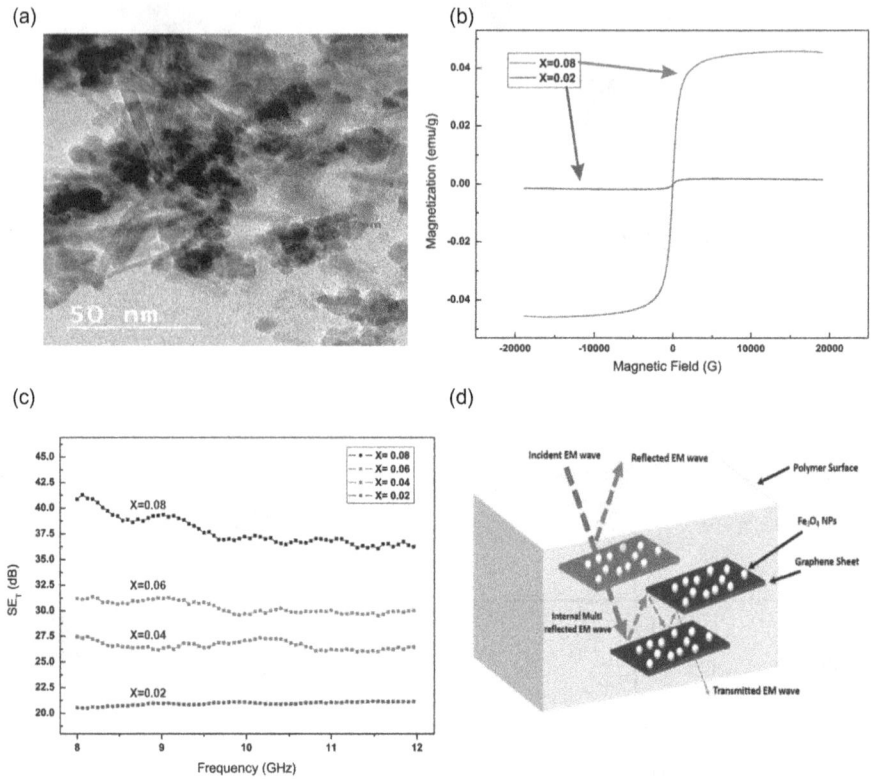

FIGURE 6.10 (a) SEM image of Gr/ Fe_3O_4 nanoparticles. (b) Hysteresis loop of PVA/Gr(-x)/Fe3O4 with different graphene concentration. (c) Total shielding effectiveness of PVA/Gr(-x)/Fe_3O_4. (d) EM shielding mechanism. Adapted from Khodiri 2020, with permission from Elsevier.

problem, the implementation of the so-called metamaterials has been explored as an alternative to enhance the MAMs performance.

Metamaterials are 3D geometric structures, usually formed by metal resona-tor arrays at micro or nanoscale developed to attenuate electromagnetic radiation at wide frequency intervals seeking to cover multiple bands (Argyros et al. 2015). Nanostructured metamaterials, build from nanocomposites with good attenuation response, combine the large interactions between the incident wave and the mate-rial, occurring in the metamaterial 3D structures with the intrinsic electromagnetic attenuation response of magnetic nanomaterials, enhancing their implementation in practical applications. Metamaterials usually have dimensions above 50 microns, but due to their remarkable attenuation properties, the nanostructured metamaterials, build from magnetic nanocomposites are considered in this chapter.

Huang et al. report an electromagnetic composite metamaterial (ECM) with truncated pyramidal shape, build from spherical carbonyl iron (CI) nanoparticles and multiwall carbon nanotubes (MWCNT), as shown in Figure 6.11a, to adjust impedance matching with air. In the nanoscale, MWCNT provides the dielectric loss behavior by polarization loss. On the other hand, CI particles (Figure 6.11b) provide the magnetic losses due to spin rotation. Different content of MWCNT and CI particles influence the position of the reflectivity peaks due to impedance coupling. The introduction of carbonyl iron particles generates magnetic losses due to the time lapse between the incident magnetic field and the magnetization generated when penetrating the compound, in addition to the eddy currents. Traditionally metamaterials used as selective frequency surfaces have thicknesses around 50 μm. However, the absorption bandwidth increases with the thickness, due to the enhanced electrical interaction in the material. 3D structures have proven their efficiency for electromagnetic attenuation by combining bulk, micro-, and nano-level attenuation mechanisms (W. Li et al. 2017). The response of the truncated

(a)

(b)

(c)

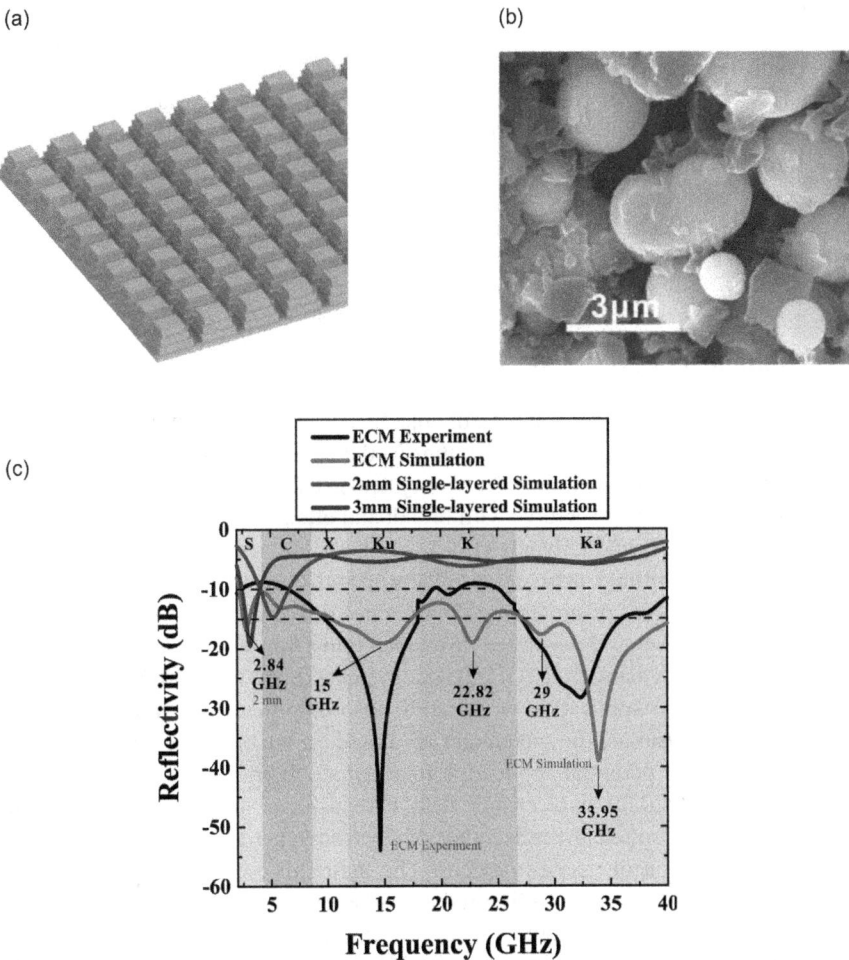

FIGURE 6.11   (a) Scheme of pyramidal ECM. (b) Detail of CI sphere particles coated by epoxy resin. (c) Experimental vs simulated reflectivity of EC with different thickness. Adapted from (Huang et al. 2018), with permission from Elsevier.

pyramidal ECM was evaluated in the frequency range of 2–40 GHz showing five absorption peaks, including 2.84, 15, 22.82, 29, and 33.95 GHz, as shown in Figure 6.11c. In this way, the combination of dielectric and magnetic loss materials with a periodic 3D structure with tunable attenuation and frequency range, was demonstrated (Huang et al. 2018).

On the other hand, *in-silico* studies have been performed to evaluate the attenuation response of metamaterials based on superparamagnetic nanocomposites. An in-silico study of chessboard-like structures with panels of different thicknesses made from a ternary $rGO/Fe_3O_4/PPy$ nanocomposite was done to optimize the thickness ratio to get a maximum attenuation response (Ruiz-Perez, Estrada, and Briones 2022). In this composite, reduced graphene oxide was decorated with magnetite nanoparticles below the superparamagnetic threshold. Figure 6.12a shows the 5×5 chessboard-like surface with different heights which were varied as shown in Figure 6.12b. As shown in Figure 6.12c, the attenuation result of single-layer composites shows a maximum response of −37 dB at 11.8 GHz. On the other hand, the response of the metamaterial can be modulated in the X-band (8–12 GHz) and has an increase in the attenuation bandwidth. These results demonstrate the synergetic effects of the nanoscale properties of the composites, combined with the metamaterial 3D structuring. Additionally, with the use of metamaterials, additive manufacturing can be implemented to enhance the applicability of magnetic nanocomposites as electromagnetic attenuation materials.

6.6 APPLICATIONS

In the military detection field, the main objective of electromagnetic absorption is to reduce the radar cross-section (RCS), *that is*, the signal reflected to the antenna by the target, to avoid its detection by radars systems. The development and implementation of materials and structures with broadband attenuation at different frequency ranges such as L, S, and C bands (1–2, 2–4, 4–8 GHz, respectively), which are of great interest on military applications (Panwar and Lee 2019; J. Li et al. 2021). In addition, with the goal of protecting the operator of radar detection systems, wearable textiles carbon-based have been developed into a non-woven matrix with potential application in X-band (8–12 GHz) radar systems (Song et al. 2017). Studies have shown the potential application of radiation-absorbing materials and structures. In communication systems, the metamaterials based on nanomagnetic materials have demonstrated their potential application to attenuate the radiation used in 5G technology (Yan et al. 2022). In electronic field, carbon-based materials have been studied to reduce or eliminate the reflectance of electromagnetic radiation to protect and ensure the operation of various devices. The materials were evaluated in the frequency range of 100 MHz to 10 GHz, reducing the reflectance by up to 30% (Kubacki et al. 2022). Another development was a wearable textile based on magnetic and conductive materials that, in combination with meta-structure arrays attenuate radiation at 2.4 GHz making it a potential candidate for the creation of devices with Wi-Fi technology such as medical monitoring devices (El Atrash, Abdalla, and Elhennawy 2021).

(a)

(b)

H (mm)	h (mm)	h/H
5	4	0.8
5	2.8	0.56
5	2.7	0.54
5	2.5	0.5
5	2.4	0.48
5	2	0.4
4	2.8	0.7
4	2.7	0.675
4	2.6	0.65
4	2.5	0.625
4	2.1	0.525

(c)

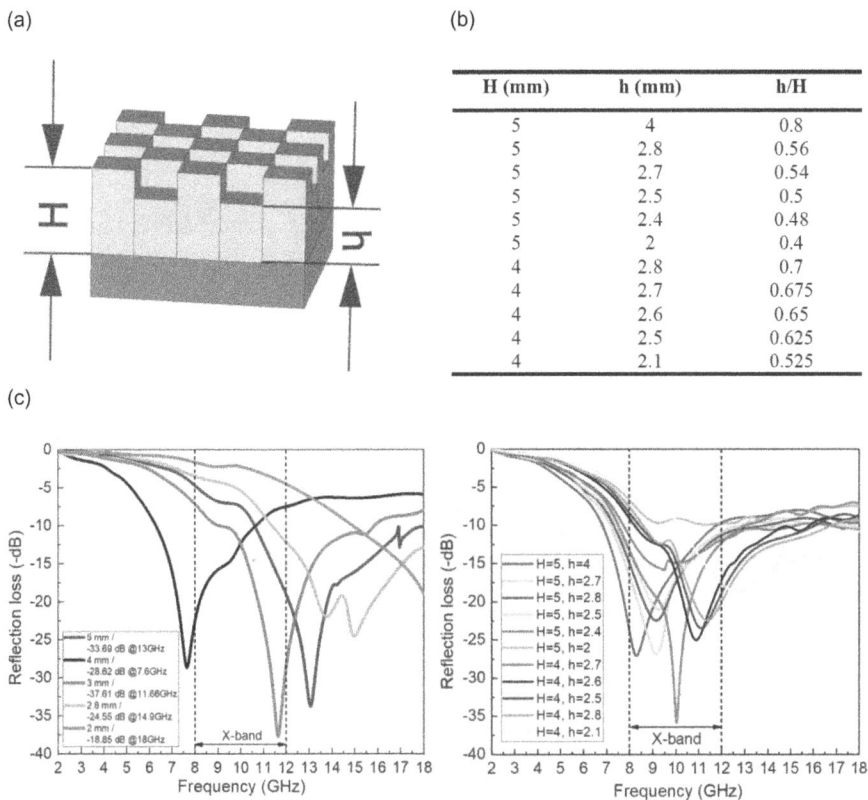

FIGURE 6.12 (a) Chessboard-like metamaterial. (b) Height setup and height ratio. (c) Reflection loss response: single-layer rGO/Fe$_3$O$_4$/PPy (left), chessboard (right). Adapted from (Ruiz-Perez, Estrada, and Briones 2022), with permission from IEEE.

6.7 CONCLUDING REMARKS

Nanomaterials have remarkable properties in comparison with their bulk counterparts. Magnetic nanomaterials have demonstrated their potential to improve the response of MAMs. Its characteristic properties, together with different carbon-based materials allowed the development of innovative technologies and methods that cover the protection needs in electronics, screening, and health care, at wide frequency intervals, as well as its application in military fields such as radar target hiding or health care of military personal. In addition, these materials overcome the disadvantages of conventional electromagnetic blocking methods because they are lightweight, flexible, anti-corrosive, and applicable in wearables and textiles. On the other hand, nanostructured metamaterials, prepared from nanocomposite radiation absorbers have a synergetic effect which has allowed to generate more efficient attenuation devices, with the possibility of being prepared by additive manufacturing. However, despite the results obtained, it is necessary to continue the developing of novel nanomaterials and composites with higher magnetic response, explore more applications in wearable technologies, and delve more into the study of the mechanisms of attenuation to better understand its operation, looking for faster product development.

ACKNOWLEDGMENTS

This work was financed by SIP-IPN under 2022–0834 grant. FRP is financed by CONACYT and BEIFI-IPN grants.

REFERENCES

Argyros, A., A. Tuniz, S.C. Fleming, and B.T. Kuhlmey. 2015. "Drawn Metamaterials." In *Optofluidics, Sensors and Actuators in Microstructured Optical Fibers*, 29–54. Elsevier. https://doi.org/10.1016/B978-1-78242-329-4.00002-3.

Bertotti, G., and F. Fiorillo. 2016. "Magnetic Losses." In *Reference Module in Materials Science and Materials Engineering*. Elsevier. https://doi.org/10.1016/B978-0-12-803581-8.02799-5.

Bi, Song, Jin Tang, Dian-jie Wang, Zheng-an Su, Gen-liang Hou, Hao Li, and Jun Li. 2019. "Lightweight Non-Woven Fabric Graphene Aerogel Composite Matrices for Assembling Carbonyl Iron as Flexible Microwave Absorbing Textiles." *Journal of Materials Science: Materials in Electronics* 30(18): 17137–17144. https://doi.org/10.1007/s10854-019-02060-y.

Canadian Nuclear Safety Commision. n.d. "Radiation Doses." Accessed August 4, 2022. http://nuclearsafety.gc.ca/eng/resources/radiation/introduction-to-radiation/radiation-doses.cfm.

Chung, D.D.L. 2020. "Materials for Electromagnetic Interference Shielding." *Materials Chemistry and Physics* 255 (November): 123587. https://doi.org/10.1016/j.matchemphys.2020.123587.

Commision, European. n.d. "How Is Radiation Exposure Measured and Assessed? - European Commission." Accessed August 4, 2022. https://ec.europa.eu/health/scientific_committees/opinions_layman/security-scanners/en/l-3/4-radiation-exposure.htm.

El Atrash, Mohamed, Mahmoud A. Abdalla, and Hadia M. Elhennawy. 2021. "A Compact Flexible Textile Artificial Magnetic Conductor-Based Wearable Monopole Antenna for Low Specific Absorption Rate Wrist Applications." *International Journal of Microwave and Wireless Technologies* 13 (2): 119–125. https://doi.org/10.1017/S1759078720000689.

El Khaled, Dalia, Nuria Castellano, Jose Gázquez, Alberto-Jesus Perea-Moreno, and Francisco Manzano-Agugliaro. 2016. "Dielectric Spectroscopy in Biomaterials: Agrophysics." *Materials* 9 (5): 310. https://doi.org/10.3390/ma9050310.

Gang, Qing, Majid Niaz Akhtar, and Reza Boudaghi. 2021. "Development of High-Efficient Double Layer Microwave Absorber Based on Fe3O4/Carbon Fiber and Fe3O4/RGO." *Journal of Magnetism and Magnetic Materials* 537 (November): 168181. https://doi.org/10.1016/j.jmmm.2021.168181.

Green, Michael, and Xiaobo Chen. 2019. "Recent Progress of Nanomaterials for Microwave Absorption." *Journal of Materiomics* 5 (4): 503–541. https://doi.org/10.1016/j.jmat.2019.07.003.

Gupta, Tejendra K., Bhanu P. Singh, Vidya Nand Singh, Satish Teotia, Avanish Pratap Singh, Indu Elizabeth, Sanjay R. Dhakate, S. K. Dhawan, and R. B. Mathur. 2014. "MnO2 Decorated Graphene Nanoribbons with Superior Permittivity and Excellent Microwave Shielding Properties." *Journal of Materials Chemistry A* 2 (12): 4256. https://doi.org/10.1039/c3ta14854h.

Hou, Mingming, Zuojuan Du, Yu Liu, Zhizhao Ding, Xiaozhong Huang, Ailiang Chen, Qiancheng Zhang, Yutian Ma, and Sujun Lu. 2022. "Reduced Graphene Oxide Loaded with Magnetic Nanoparticles for Tunable Low Frequency Microwave Absorption." *Journal of Alloys and Compounds* 913 (August): 165137. https://doi.org/10.1016/j. jallcom.2022.165137.

Huang, Qianqian, Gehuan Wang, Ming Zhou, Jing Zheng, Shaolong Tang, and Guangbin Ji. 2022. "Metamaterial Electromagnetic Wave Absorbers and Devices: Design and 3D Microarchitecture." *Journal of Materials Science & Technology* 108 (May): 90–101. https://doi.org/10.1016/j.jmst.2021.07.055.

Huang, Yixing, Wei-Li Song, Changxian Wang, Yuannan Xu, Weiyi Wei, Mingji Chen, Liqun Tang, and Daining Fang. 2018. "Multi-Scale Design of Electromagnetic Composite Metamaterials for Broadband Microwave Absorption." *Composites Science and Technology* 162 (July): 206–214. https://doi.org/10.1016/j.compscitech.2018.04.028.

Huo, Jia, Li Wang, and Haojie Yu. 2009. "Polymeric Nanocomposites for Electromagnetic Wave Absorption." *Journal of Materials Science* 44 (15): 3917–3927. https://doi. org/10.1007/s10853-009-3561-1.

Jamkhande, Prasad Govindrao, Namrata W. Ghule, Abdul Haque Bamer, and Mohan G. Kalaskar. 2019. "Metal Nanoparticles Synthesis: An Overview on Methods of Preparation, Advantages and Disadvantages, and Applications." *Journal of Drug Delivery Science and Technology* 53 (October): 101174. https://doi.org/10.1016/j.jddst.2019.101174.

Jauchem, James R. 2008. "Effects of Low-Level Radio-Frequency (3kHz to 300GHz) Energy on Human Cardiovascular, Reproductive, Immune, and Other Systems: A Review of the Recent Literature." *International Journal of Hygiene and Environmental Health* 211 (1–2): 1–29. https://doi.org/10.1016/j.ijheh.2007.05.001.

Khodiri, Ahmed A., Magdy Y. Al-Ashry, and Ahmed G. El-Shamy. 2020. "Novel Hybrid Nanocomposites Based on Polyvinyl Alcohol/Graphene/Magnetite Nanoparticles for High Electromagnetic Shielding Performance." *Journal of Alloys and Compounds* 847 (December): 156430. https://doi.org/10.1016/j.jallcom.2020.156430.

Kim, Jonghoon, Hongseok Kim, Chiuk Song, In-Myoung Kim, Young-il Kim, and Joungho Kim. 2014. "Electromagnetic Interference and Radiation from Wireless Power Transfer Systems." In *2014 IEEE International Symposium on Electromagnetic Compatibility (EMC)*, 171–176. IEEE. https://doi.org/10.1109/ISEMC.2014.6898964.

Kubacki, Roman, Ludwika Lipińska, Rafał Przesmycki, and Dariusz Laskowski. 2022. "The Comparison of Microwave Reflectance of Graphite and Reduced Graphene Oxide Used for Electronic Devices Protection." *Energies* 15 (2): 651. https://doi. org/10.3390/en15020651.

Li, Jing, Di Zhou, Peng-Jian Wang, Chao Du, Wen-Feng Liu, Jin-Zhan Su, Li-Xia Pang, Mao-Sheng Cao, and Ling-Bing Kong. 2021. "Recent Progress in Two-Dimensional Materials for Microwave Absorption Applications." *Chemical Engineering Journal* 425 (December): 131558. https://doi.org/10.1016/j.cej.2021.131558.

Li, Weiwei, Mingji Chen, Zhihui Zeng, Hao Jin, Yongmao Pei, and Zhong Zhang. 2017. "Broadband Composite Radar Absorbing Structures with Resistive Frequency Selective Surface: Optimal Design, Manufacturing and Characterization." *Composites Science and Technology* 145 (June): 10–14. https://doi.org/10.1016/j.compscitech.2017.03.009.

Micheli, D., C. Apollo, R. Pastore, and M. Marchetti. 2010. "X-Band Microwave Characterization of Carbon-Based Nanocomposite Material, Absorption Capability Comparison and RAS Design Simulation." *Composites Science and Technology* 70 (2): 400–409. https://doi.org/10.1016/j.compscitech.2009.11.015.

NASA. 2010. "Introduction to the Electromagnetic Spectrum." National Aeronautics and Space Administration, Science Mission Directorate. https://science.nasa.gov/ems/01_intro.

Nguyen, Minh Dang, Hung-Vu Tran, Shoujun Xu, and T. Randall Lee. 2021. "Fe3O4 Nanoparticles: Structures, Synthesis, Magnetic Properties, Surface Functionalization, and Emerging Applications." *Applied Sciences* 11 (23): 11301. https://doi.org/10.3390/ app112311301.

Niu, Yuting, Yufei Chen, Wenjing Li, Ruiqin Xie, and Xuliang Deng. 2020. "Electromagnetic Interference Effect of Dental Equipment on Cardiac Implantable Electrical Devices: A Systematic Review." *Pacing and Clinical Electrophysiology* 43 (12): 1588–1598. https://doi.org/10.1111/pace.14051.

Pandey, Rachit, Sravya Tekumalla, and Manoj Gupta. 2020. "EMI Shielding of Metals, Alloys, and Composites." In *Materials for Potential EMI Shielding Applications*, 341–355. Elsevier. https://doi.org/10.1016/B978-0-12-817590-3.00021-X.

Panwar, Ravi, and Jung Ryul Lee. 2019. "Recent Advances in Thin and Broadband Layered Microwave Absorbing and Shielding Structures for Commercial and Defense Applications." *Functional Composites and Structures* 1 (3): 032001. https://doi.org/10.1088/2631-6331/ab2863.

Qin, F., and C. Brosseau. 2012. "A Review and Analysis of Microwave Absorption in Polymer Composites Filled with Carbonaceous Particles." *Journal of Applied Physics* 111 (6): 061301. https://doi.org/10.1063/1.3688435.

Qu, Zhongji, Yu Wang, Wei Wang, and Dan Yu. 2021. "Robust Magnetic and Electromagnetic Wave Absorption Performance of Reduced Graphene Oxide Loaded Magnetic Metal Nanoparticle Composites." *Advanced Powder Technology* 32 (1): 194–203. https://doi.org/10.1016/j.apt.2020.12.002.

Ranjkesh, Zahra, and Komeil Nasouri. 2022. "Facile Synthesis of Novel Porous Nickel/Carbon Fibers Obtained from Cigarette Butts for High-Frequency Microwave Absorption." *Journal of Environmental Chemical Engineering* 10 (1): 106969. https://doi.org/10.1016/j.jece.2021.106969.

Ruiz-Perez, F., S. M. Lspez Estrada, and F. Caballero Briones. 2022. "Tunable, Wideband X-Band Microwave Absorbers Using Variable Chessboard Surfaces." *IEEE Letters on Electromagnetic Compatibility Practice and Applications*, 1. https://doi.org/10.1109/LEMCPA.2022.3141775.

Ruiz-Perez, F., S.M. López-Estrada, R.V. Tolentino-Hernández, and F. Caballero-Briones. 2022. "Carbon-Based Radar Absorbing Materials: A Critical Review." *Journal of Science: Advanced Materials and Devices* 7 (3): 100454. https://doi.org/10.1016/j.jsamd.2022.100454.

Safari Variani, Ali, Somayeh Saboori, Saeed Shahsavari, Saeed Yari, and Vida Zaroushani. 2019. "Effect of Occupational Exposure to Radar Radiation on Cancer Risk: A Systematic Review and Meta-Analysis." *Asian Pacific Journal of Cancer Prevention* 20 (11): 3211–3219. https://doi.org/10.31557/APJCP.2019.20.11.3211.

Soler, M.A.G., and L.G. Paterno. 2017. "Magnetic Nanomaterials." In *Nanostructures*, 147–186. Elsevier. https://doi.org/10.1016/B978-0-323-49782-4.00006-1.

Song, Wei-Li, Li-Zhen Fan, Zhi-Ling Hou, Kai-Lun Zhang, Yongbin Ma, and Mao-Sheng Cao. 2017. "A Wearable Microwave Absorption Cloth." *Journal of Materials Chemistry C* 5 (9): 2432–2441. https://doi.org/10.1039/C6TC05577J.

Sun, Keju, Masanori Kohyama, Shingo Tanaka, and Seiji Takeda. 2015. "Understanding of the Activity Difference between Nanogold and Bulk Gold by Relativistic Effects." *Journal of Energy Chemistry* 24 (4): 485–89. https://doi.org/10.1016/j.jechem.2015.06.006

Upadhyay, Sneha, Kinnari Parekh, and Brajesh Pandey. 2016. "Influence of Crystallite Size on the Magnetic Properties of Fe3O4 Nanoparticles." *Journal of Alloys and Compounds* 678 (September): 478–485. https://doi.org/10.1016/j.jallcom.2016.03.279.

(U.S.NRC), United States Nuclear Regulatory Commision. n.d. "Subpart C—Occupational Dose Limits." Accessed August 4, 2022. https://www.nrc.gov/reading-rm/doc-collections/cfr/part020/part020-1201.html.

Wang, Baolei, Qian Wu, Yonggang Fu, and Tong Liu. 2021. "A Review on Carbon/Magnetic Metal Composites for Microwave Absorption." *Journal of Materials Science & Technology* 86 (September): 91–109. https://doi.org/10.1016/j.jmst.2020.12.078.

Wang, Ding-Chuan, Yu Lei, Wei Jiao, Yi-Fan Liu, Chun-Hong Mu, and Xian Jian. 2021. "A Review of Helical Carbon Materials Structure, Synthesis and Applications." *Rare Metals* 40 (1): 3–19. https://doi.org/10.1007/s12598-020-01622-y.

Wang, Yan, Xiang Gao, Xinming Wu, and Chunyan Luo. 2020. "Facile Synthesis of Mn3O4 Hollow Polyhedron Wrapped by Multiwalled Carbon Nanotubes as a High-Efficiency Microwave Absorber." *Ceramics International* 46 (2): 1560–1568. https://doi.org/10.1016/j.ceramint.2019.09.124.

Wdowiak, Artur, Paweł A. Mazurek, Anita Wdowiak, and Iwona Bojar. 2017. "Effect of Electromagnetic Waves on Human Reproduction." *Annals of Agricultural and Environmental Medicine* 24 (1): 13–18. https://doi.org/10.5604/12321966.1228394.

Yahyaei, Hossein, and Mohsen Mohseni. 2018. "Nanocomposites Based EMI Shielding Materials." In *Advanced Materials for Electromagnetic Shielding*, 263–288. John Wiley & Sons, Inc. https://doi.org/10.1002/9781119128625.ch12.

Yan, Dexian, Erping Li, Qinyin Feng, Xiangjun Li, and Shihui Guo. 2022. "Design and Analysis of a Wideband Microwave Absorber Based on Graphene-Assisted Metamaterial." *Optik* 250 (January): 168310. https://doi.org/10.1016/j.ijleo.2021.168310.

Yoo, Young Joon, Sanghyun Ju, Sang Yoon Park, Young Ju Kim, Jihye Bong, Taekyung Lim, Ki Won Kim, Joo Yull Rhee, and YoungPak Lee. 2015. "Metamaterial Absorber for Electromagnetic Waves in Periodic Water Droplets." *Scientific Reports* 5 (1): 14018. https://doi.org/10.1038/srep14018.

Yu, Di, Gui-Mei Shi, Fa-Nian Shi, Xiu-Kun Bao, Shu-Tong Li, and Qian Li. 2022. "N-Doped Carbon Nanofiber Embedded with TiN Nanoparticles: A Type of Efficient Microwave Absorbers with Lightweight and Wide-Bandwidth." *Journal of Alloys and Compounds* 920 (November): 165791. https://doi.org/10.1016/j.jallcom.2022.165791.

Zeng, Xiaojun, Xiaoyu Cheng, Ronghai Yu, and Galen D. Stucky. 2020. "Electromagnetic Microwave Absorption Theory and Recent Achievements in Microwave Absorbers." *Carbon* 168: 606–623. https://doi.org/10.1016/j.carbon.2020.07.028.

Zhou, Jin, Zhengqi Liu, Xiaoshan Liu, Guolan Fu, Guiqiang Liu, Jing Chen, Cong Wang, Han Zhang, and Minghui Hong. 2020. "Metamaterial and Nanomaterial Electromagnetic Wave Absorbers: Structures, Properties and Applications." *Journal of Materials Chemistry C* 8 (37): 12768–12794. https://doi.org/10.1039/D0TC01990A.

Zhu, Yanwu, Shanthi Murali, Weiwei Cai, Xuesong Li, Ji Won Suk, Jeffrey R. Potts, and Rodney S. Ruoff. 2010. "Graphene and Graphene Oxide: Synthesis, Properties, and Applications." *Advanced Materials* 22 (35): 3906–3924. https://doi.org/10.1002/adma.201001068.

7 Functionalized Magnetic Nanoparticles for Photocatalytic Applications

Muhammad Munir Sajid and Haifa Zhai
Henan Normal University

Ali Raza Ishaq
Hubei University

Thamer Alomayri
Umm Al-Qura University

Nadia Anwar and Muqarrab Ahmed
Tsinghua University

CONTENTS

7.1 INTRODUCTION

It has long been understood that a structure's properties, behavior, and potential applications are intensely affected by the number of atoms in that structure [1]. The reduction of crystalline structures suggests that the characteristics of individual atoms are entirely different than in the bulk of the same material. The recent size decrease of device components, mostly electronic components, has achieved great interest in understanding the peculiar behavior of very small structures [2]. The importance of the length scale of nanometers in terms of size effects increases when at least one

DOI: 10.1201/9781003335580-7

dimension of a crystal is reduced to the order of hundreds of atoms. As a result, the field of "nanotechnology" has grown quickly, and nanotechnology is the study and creation of matter on a microscopic scale [3]. In general, structures between 1 and 100 nm in at least one dimension are of interest to nanotechnology, which involves the creation of materials or technologies within that range. Richered Feyman was the first scientist who used the word nanotechnology in his speech "There's Plenty of Room at the Bottom," which he delivered on December 29, 1959 [4, 5]. Feynman believed that direct atom manipulation has a more effective kind of synthetic chemistry than bulk methods [6]. In 1974, at Tokyo University of Sciences professor Norio Taniguchi invented the term "nanotechnology". Due mostly to the development of STM (scanning tunneling microscope) and the emergence of cluster science, nanotechnology gained prominence in the 1980s. The discovery and use of fullerenes came next in the middle of the 1980s, semiconductor nanocrystal discovery also supported its development [7, 8].

A nanomaterial is a physical substance having at least one characteristic dimension ranging from 1 to 100 nm [9]. The characteristics of nanomaterials can differ from those of materials of micron or millimeter dimensions. Surface-related effects and quantum confinement effects are two main categories into which the unique properties of nanomaterials can be classified as indicated in Figure 7.1. Atoms on a material surface have a different bonding structure from those below the surface, and they could be exposed to a different chemical environment, such as oxide or hydrogen termination forms [10, 11]. These atoms act differently from other atoms in a crystal as a result, and events occurring at a material surface are typical of little consequence to the material's overall behavior since the surface atoms to bulk atoms ratio is negligible in bulk materials. However, because creation at the nanoscale involves such a high ratio of surface to volume, surface effects are unavoidable. Additionally, substantial portion of surface atoms may cause the solid-state structure to change slightly, decreasing overall stability. Quantum effects happen when an electron wavelength in a material is on level with the material dimension. The substance becomes quantized in that restricting dimension due to the restriction on an electron path of travel. The density of states is then dependent on the number of dimensions in which electrons are quantized.

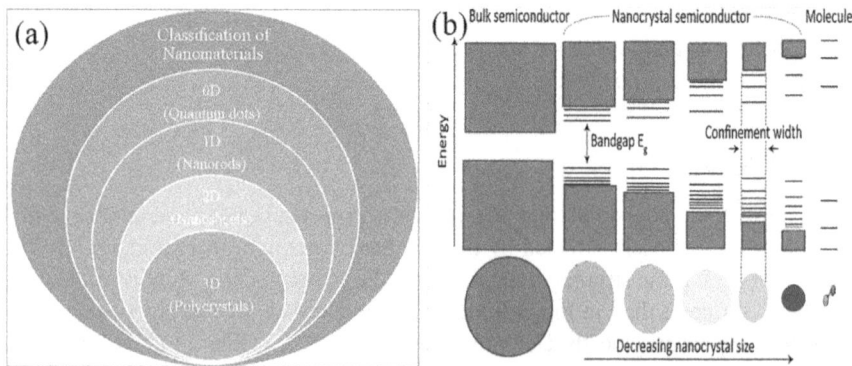

FIGURE 7.1 (a) Classification of nanomaterials. (b) Effect of size on band gaps.

Furthermore, when compared to macroscopic systems, many physical (mechanical, electrical, optical, etc.) attributes alter. The rise in energy difference between energy levels and bandgap is caused by quantum confinement. When the restricting dimension is enormous in comparison to the particle wavelength, the particle acts as if it were free. Because of the continuous energy state, the bandgap stays at its initial energy during this condition. The energy spectrum becomes distinct when the limiting dimension lowers and reaches a particular limit, often at the nanoscale. As a result, the bandgap now depends on the size [12, 13].

Several magnetic nanoparticles, for example, iron, nickel, aluminum, gold, and cobalt, are used as a photocatalyst, and much research into such materials has been conducted. Functionalized magnetic particles with certain structural arrangements are prepared by coprecipitation, thermal decomposition, microemulsion, hydrothermal, sol-gel, microwave, flame spray, laser ablation, ultrasonic-assisted, and sonochemical techniques. Aside from photocatalysis, these magnetic nanomaterials can be employed in lithium-ion batteries and solar panel sensors. As shown in Figure 7.2, several methods have been employed to synthesize magnetic nanomaterials.

These methods are mainly based on two approaches, namely the "top-down approach" as well as "bottom-up approach." Large materials (bulk materials) are broken down into little fragments of material in the top-down technique [14]. This

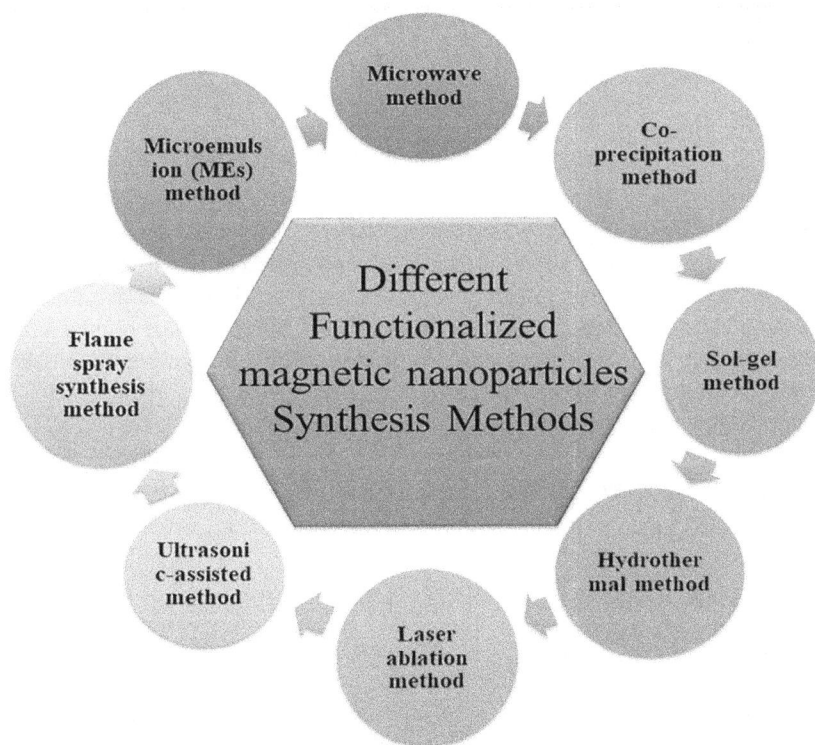

FIGURE 7.2   Different synthesis methods for functionalized magnetic nanoparticles.

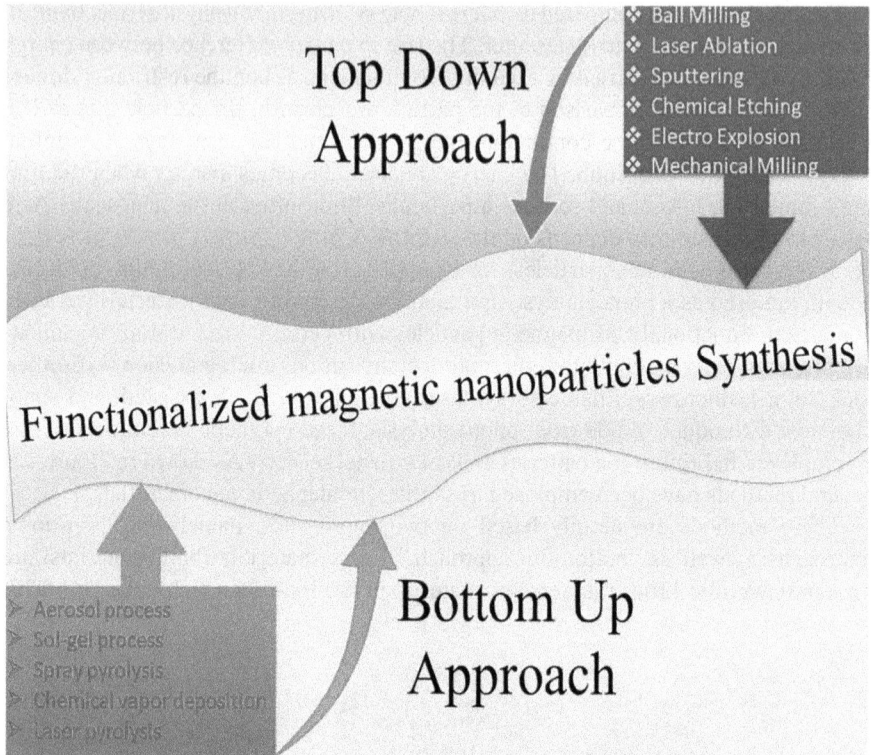

FIGURE 7.3 Various techniques to synthesize metallic nanoparticles.

method is commonly utilized when external control is required to get the desired form and morphologies. This technique is appealing because of its application in metals, semiconductors, large-scale production, lesser precision accuracy, and regulated parameters [15]. Bottom-up approaches are used to create more refined materials by combining simple atomic-level components while managing process factors like pH, size, morphology, pattern, and well-arrangement. Bottom-up approaches are commonly employed because they are speedier, have large-scale manufacturability, have extremely little contamination, are more cost-effective, have high quality, and have an aligned structure. A schematic for bottom-up and top-down approaches is illustrated in Figure 7.3.

7.2 PHOTOCATALYTIC DEGRADATION (PCD)

One of the first to perform tests to see whether "light and light alone" would assist chemical reactions was chemist Giacomo Ciamician in 1901 [16, 17]. He conducted tests with blue as well as red lights and then discovered that only blue light had a chemical reaction. He was careful enough to rule out the concept that thermal heating brought on by light was what was driving these reactions. The term "photocatalysis" first appears in the scientific literature in 1911. In 1932, it was reported that TiO_2

and Nb_2O_5 were the catalysts for photocatalytic reductions of $AgNO_3$ to Ag and then $AuCl_3$ to Au. Following that, in 1938, TiO_2 was studied in the presence of O_2 as a photosensitizer to bleach dyes. Nevertheless, the lack of widely used practical applications meant that interest in photocatalysis continued as a hobby. Early in the 1970s, things started to shift for two reasons. First, the "oil crisis" inspired scientists to look for fossil fuel alternatives. Second, the search for renewable energy sources was generated by the researcher about the effects of large-scale industrial operations on the environment. Several significant papers were published at this time. Bell Lab researchers published the first study on O_2 evolution from TiO_2 in 1968 [18], using TiO_2 electrode and UV-light irradiation, Fujishima, as well as Honda, stated photo-assisted H_2O oxidation with H_2 generation in 1972 [19]. In 1979, investigations on photocatalytic CO_2 reduction retaining several inorganic semiconductors as photocatalysts were published by Fujishima et al. [20]. In the 1980s, research on analogous processes utilizing, in particular, TiO_2 nanoparticles as photocatalysts received substantial interest as a result of these early efforts to expand the uses of photocatalysis. Since then, research has focused on realizing the underlying ideas, improving photocatalytic effectiveness, looking for new photocatalysts as well as broadening the scope of reaction [21]. For example, in 1997, the discovery of the TiO_2 photo-induced super-hydrophilicity effect. TiO_2, which possesses self-cleaning also anti-fogging properties, had thus been used in construction materials. Various candidates with greater photocatalytic activity than TiO_2 had been examined in the creation of novel photocatalysts; the majority of these candidates have large band gaps also are only active under UV rays. Photocatalysts that absorb visible light have also been studied in parallel for improved efficiency.

Bokare et al. investigated the use of Fe/Ni bimetallic nanoparticles to remove the pollutant organic dye Orange G from an aqueous solution [22]. Fe/Pd bimetallic nanoparticle dechlorinate 5 mg/L of lindane (gamma-hexachlorocyclohexane) regulated within 5 minutes at a catalyst stocking of 0.5 g/L, according to research on the Fe/Pd nanoparticles degrade lindane 120 (gamma-hexachlorocyclohexane) in an aqueous solution [23]. According to Mohamed et al. [28], there are double primary categories of photocatalysis processes: (i) catalyzed photoreactions also (ii) sensitized photoreactions. The first category is catalyst-driven photoreaction. A ground-state molecule is then given an electron via a photo-excited catalyst. When dye molecules undergo sensitized photo-excitation and relate to the ground-state catalyst, early light excitation results [24].

According to Qu et al. [25], photocatalysis has exhibited tremendous potential since it is inexpensive, environmentally friendly, and supportability in water degradation technology. The exploitation of nanoparticles in this remedy method, in particular, is of critical concern. Nanoscale semiconductors including nano-TiO_2, ZnO, CdS, $BiVO_4$, and WO_3 are among the mainly utilized photocatalytic nanocatalyst in wastewater remedies. TiO_2 is furthermost preferred because of its low toxicity, chemical strength, and low cost, as well as wide obtainability as a material. Though, the actual application of TiO_2 as a photocatalyst has been disadvantaged by various restrictions, such as the requirement for employing aqueous solutions limits its broad practical usage and challenges its residue separation and recycling of the photocatalyst. Another issue is photocatalyst optimization to take advantage of visible light from solar radiation which is 45% of the solar spectrum. The latter difficulty, fortunately, may be handled by doping metal or anion-impurities into nano-TiO_2, utilizing dye sensitizers, or combining nano-TiO_2 with the narrow bandgap semiconductor to generate hybrid nano-composites [25].

7.2.1 Basic Principle of a Photocatalytic Oxidation Process

In continuous semiconductor photocatalysis, photocatalysis is an advanced oxidation process (AOP) for eliminating organic contaminants (e.g., TiO_2, $BiVO_4$, ZnO, FeO), a light source as well as an oxidizing facilitator [26]. When a semiconducting catalyst is permitted to be lit by a photon of energy equal to or higher than the bandgap energy (E_g) of the semiconductor, an electron from the valence band to the conduction band is accelerated. Simultaneously, it created an electron vacancy, also known as a hole (h), in the valence band within a semiconductor [27].

Mohamed et al. [28] explore how photogenerated positive holes oxidize with water to form hydroxyl radicals ($OH^•$), whereas electrons produce superoxide radicals via the oxygen reaction. Both of these major products, specifically ($OH^•$), are long-lasting oxidizing types that produce organic pollutants [28].

According to Ahmed and Khurshid [26], the rate of oxygen breakdown and contaminant oxidation must be consistent during the photocatalytic removal of carbon-based contaminants. As a result, the photocatalytic oxidation process will be slowed down and the rate of electron–hole pair (EHP) recombination will increase (usually if there is not enough oxygen in the solution). However, intrinsic catalyst characteristics including crystal structure, surface area, particle size, band gap, and outward hydroxyl compactness play a significant role in photocatalytic activities [26].

When the catalyst is illuminated, EHPs are produced to the valence electron promotion to the conduction band. The water molecule is conquered by the hole created on the catalyst and transformed into a hydroxyl radical, which possesses the same significant oxidizing exponents. When organic pollutants are present, they react with these hydroxyl groups, which causes them to break up. If the entire process is carried out in the presence of oxygen, a chain reaction between the constituent molecules intervening in radicals and oxygen atoms occurs. When an organic pollutant is produced, carbon dioxide and water are the consequences. A schematic diagram for photocatalytic procedure is illustrated in Figure 7.4.

The copulating reaction is one way that atmospheric oxygen is reduced. Because oxygen is easily reduced, it has a step-down option for producing hydrogen. Conduction band electrons react with oxygen to create a superoxide anion. The ion can also form a link with the mediator of an oxidation chemical process, which produces peroxides before turning them into water. Similar to water, organic matter can quickly experience cognitive reduction. This is why, as shown in Figure 7.5, a high organic matter concentration increases the photocatalytic response by increasing the likelihood of the number of holes, which ultimately deoxidizes the recombination value of carriers.

Oxidation and reduction responses occur simultaneously in a photocatalytic process. For photocatalytic response, we need specific materials or catalysts that accompany particular oxidation and reduction reactions. Materials are generally

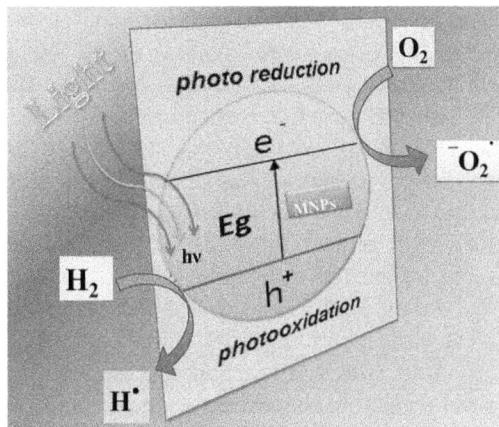

FIGURE 7.4   Schematic representation of photocatalytic procedures (Mohamed and Bahnemann, 2012).

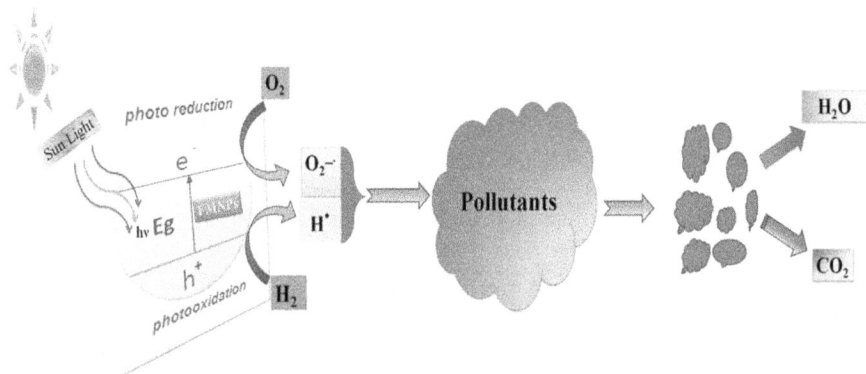

FIGURE 7.5   Schematic presentation of photocatalytic basic principle.

divided into three primary classes: conductor, dielectric, and semiconducting material. This division is based on electronic characteristics. Convergence between the valence and conduction band occurs in a conductor. Essential to a photocatalytic reaction is simultaneous oxidation and reduction, but only free electrons may be used in conductivity. We only do one oxidation reaction at a time on a conductor, not both reactions at once. Alkaline earth metals, transition metals, and alkalis are the best conductors. They either lack a suitable band gap or, more commonly, their conduction and valence bands intersect. They weren't favored for catalytic activity in reactions. Dielectrics have a large band gap and a high energy threshold for carrying out oxidation and reduction reactions. We cannot use dielectric as a catalyst to split apart water molecules. We need an eccentric catalyst that ignites in either the visible or ultraviolet illumination spectrum. Additionally, insulators lack free electrons, which prevents oxidation from occurring, that is why they are not suitable for the photolytic process. Being dielectric, all gases weren't suitable for the photocatalytic reaction. When it comes to semiconducting materials, oxidation and reduction have the capacity to occur simultaneously and have tunable band gaps. Free EHPs are produced as the light falls. The low recombination value is an essential need for a semiconducting substance to function as a photocatalyst. Additionally, semiconductors with a band gap between 1.5 and 3.5 eV or an unnoticeable domain between 420 and 800 nm are appropriate for photocatalytic activity. Since they exhibit catalytic activity in visible light. As a photocatalyst for the UV–visible range, we expect a semiconductor with a band gap of only 1.5–3.5 eV, which is unusual for semiconductors. Most metallic oxides belong to this category, they also possess additional qualities that create them appropriate as photocatalysts, which are mentioned lower. Metal oxides are crucial in photocatalysis and several electrical applications. They completely meet our necessities for a photocatalyst. Metal oxides are suitable for this assumption because of its favorable band gap, electrical structure, light absorption, and carrier transportation. A band gap is the most fundamental and crucial dimension that a photocatalyst should possess. Band gap should be within the UV–visible range are affordable, the stability of the building is one of the other properties. Such as morphology, reusability, and large area. These characteristics are shared by a variety of metal oxides, including the oxides of titanium, vanadium, cerium, zinc, and chromium. They are used as photocatalysts as a result when unprotected from visible light, the photocatalyst absorbs a photon and also activates the higher conduction band, producing an EHP. This EHP causes oxidation–reduction reactions, which break down contaminants, to occur along the surface of metal oxide. Because of this, metal oxide is often used as a photocatalyst.

The general mechanism for a photocatalytic chemical reaction involves five basic steps: reactant diffusion on the surface of the catalyst in step 1, reactant adsorption on the catalyst surface in step 2, basic reaction on the catalyst surface in step 3, production desorption in step 4, and product diffusion on the catalyst surface in step 5. General mechanism of organic pollutant degradation is in the steps that follow:

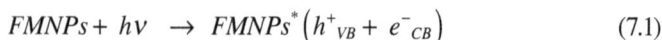

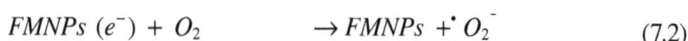

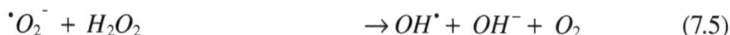

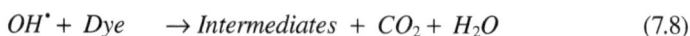

$$FMNPs + hv \rightarrow FMNPs^* \left(h^+_{VB} + e^-_{CB} \right) \tag{7.1}$$

$$FMNPs\ (e^-) + O_2 \rightarrow FMNPs +^{\bullet} O_2^- \tag{7.2}$$

$$O_2 + 2H^+ + e^- \rightarrow HO_2^{\bullet} \tag{7.3}$$

$$HO_2^{\bullet} + HO_2^{\bullet} \rightarrow H_2O_2 + O_2 \tag{7.4}$$

$$^{\bullet}O_2^- + H_2O_2 \rightarrow OH^{\bullet} + OH^- + O_2 \tag{7.5}$$

$$FMNPs\ (h^+) + OH^- / H_2O \rightarrow FMNPs + OH^{\bullet} \tag{7.6}$$

$$H_2O_2 + hv \rightarrow 2OH^{\bullet} \tag{7.7}$$

$$OH^{\bullet} + Dye \rightarrow Intermediates + CO_2 + H_2O \tag{7.8}$$

When comparing wastewater handling and photocatalytic activity, different factors play an essential role.

EHP detachment rate is one important component. Semiconducting material's EHPs have very short lifetimes. The use of a catalyst in a subordinate response before recombination must be permanent. When compared to a conventional hydrogen electrode, the redox potential for the hole also electron in TiO_2 is, respectively, 1.0–3.5 and 0.5–1.5 V. The hole is a good oxidizing agent in the valance band. A great photocatalyst should have a broad band gap as well as least amount of electron–hole recombination [29].

The catalyst structure affect the photocatalytic activity. For instance, rutile, brookite, and anatase are the three different forms of titanium dioxide (TiO_2). Anatase is the most effective of these three due to its well-balanced structure, exceptional adsorption capacity, and appropriate conduction band position. A key component of the corrosion process is morphology. Photocatalyst is used in nano form rather than bulk form because we understand that on the nanoscale, materials bear tiny size and also higher expanse attainable for chemical reaction, increasing reaction range. Because it has a greater specific surface area, zinc oxide sample distribution with a spherical shape performs more effectively than rod-like structures. To put it another way, we assume that surface to volume ratio is modified, allowing multiple reactants to respond to the catalyst at the same time, for example, nano-sized TiO_2 is stronger than bulk TiO_2 [30].

The pH of solution also plays a crucial role in effectiveness of a photocatalyst. By altering the surface charges of the photocatalyst, pH affects its activity. It can be explained by electrostatic interaction between pollution atoms and charged particles. Here, results of photocatalytic activity for dye versus photocatalyst in metallic element oxide semiconducting materials under various pH scales are discussed. According to published research, photocatalyst performance is lower for pH scales lower than 5, or moderately acidic conditions. This is because the proton concentration is higher at these pH levels, which causes slower dye debasement. After all, there are fewer available $OH^{\cdot}$ molecules. The photo activity of the catalyst increase for pH scale ranges of 5 to 10, and for pH 10 (alkaline medium), proficiency is at its highest level because the $OH^{\cdot}$-generated reaction with organic pollutant also degrades organic pollutants. Due to the extremely abundant hydroxyl radical immersion in the pH range of 11–14, there is a decline in efficiency. They are not allowed to react with the dye. As a result, this pH array's efficiency is very poor.

The amount of catalyst directly correlates with the amount of photocatalytic action, that is, increasing the amount of catalyst will increase the number of free radicals required for degradation, which in turn will raise the rate of the chemical reaction. But after an optimum limit is reached, catalyst quantity doesn't favor the reaction because light can't travel through easily, therefore the majority of the catalyst surfaces remain passive [31].

The rate of photocatalytic degradation increases with omissible light intensity. The quantum yield increases when the light power is increased. The proportion of response rate to absorption rate is known as quantum yield. The effectiveness of the response is also influenced by wavelength. The bandgap of Fe_2O_3 is 3.2 eV; UV light is absorbed by it. The degradation rate rises for the light intensity between 0 and

20 mW/cm². This is a half-order response, meaning that the rate is dependent on the square root of incident light. Prior to the intermediate intensity, or 25 mW/cm², the response rate falls as the recombination of EHPs is accelerated by intense intensity light.

Numerous researchers have looked into how the efficiency of the photocatalytic rate depends on response temperature. Due to an increase in the rate of electron and hole recombination and then a drop in the absorption rate of the catalyst at this temperature, the reaction rate for TiO_2 is reduced at 80°C. We achieve a maximum reaction rate of 20°C–800°C, which is the ideal temperature.

7.3   WHY MAGNETIC NPs ARE SO IMPORTANT FOR PCD?

The purpose of magnetic photocatalyst is to obtain better photocatalytic performance regarding to its stability and activity during the catalytic process. In general, there are several cases of modified photocatalysts based on physical and chemical interactions. Although those methods and techniques could maintain good stability, two critical problems occur: (i) recycle and reuse of the photocatalyst become difficult when solid impurities exist in the catalytic system and (ii) separation of photocatalyst material from the supernatant. Recently, the applications of MNPs photocatalysts have been proposed, which could effectively solve those above-mentioned problems. First, the magnetic material could be easily separated by applying an external magnetic field. Thus, recycling and reuse could be simplified and be conducted for several cycles without influencing the process, while the cost is also slowed down (Figure 7.6).

7.4   PHOTOCATALYTIC ACTIVITY OF FUNCTIONALIZED MAGNETIC IRON OXIDE AT NANOSCALE

The researcher has tried to shrink the size of photocatalytic materials and then also improve the photocatalytic capabilities of these materials by manufacturing photo-catalysts in a nanoscale powder form. The surface area plays a significant role in influencing the photocatalytic activity of materials. To explore the magnetic and photocatalytic properties of uniformly sized α-Fe_2O_3 nanoparticles, S Yang et al.

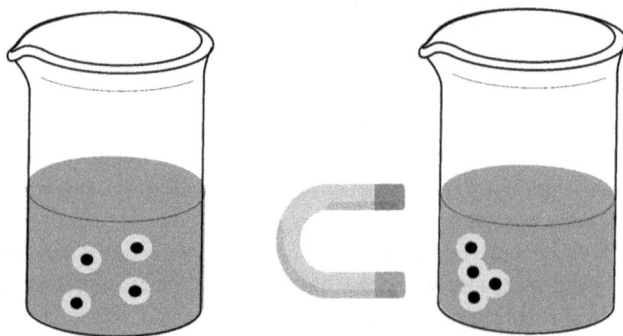

FIGURE 7.6   Schematic separation of functionalized magnetic nanoparticles by magnetic.

produced α-Fe$_2$O$_3$ hydrothermally; in comparison to powders with larger crystallite sizes, α-Fe$_2$O$_3$ powders with lesser crystallite sizes exhibit the best photocatalytic degradation proficiency. Furthermore, compared to Degussa P25, which is readily available commercially, all of the compounds demonstrated greater dye degradation efficiency. Powders of nanostructured α-Fe$_2$O$_3$ measuring 25–55 nm in size [32]. Hydrogen sulfide (H$_2$S) gas was broken down using the photocatalytic activity of the synthetic material by Apte et al. The necked structures of α-Fe$_2$O$_3$ showed good photocatalytic abilities and H$_2$ generation. By thermal dehydration, Zhou et al. produced nanorods of α-Fe$_2$O$_3$ and compared their photocatalytic activity to that of micro-rods. According to scientists, nano-dimensional α-Fe$_2$O$_3$ degrades Rhodamine B (RhB) more quickly than the matching micron-size rods. One of the main explanations for the increased photocatalytic activity was suggested to be a higher Fe-O bond stretching frequency. Another important aspect that affects a material's ability to operate as a photocatalyst is the size, composition, porosity, then local structures of particles [33]. Troy et al. examined the photocatalytic activity of three distinct types of bulk (crystallite size 120 nm), ultrasonicated (crystallite size 40 nm), and nano α -Fe$_2$O$_3$ powders (crystallite size 5.4 nm). They discovered that the rate of oxygen evolutions increases as crystallite size decreases, with α-Fe$_2$O$_3$ nano-powders reporting the greatest rate (1072 mol/g). Because the hole diffusion length in α-Fe$_2$O$_3$ nano-powders is equivalent to the crystallite size, there are more holes available for water-reactive reactions [34]. Jin et al. reported the impact on the catalytic characteristics of calcination temperature, reaction temperature, catalyst quantity, also reaction time. According to their findings, photocatalytic activity increases with raising calcination temperature, reaction temperature as well as catalytic-quantity up to a point before declining [35]. Additionally, similar results were reported by W Dong; for α-Fe$_2$O$_3$ nanoparticles produced using the sol-gel method and then heated at various calcination temperatures. The efficiency of the catalyst was examined for a number of experimental factors, including calcination temperature, pH, light intensity, and dye also catalyst concentration. The maximum photocatalytic activity can be observed in samples that have been calcined at 600°C due to the creation of the more prevalent α-Fe$_2$O$_3$ phase. The reaction at higher pH settings demonstrated superior photocatalytic characteristics, according to an analysis of the photocatalytic properties for the pH range of 3 to 10. Electrostatic abstractive properties among negatively charged surfaces of α-Fe$_2$O$_3$ also cationic malachite green dye increase at basic pH settings, favoring the formation of the OH$^\bullet$ radical and increasing the likelihood of dye degradation. The photocatalytic properties of α-Fe$_2$O$_3$ are linearly influenced by light intensity since there are more photons available for reaction. Additionally, similar results were investigated by Liu et al. for nanorods made of α-Fe$_2$O$_3$. The authors looked at the influence of initial dye concentration as well as catalyst concentration on photocatalytic characteristics. According to studies, 50 mg/L of catalyst is the ideal concentration to obtain the greatest photocatalytic activity. However, as dye concentration rises, the photocatalytic activity declines. After a certain value, this effect was explained by a decrease in transparency with an increase in dye and then catalyst concentration. The catalyst porosity significantly contributes to the photocatalytic properties of a photoreaction [36]. Sundarmurthy et al. study the photocatalytic characteristics, electrospun 1-dimensional Fe_2O_3 nanobraids and nanoporous structures created. The nanostructures display enhanced photocatalytic activity for CR degradation in a short amount of time because of their porous surface and nano-sized crystallites of α -Fe_2O_3, which offer more active catalytic centers and allow efficient interaction between organic dye and α -Fe_2O_3 [37]. Porous structures made of Fe_2O_3 were reported by Zhang et al. also photocatalytic activity was analyzed by degradation of methylene blue (MB). They examined the impact of catalyst concentration also porosity on photocatalytic activity. According to reports, the optimal amount of catalyst (20 mg) is needed to achieve the highest rate of MB degradation; using fewer or additional catalysts results in lessened photocatalytic activity. Smaller illumination is the result of using more catalysts, and when there aren't enough active sites to break down the organic dye, there aren't enough active sites either [38]. Geng et al. followed a Ni^{+2}/surfactant system method to create porous and rough-surfaced α-Fe_2O_3, which exhibits superior photocatalytic capabilities than α-Fe_2O_3 nanoparticles in the destruction of MB due to their greater surface area [39]. Mishra et al; Fe_2O_3 micro/nano-spheres that have been generated using hydrothermal synthesis and thermal treatment. The dye degradation efficiency of the micro/nano-spheres is higher than that of nano-powders. The calculated response rate for spherical structures is more than twice that of nano-powders and 12 times that of powders with a particle size of less than one micron. The greater specific surface area also porous structure are responsible for improved photocatalytic activity. Bharathi et al; investigated a correlation between the photocatalytic activity of α-Fe_2O_3 and its surface shape. RhB degradation in presence of the catalyst was used to assess the photocatalytic activity of α-Fe_2O_3 nanostructures of various morphologies, including micro flowers, nanospindles, nanoparticles, and nanorhombohedra. The samples with the highest surface area and porosity showed the best photocatalytic activity [40]. Additionally, similar surface area influences were investigated by Cheng et al; for flower-like biphasic interfacial reaction-produced α-Fe_2O_3 nanostructures. RhB degradation was used to measure the photocatalytic abilities of α-Fe_2O_3. Nano-flowers were discovered to have better photocatalytic properties than commercial α-Fe_2O_3 powders when findings were compared. According to the result of other authors for TiO_2 and Fe_2O_3, the augmentation was associated with an increase in crystallinity also surface area [41]. Additionally, similar surface area influences were investigated by Cao et al. [42], and Lili et al. [43] synthesized hollow microspheres of Fe_2O_3 by solvothermal as well as hydrothermal method for the breakdown of salicylic acid was used to gauge the photocatalytic activity. In comparison to α-Fe_2O_3 nanoparticles, hollow spheres linked with nanosheets exhibit greater photocatalytic activity. The same outcomes were also reported by Wei et al.; Due to their greater surface area, α-Fe_2O_3 powder made at 500°C exhibit stronger photocatalytic activity for the degradation of rose Bengal dye than powders prepared at 600°C and commercially available TiO_2 (Degussa-25) [44]. Hollow spindles and spheres made of Fe_2O_3 prepared by Wei et al. [45] and Xu et al. [46], correspondingly. These authors noted that an increase in specific surface area, which exposes more unsaturated surface coordination sites to solution, managed to an improvement in photocatalytic degradation efficiency. More electron–hole transport is made possible by the hollow microsphere, which also slows down recombination. Hollow microspheres enable various visible light reflections within the interior, which promotes more effective utilization of the light source and improves light harvesting, increasing the amount of OH· that is accessible to take part in photocatalytic reactions. Additionally, the hollow spheres offer the dye molecules optimal routes and raise the likelihood of contact. In addition to crystallite size, crystallite orientation has a significant impact on the development of photocatalytic capabilities. This effect has been stated by Wei et al. [47] who produced α-Fe$_2$O$_3$ nanocubes using a solvothermal process and described that the samples with (104) planes had a stronger photocatalytic property than those with (012) planes. Fenton's reaction plays a role in the photocatalytic qualities. The Fenton reaction, which produces hydroxyl radicals (OH·) when Fe^{3+} is reduced to Fe^{2+}, depends heavily on the the amount of Fe^{3+} on the catalyst's surface. It has been stated by Lv et al.; specifies the exposed Fe^{3+} ion density in the (104) planes of α-Fe$_2$O$_3$ is 10.3 atoms/nm^2, while the exposed Fe^{3+} ion density in (012) planes is 7.33 atoms/nm^2. This explains why (104) planes react more quickly than (012) planes. Oxygen pressure and catalyst quantity, in addition to surface shape, are important factors in improving the photocatalytic capabilities [48]. Isaev et al. stated up to a certain point, there was an increase in photocatalytic activity with an increase in the amount of Fe_2O_3, but after that, the photocatalytic activity dropped. Similar to this, the scientists noted that higher oxygen levels accelerated dye breakdown. The creation of extra oxygen-containing active species, such as OH·, O_2, also H_2O oxidizing species, is what causes the improvement in photocatalytic behavior [49].

7.5   CONCLUSIONS AND FUTURE PROSPECTS

Nanotechnology provides a cost-effective solution to cleanse water while also lowering overall expenses by removing hazardous and undesired pollutants. Several studies have been conducted that show versatile NPs can remove multiple pollutants in water at the same time. This study sought to eradicate inorganic chemicals, and environmental toxins from wastewater by using FMNPs. An advanced effort in the future is required to initiate the metal oxide reaction in visible light. Population increase, along with rapid resource consumption also continuous industrial and agricultural progress has resulted in surplus effluent with integrative changes, texture, perniciousness, also harmfulness owing to various types of pollutants found in wastewater. Because of inherent nanoscale dimensions, the internalization of nanomaterials for remedy of decontamination and toxins brings large benefits, such as the ability to stimulate information at atomic solvents, improve inferior signal charges, accelerate reception times, and incessantly influence changes. Developing nanoscale instruments for an environmentally friendly atmosphere necessitates technical advancements in recognizing and differentiating processes and configurations for future rising requirements. The development of novel photocatalysts with favored properties to promote the oxidization process of constitutive substrates or decrease of pollutants under visible light irradiation should be accelerated. Currently, challenges associated with wastewater management are primarily related to the complexity of the wastewater structure, which causes inconveniences in handling progress by requiring distinct equipment and advanced technologies, resulting in longer treatment periods as well as higher working expenses.

REFERENCES

1. Busi, K.B., et al., Engineering colloidally stable, highly fluorescent and nontoxic Cu nanoclusters via reaction parameter optimization. *RSC Advances*, 2022. **12**(27): p. 17585–17595.
2. Gasparini, N., et al., The role of the third component in ternary organic solar cells. *Nature Reviews Materials*, 2019. **4**(4): p. 229–242.
3. Bayda, S., et al., The history of nanoscience and nanotechnology: from chemical–physical applications to nanomedicine. *Molecules*, 2019. **25**(1): p. 112.
4. Talebian, S. and J. Conde, Why go NANO on COVID-19 pandemic? *Matter*, 2020. **3**(3): p. 598–601.
5. Srivastava, A., A. Seth, and K. Katiyar, Microrobots and nanorobots in the refinement of modern healthcare practices, in *Robotic Technologies in Biomedical and Healthcare Engineering*. 2021, CRC Press. p. 13–37.
6. Ji, S., et al., Chemical synthesis of single atomic site catalysts. *Chemical Reviews*, 2020. **120**(21): p. 11900–11955.
7. Gross, J.H., From the discovery of field ionization to field desorption and liquid injection field desorption/ionization-mass spectrometry—A journey from principles and applications to a glimpse into the future. *European Journal of Mass Spectrometry*, 2020. **26**(4): p. 241–273.
8. Prasanth, R., et al., *The High-Impact Engineering Materials of the Millennium.*
9. Sahoo, M., et al., Nanotechnology: Current applications and future scope in food. *Food Frontiers*, 2021. **2**(1): p. 3–22.
10. Xiong, J., et al., Surface defect engineering in 2D nanomaterials for photocatalysis. *Advanced Functional Materials*, 2018. **28**(39): p. 1801983.
11. Lotfi, R., et al., A comparative study on the oxidation of two-dimensional Ti 3 C 2 MXene structures in different environments. *Journal of Materials Chemistry A*, 2018. **6**(26): p. 12733–12743.
12. Premaratne, M. and G.P. Agrawal, *Theoretical Foundations of Nanoscale Quantum Devices*. 2021: Cambridge University Press.
13. Autere, A., et al., Nonlinear optics with 2D layered materials. *Advanced Materials*, 2018. **30**(24): p. 1705963.
14. Usman, K.A.S., et al., Downsizing metal–organic frameworks by bottom-up and top-down methods. *NPG Asia Materials*, 2020. **12**(1): p. 1–18.
15. Gregorczyk, K. and M. Knez, Hybrid nanomaterials through molecular and atomic layer deposition: Top down, bottom up, and in-between approaches to new materials. *Progress in Materials Science*, 2016. **75**: p. 1–37.
16. Zhu, S. and D. Wang, Photocatalysis: basic principles, diverse forms of implementations and emerging scientific opportunities. *Advanced Energy Materials*, 2017. **7**(23): p. 1700841.
17. Coronado, J.M., A historical introduction to photocatalysis, in *Design of Advanced Photocatalytic Materials for Energy and Environmental Applications*. 2013, Springer. p. 1–4.
18. Tahir, M.B. and K.N. Riaz, *Nanomaterials and Photocatalysis in Chemistry [M]*. 2021.
19. Singh, R. and S. Dutta, A review on H_2 production through photocatalytic reactions using TiO2/TiO2-assisted catalysts. *Fuel*, 2018. **220**: p. 607–620.
20. Yui, T., et al., Photocatalytic reduction of CO_2: from molecules to semiconductors. *Topics in Current Chemistry*, 2011. **303**: p. 151–184.

21. Jiang, D., S. Zhang, and H. Zhao, Photocatalytic degradation characteristics of different organic compounds at TiO2 nanoporous film electrodes with mixed anatase/rutile phases. *Environmental Science & Technology*, 2007. **41**(1): p. 303–308.

22. Bokare, A.D., et al., *Iron-nickel bimetallic nanoparticles for reductive degradation of azo dye Orange G in aqueous solution.* Applied Catalysis B: Environmental, 2008. **79**(-3): p. 270–278.

23. Murugesan, K., et al., Effect of Fe–Pd bimetallic nanoparticles on Sphingomonas sp. PH-07 and a nano-bio hybrid process for triclosan degradation. *Bioresource Technology*, 2011. **102**(10): p. 6019–6025.

24. Mohamed, H.H. and D.W. Bahnemann, *The role of electron transfer in photocatalysis: Fact and fictions. Applied Catalysis B: Environmental*, 2012. **128**: p. 91–104.

25. Qu, X., P.J. Alvarez, and Q. Li, Applications of nanotechnology in water and wastewater treatment. *Water Research*, 2013. **47**(12): p. 3931–3946.

26. Ahmed, M. and Y. Khurshid, Does silicon and irrigation have impact on drought tolerance mechanism of sorghum? *Agricultural Water Management*, 2011. **98**(12): p. 1808–1812.

27. Humayun, M., C. Wang, and W. Luo, Recent progress in the synthesis and applications of composite photocatalysts: a critical review. *Small Methods*, 2022. **6**(2): p. 2101395.

28. Mohamed, R., D. McKinney, and W. Sigmund, Enhanced nanocatalysts. *Materials Science and Engineering: R: Reports*, 2012. **73**(1): p. 1–13.

29. Sun, M., et al., Enhanced visible light photocatalytic activity in BiOCl/SnO$_2$: heterojunction of two wide band-gap semiconductors. *RSC Advances*, 2015. **5**(29): p. 22740–22752.

30. Hao, Z., A. AghaKouchak, and T.J. Phillips, Changes in concurrent monthly precipitation and temperature extremes. *Environmental Research Letters*, 2013. **8**(3): p. 034014.

31. Rauf, M. and S.S. Ashraf, Fundamental principles and application of heterogeneous photocatalytic degradation of dyes in solution. *Chemical Engineering Journal*, 2009. **151**(1–3): p. 10–18.

32. Yang, S., et al., Size-controlled synthesis, magnetic property, and photocatalytic property of uniform α-Fe$_2$O$_3$ nanoparticles via a facile additive-free hydrothermal route. *CrystEngComm*, 2012. **14**(23): p. 7915–7921.

33. Zhou, X., et al., Visible light induced photocatalytic degradation of rhodamine B on one-dimensional iron oxide particles. *The Journal of Physical Chemistry C*, 2010. **114**(40): p. 17051–17061.

34. Townsend, T.K., et al., Photocatalytic water oxidation with suspended alpha-Fe 2 O 3 particles-effects of nanoscaling. *Energy & Environmental Science*, 2011. **4**(10): p. 4270–4275.

35. Jin, Y., et al., Synthesis of unit-cell-thick α-Fe2O3 nanosheets and their transformation to γ-Fe2O3 nanosheets with enhanced LIB performances. *Chemical Engineering Journal*, 2017. **326**: p. 292–297.

36. Liu, Y., et al., Fast degradation of methylene blue with electrospun hierarchical α-Fe2O3 nanostructured fibers. *Journal of Sol-Gel Science and Technology*, 2011. **58**(3): p. 716–723.

37. Sundaramurthy, J., et al., Superior photocatalytic behaviour of novel 1D nanobraid and nanoporous α-Fe 2 O 3 structures. *RSC Advances*, 2012. **2**(21): p. 8201–8208.

38. Zhang, J., et al., Fabrication of carbon quantum dots/TiO2/Fe2O3 composites and enhancement of photocatalytic activity under visible light. *Chemical Physics Letters*, 2019. **730**: p. 391–398.

39. Geng, W., et al., Volatile organic compound gas-sensing properties of bimodal porous α-Fe2O3 with ultrahigh sensitivity and fast response. *ACS Applied Materials & Interfaces*, 2018. **10**(16): p. 13702–13711.

40. Bharathi, S., et al., Highly mesoporous α-Fe2O3 nanostructures: preparation, characterization and improved photocatalytic performance towards Rhodamine B (RhB). *Journal of Physics D: Applied Physics*, 2009. **43**(1): p. 015501.

41. Cheng, X.-L., et al., Controlled synthesis of novel flowerlike α-Fe 2 O 3 nanostructures via a one-step biphasic interfacial reaction route. *CrystEngComm*, 2012. **14**(22): p. 7701–7708.

42. Cao, S.-W., et al., Preparation and photocatalytic property of α-Fe2O3 hollow core/shell hierarchical nanostructures. *Journal of Physics and Chemistry of Solids*, 2010. **71**(12): p. 1680–1683.

43. Li, L., Y. Chu, and Y. Liu, Synthesis and characterization of ring-like α-Fe2O3. *Nanotechnology*, 2007. **18**(10): p. 105603.

44. Wei, Q., et al., Metal-organic frameworks-derived porous α-Fe2O3 spindles decorated with Au nanoparticles for enhanced triethylamine gas-sensing performance. *Journal of Alloys and Compounds*, 2020. **831**: p. 154788.

45. Wei, Z., et al., Facile template-free fabrication of hollow nestlike α-Fe2O3 nanostructures for water treatment. *ACS Applied Materials & Interfaces*, 2013. **5**(3): p. 598–604.

46. Xu, J.-S. and Y.-J. Zhu, α-Fe 2 O 3 hierarchically hollow microspheres self-assembled with nanosheets: surfactant-free solvothermal synthesis, magnetic and photocatalytic properties. *CrystEngComm*, 2011. **13**(16): p. 5162–5169.

47. Wei, Q., et al., MOF-derived α-Fe2O3 porous spindle combined with reduced graphene oxide for improvement of TEA sensing performance. *Sensors and Actuators B: Chemical*, 2020. **304**: p. 127306.

48. Lv, H., et al., Coin-like α-Fe2O3@ CoFe2O4 core–shell composites with excellent electromagnetic absorption performance. *ACS Applied Materials & Interfaces*, 2015. **7**(8): p. 4744–4750.

49. Isaev, A., et al., Electrochemical synthesis and photocatalytic properties of α-Fe2O3. *Journal of Nanoscience and Nanotechnology*, 2017. **17**(7): p. 4498–4503.

8 Recent Progress in Green Synthesis of Functionalized Magnetic Nanoparticles as Retrievable Photocatalyst

M. Shalini, G. Murali Manoj,
R. Subramaniyan Raja, and H. Shankar
KPR Institute of Engineering and Technology

R. Saidur
Sunway University

CONTENTS

DOI: 10.1201/9781003335580-8

8.1 INTRODUCTION

Water scarcity, energy demand, and increasing CO_2 concentration in the atmosphere are critical global challenges for the next decade, and researchers are being directed to address these environmental issues in a scientific and eco-friendly manner (Malakootian et al. 2019; Mehdizadeh et al. 2020; Atrak, Ramazani, and Taghavi Fardood 2018; Adarsha et al. 2021). Increasing human needs lead to the industries such as textile, plastic, food, cosmetics, and paper to produce a large amount of dye wastewater annually (Helmiyati et al. 2022; Madhukara Naik, Bhojya Naik, Nagaraju, Vinuth, Raja Naika, et al. 2019). Industrial dyes are highly toxic and tough to degrade naturally, and they affect the soil and plant's ability to photosynthesize. Various methods have been used in the past to remove colors and hazardous compounds from the industrial effluents (Lohrasbi et al. 2019; Helmiyati et al. 2022; Atrak, Ramazani, and Taghavi Fardood 2019). Photocatalysis is one of the advanced oxidation processes that can convert solar energy into chemical and electrical energy with the aid of strong radicals (OH and O_2) (Somanathan et al. 2019; Mehdizadeh et al. 2020). During the photocatalytic process, a light photon is absorbed by the semiconductor when the wavelength of the light is equal to or greater than equal to the band gap. The electron–hole pair is created in the valence and conduction band of the semiconductor. Photoexcited electrons and holes undergo a redox reaction at the surface of photocatalytic material and degrade the organic dye from the wastewater and also convert H_2O, CO_2, and N_2 into the organic fuels such as H_2 and CH_4 (Irshad et al. 2022). Nanomaterials with semiconducting property that have an advantage such as a high surface-to-volume ratio (Shaker Ardakani et al. 2021) produce more charge carriers and light absorption, etc. Nanomaterials are more effective at eliminating the dye from the industrial waste than the bulk metal oxide semiconductor (Sarkar, Sarkar, and Bhattacharjee 2019; Kamaraj et al. 2019).

Metal oxide semiconductor (ZnO, TiO_2, $SrTiO_3$, etc.) and metal (Ag, Au, Pd, etc.) nano-photocatalysts have photocatalytic activity in the presence of the UV region, and are nontoxic, wide band gap, economical and chemically stable (Surendra 2018; Indriyani et al. 2021). These semiconducting nanomaterials show better photocatalytic performance and photocorrosion resistance. However, (i) The rapid recombination of the photoexcited charge carrier (Indriyani et al. 2021), (ii) Separation of the photocatalysts after the degradation studies from the reaction mixture (Malakootian et al. 2019) and (iii) Utilization of the complete solar spectrum are being the major challenges in employing these nano-photocatalysts for real time applications. To address the aforementioned challenge, magnetic nanomaterials (MNs) such as Fe_3O_4, FeO, Fe_2O_3 (Madubuonu et al. 2019), M Fe_2O_4 (M= Ni, Mg, Zn, Co, etc.) (Atrak, Ramazani, and Taghavi Fardood 2019), and M FeO_3 (M= Bi, Mn, Tb, etc.) (Abdul Satar et al. 2019) have gained more attention among researchers in recent years due to their superior properties: photoresponse in the UV and visible range, magnetically recoverable, and ability to reuse. They are used in various applications such as dye degradation, water splitting, removal of heavy metals, and antibiotics (Indriyani et al. 2021; Vasantharaj et al. 2019). The performance of the magnetic photocatalyst is unsatisfactory due to its stability and fast recombination of the charge carrier. Functionalization is one method for improving the stability and performance of the photocatalysts, and it can be applied to increase more surface-active sites and also to control the morphology and electronic and optical properties of the materials. Synthesis strategies for functionalizing the magnetic nanomaterials include covalent and noncovalent interactions, ligand exchange, and in situ surface engineering (defect engineering, doping, heteroatom incorporation) (John, Nair, and Vinod 2021; Kumar and Sinha Ray 2018; Bhatia et al. 2021). So far, various conventional synthesis techniques such as coprecipitation, microemulsion, thermal decomposition, solvothermal, and chemical reduction are used to prepare magnetic nanoparticles (Faraji, Yamini, and Rezaee 2010; Somanathan et al. 2019). However, harmful chemicals are used as reducing agents, capping agents, and surfactants to prepare the magnetic nanoparticle by chemical and physical methods, which affect the human health and environment. Green synthesis is one of the best alternatives to conventional synthesis methods to overcome these issues. The green synthesis method is economical, nontoxic, eco-friendly, and biocompatible and also the source used for preparing the nanomaterial is renewable to control the various parameters of the materials during synthesis (Vasantharaj et al. 2019; Madubuonu et al. 2019; Qasim et al. 2020; Bishnoi, Kumar, and Selvaraj 2018). Biological sources such as microorganisms (bacteria, yeast, fungi, algae) and plant extracts (leaf, fruit, stem, seed, peel) (Indriyani et al. 2021) are used in green synthesis methods as reducing agents, capping agents, surfactants, and stabilizing agents to replace the toxic chemicals NaOH, NH$_4$OH, and KOH (Bhuiyan et al. 2020; Abid and Kadhim 2020; Indriyani et al. 2021). Biosynthesis (using microorganisms) and phytosynthesis (using plant extracts) are the two roots of the green synthesis method for preparing nanomaterials. Functionalization of green-synthesized magnetic nanoparticles shows the desired particle size, better photocatalytic performance in the visible region, shorter degradation time, and easy recoverability (Surendra 2018; Abdul Satar et al. 2019; Atrak, Ramazani, and Taghavi Fardood 2018).

This chapter focuses on the various synthesis methods of green-synthesized magnetic particles including hydrothermal method, sol-gel method, microwave-assisted synthesis, and biosynthesis methods. Also, various synthesis strategies, various plant extracts, and functionalization of MNs (by forming metal nanocomposite, metal oxide nanocomposite, and carbon-based nanocomposite) and their photocatalytic applications have been summarized.

8.2 GREEN SYNTHESIS

The alternative choice to physical and wet chemical synthesis for the preparation of nanomaterials is green synthesis and it is a nontoxic, eco-friendly, cost-effective and renewable process. During this synthesis method, metal salts and plant extracts or microorganisms are used as precursors and reducing agents. Various parameters such as morphology, structure, size, and properties of materials depend on the concentration of precursors and reducing agent, temperature, and pH at the time of the synthesis process. Various types of nanoparticle synthesis and green synthesis methods are shown in Figures 8.1 and 8.2. In general, biosynthesis and photosynthesis are the two types of green synthesis processes used to prepare the nanomaterial. The plant extract aided synthesis process is facile and most preferable because microbial

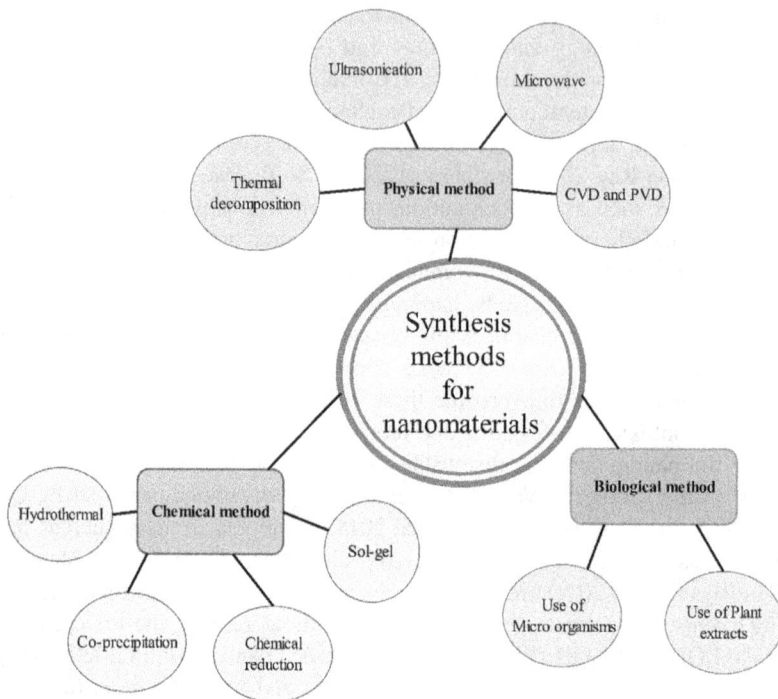

FIGURE 8.1 Various synthesis methods for preparation of nanomaterials.

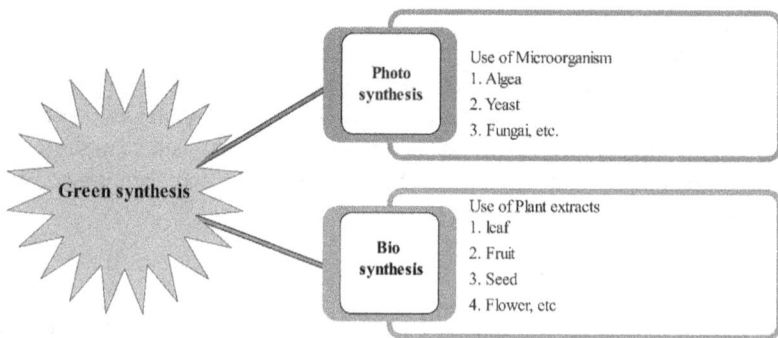

FIGURE 8.2 Types of green synthesis methods.

cell culture is required and the biohazard problem for the biosynthesis method is not present (Pal, Rai, and Pandey 2019). Plant extract consists of phytochemicals like proteins, carbohydrates, glucose, polyphenols, terpenoids, etc., which are used to reduce and stabilize the metal precursor into a metal ion during the synthesis process (Lohrasbi et al. 2019; Liaskovska et al. 2019; Sarkar, Sarkar, and Bhattacharjee 2019; Vasantharaj et al. 2019; Bhuiyan et al. 2020) (Figures 8.1 and 8.2).

8.3 MAGNETIC NANOMATERIAL AS PHOTOCATALYST

In photocatalytic degradation, recovery and recyclability of nano semiconductor photocatalyst after the completion of reaction are the challenging tasks. Figure 8.3. shows several forms iron oxide nanomaterials such as magnetite (Fe_3O_4), maghemite (γ-Fe_2O_3 and β-Fe_2O_3), hematite (α-Fe_2O_3), and Ferrites (MFe_2O_4, M = Mn, Ba and Ni) materials having magnetic properties, a larger active surface area, good chemical stability, are nontoxic, are active in UV and visible light spectrum range, are easily recoverable and reusable (Malakootian et al. 2019; Vasantharaj et al. 2019; Arularasu, Devakumar, and Rajendran 2018; Roy, Das, and Dhar 2021). Magnetic nanoparticles are used in the application of magnetic drug delivery, sensors, catalysis, tissue repair, and magnetic storage media (Hedayati, Joulaei, and Ghanbari 2020; Shaker Ardakani et al. 2021; Saeid Taghavi Fardood, Farzaneh Moradnia, Miad Mostafaei, Zolfa Afshari, Vahid Faramarzi 2019). Jianmin Gu et al. reported single crystalline α-Fe_2O_3 with hierarchical structures (micro-pine, snowflake and bundle) prepared by hydrothermal method and shows good photocatalytic properties of degradation of

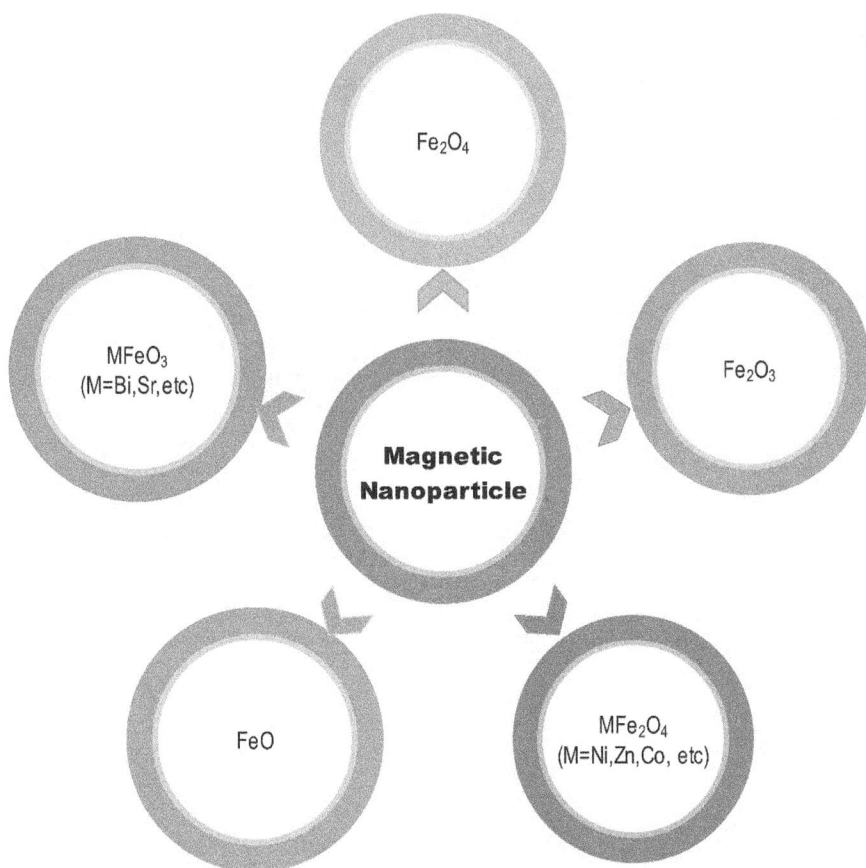

FIGURE 8.3 Various types of magnetic nanoparticles.

salicyclic acid (Gu et al. 2009). Khedr has prepared nano sized iron oxide material by thermal evaporation and coprecipitation method. Prepared iron oxides with different average crystallite size of 35, 100, and 150 nm were found to completely decompose the Congo red dye. The degradation rate of Congo red dye with respect to the particle was observed to be in the order of 35 nm > 150 nm > 100 nm. Degradation decreases at a particle size of 100 nm as large number of by-products are attached to the surface of the material (Khedr, Abdel Halim, and Soliman 2009). Mishra et al. have reported poor photocatalytic activity of iron oxide magnetic semiconductor due to a fast electron–hole recombination rate and they were found to be less stable due to low bandgap energy and easy aggregation (Mishra, Patnaik, and Parida 2019; Mamba and Mishra 2016).

8.4   GREEN SYNTHESIS OF MAGNETIC NANOPARTICLES

The synthesis of magnetic nanoparticle carried out with different synthesis routes is shown in Figure 8.4.

8.4.1   Biosynthesis Method

N.S. Abdul Satar et al. have used facile green synthesis route for the fabrication of yttrium (Y)-doped $BiFeO_3$ nanoparticle using κ-carrageenan seaweed as a biotemplate for photodegradation of methylene blue (MB). Doping with yttrium $BiFeO_3$ was found to decrease the grain size and also increase surface area compared to undoped nanoparticle (Abdul Satar et al. 2019). K.D. Sirdeshpande et al. have reported the preparation of Fe_3O_4 magnetic nanoparticle using *Calliandra haematocephala* leaf extract as a reducing agent for degradation of malachite green dye under sunlight irradiation. Larger specific surface area and mesoporous structure with a pore radius of 34.18 A° observed from Brunauer–Emmett–Teller (BET) analysis were attributed to the photocatalytic activity of the prepared photocatalysts (Sirdeshpande et al. 2018). FeO nanoparticle was developed by Kamaraj et al., using iron nitrate, sodium

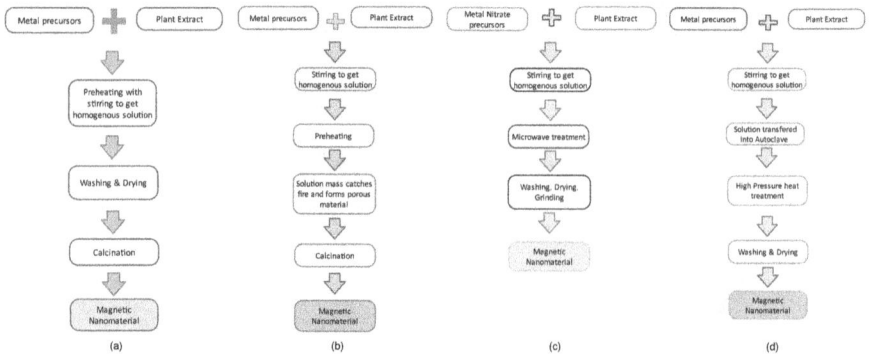

FIGURE 8.4   (a) Coprecipitation, (b) combustion, (c) microwave-assisted combustion, (d) hydrothermal method (Noroozifar, Yousefi, and Jahani 2013; Cristiani et al. 2013; Hayashi and Hakuta 2010; Zhong et al. 2017).

hydroxide, and avocado fruit extract, and the prepared catalyst was tested for the degradation of various organic dye solutions under direct sunlight irradiation (Kamaraj et al. 2019). N. Madubuonua et al. prepared FeO magnetic nanoparticle for antibacterial and photodegradation of dye solution using iron chloride and composition of leaf extracts of *Psidium guajava–Moringa oleifera*, which served as a reducing and capping agent for the iron oxide preparation (Madubuonu et al. 2019). Daphne mezereum leaf extract was used as reducing and stabilizing agents for the preparation of FeO magnetic nanoparticle without any additional reducing agent (Beheshtkhoo et al. 2018).

8.4.2 CHEMICAL REDUCTION AND COPRECIPITATION METHOD

Coprecipitation method provides inexpensive, easy control on particle size and composition, high purity, and high yield product of nanoparticle (Noroozifar, Yousefi, and Jahani 2013; Cristiani et al. 2013). Md.S.H. Bhuiyan et al. prepared α-Fe_2O_3 nanoparticles by using leaf extract of papaya plant and its photocatalytic efficiency for the degradation of remazol yellow RR was studied. The plant extract contains phytochemicals such as flavonoids, glycosides, and polyphenols, which act as reducing and stabilizing agent and react with the Fe ions and produce iron oxide nanoparticle. The functional group present in the α-Fe_2O_3 nanoparticles and papaya plant extract was confirmed by FITR analysis (Bhuiyan et al. 2020). Synthesis of Fe_2O_3 and FeO nanoparticle by using henna leaf extract by the chemical reduction method with and without pulsed laser ablation (PLA) was reported by M.A. Abid and D.A. Kadhim. Presence of gallic acid, hydroxyl, and lawsone (2-hydroxy 1,4-naphthoquinone) groups in the henna extract may be the reasons for the reduction process in the iron precursor to produce Fe_2O_3 nanoparticles. During the PLA process, the prepared Fe_2O_3 nanoparticles were changed into the formation of FeO nanoparticle. Fe_2O_3 nanoparticle shows the wurtzite structure and PLA processed FeO shows the cubic structure in the XRD analysis. The bandgap range of Fe_2O_3 and FeO nanoparticle was found to be 2.75 eV and 3.3 eV (Abid and Kadhim 2020). Iron oxide nanorods were prepared using iron chloride precursors with *Withania coagulans* fruit extract as a reducing and capping agent, and they were also compared with chemical method using iron chloride, Na_2SO_3, and ammonia as the precursors (Qasim et al. 2020). Even though this method has many advantages, it needs continuous washing, drying, and high-temperature calcination to eliminate the impurity phases.

8.4.3 SOL-GEL METHOD

Metal oxide nanoparticles and their composites were prepared by sol-gel method with desire texture and surface property at low-temperature process (Karahroudi, Hedayati, and Goodarzi 2020; Chislova et al. 2011; Kumar et al. 2017). Indriyani et al. synthesized $BiFeO_3$ nanoparticle using *Abelmoschus esculentus* leaf extract for photodegradation of MB under sunlight irradiation. The secondary metabolite flavonoids and saponins in the *A. esculentus* act as a capping agent to provide uniform shape and grain size during the formation of nanoparticle. The photocatalytic activity tests were performed under various catalyst dosages 10, 15, and 20 mg for

MB degradation, and the catalyst with 15 mg of dosage was observed to be 94.04% degradation under the visible light irradiation. Below and above the 15 mg of catalyst dosages, photocatalytic activity was less due to the low active site in the dosage of 10 mg and also due to the higher concentration of catalyst in the dosage of 20 mg. The stability and recyclability of prepared nanoparticle revealed 90.20% after the four cycles of degradation of MB under visible light, and in each cycle, a small amount of catalyst loss was observed due to loss of mass of catalyst during the reproduction (Indriyani et al. 2021). $Mg–Ni–Al–Fe_2O_4$ was synthesized by sol-gel method using tragacanth gel with different dosages of Al concentration for the degradation of Direct Blue129 dye under visible light irradiation. Bandgap of the $Mg_{0.5}Ni_{0.5}Al_xFe_{2-x}O_4$ magnetic nanoparticle increases and crystallite size decreases with increasing concentration of Al ions (Atrak, Ramazani, and Taghavi Fardood 2019). $SrFe_{12}O_{19}$–$SrTiO_3$ nanocomposite was prepared by auto-combustion sol-gel method using lemon juice as gelling and capping agent. $SrFe_{12}O_{19}$ and $SrTiO_3$ nanoparticle shows agglomeration without calcination. Due to calcination at 850°C and 500°C, agglomeration was reduced. $SrFe_{12}O_{19}$ and $SrTiO_3$ nanoparticle exhibited needle shape and spherical shape. Composition of $SrFe_{12}O_{19}$-$SrTiO_3$ shows needle and plane shapes (Karahroudi, Hedayati, and Goodarzi 2020). Hence, in sol-gel method, it is possible to change the size, shape, and particle distribution of magnetic nanoparticle prepared by varying the pH, processing time, temperature, and solvent during the synthesis process.

8.4.4 HYDROTHERMAL METHOD

The hydrothermal method provides high crystallinity and controlled particle size under the high pressure at a mild temperature. In general, metal precursors are mixed with solvent until a homogeneous solution is obtained and then the solution is transferred into an autoclave and kept in the hot air oven or furnace for some time (Hayashi and Hakuta 2010). There is only one report found on green synthesis of magnetic nanohybrid $ZnFe_2O_4$. Krishnan et al. synthesized $ZnFe_2O_4$ by using metal nitrate, GO powder, and orange peel extract as a natural surfactant. They have found a photocatalytic efficiency of 92.4% for MB degradation (Krishnan et al. 2021).

8.4.5 COMBUSTION METHOD

Synthesis of the magnetic nanoparticle by combustion method is a rapid wet chemical synthesis and exothermic reaction process. In this method, preheating and additional calcination processes are not required (Das and Kandimalla 2017). $CaFe_2O_4$ nanoparticle was prepared by combustion method using lemon juice as a capping agent (Adarsha et al. 2021). The prepared nanocatalyst has an orthorhombic phase with a grain size of 30 nm and porous with a lot of voids in the structure, which leads to high surface area and chemical reactivity and avoids aggregation $CoFe_2O_4$, and Zn-doped $CoFe_2O_4$ was prepared using curd as reducing agent (Madhukara Naik, Bhojya Naik, Nagaraju, Vinuth, Vinu, et al. 2019). Ag-doped $CuFe_2O_4$ nanoparticles were synthesized using an extract of jatropha oil by combustion method for decomposition of malachite green dye under UV-light irradiation. The prepared Ag-doped $CuFe_2O_4$ nanoparticles exhibited improved photocatalytic efficiency and reusability

(Surendra 2018). Arularasu et al. prepared Fe_3O_4 nanoparticle by the hotplate combustion method using *Kappaphycus alvarezii* seaweed plant extract as a stabilizer to study the degradation of cationic textile dye waste. Prepared magnetic nanoparticle shows ferromagnetic and is mesoporous in nature with a particle size range of 10–30 nm (Arularasu, Devakumar, and Rajendran 2018). Surendra et al. have reported the preparation of $ZnFe_2O_4$ nanoparticle via combustion method using jatropha oil extract and metal nitrate precursors for the photocatalytic application. The prepared magnetic nanoparticle has porous flakes with agglomeration structure in nature. The agglomeration occurs because of fatty acids present in the oil extract, which coordinates with Zn ion and forms a complex network with hydroxyl groups (Surendra et al. 2020).

8.4.6 MICROWAVE-ASSISTED METHOD

Microwave-assisted combustion method is another facile and rapid preparation method for synthesizing nanoparticles, and advantage of this method is no need for additional high-temperature calcination after the synthesis step (Zhong et al. 2017). Somanathana et al. reported Ce-incorporated $NiFe_2O_4$ magnetic nanoflakes using *Calotropis gigantean* plant leaf extract for the removal of MB dye effectively. *C. gigantean* plant leaf extract acts as a natural reducing and capping agent at the time of the synthesis process to reduce the metal nitrate to form the Ce-incorporated $NiFe_2O_4$ nanoflakes (Somanathan et al. 2019). $ZnFe_2O_4$ nanoparticles were prepared through a microwave-assisted method by Madhukara Naik et al., using *Limonia acidissima* juice extract as reducing agent and metal precursors. Uniform particle distribution with agglomeration was noticed in the SEM micrograph. But, agglomeration was found in the SEM images, due to the high energy produced by the microwave oven at the time of synthesis (Madhukara Naik, Bhojya Naik, Nagaraju, Vinuth, Raja Naika, et al. 2019).

8.4.7 SONICATION METHOD

The advantage of sonication method is, while using the ultrasonic wave, the metal precursors and surfactants are mixed well, and then after the calcination, desired structure of nanoparticle can be easily obtained (Alizadeh, Fallah, and Nikazar 2019). $TbFeO_3$ magnetic nanoparticle was prepared by Mehdizadeh et al., using various synthetic (sodium dodecyl sulfate (SDS) and cetyltrimethylammonium bromide (CTAB)) and natural surfactants (cherry and orange juices) via ultrasonication method. Orange juice used $TbFeO_3$ nanoparticle has high crystallinity and no impurity peak was observed from XRD pattern. Uniform particle distribution was occurred while using CTAB and orange juice as shown in the SEM image (Figure 8.5). Compared with other surfactants, orange juice used $TbFeO_3$ nanoparticle has smaller particle size (Mehdizadeh et al. 2020).

8.5 FUNCTIONALIZATION OF MAGNETIC NANOMATERIALS

Functionalization is a process in which functional groups (carboxylic, hydroxyl, and amine groups) and active sites can be created on the surface of the nanomaterials at

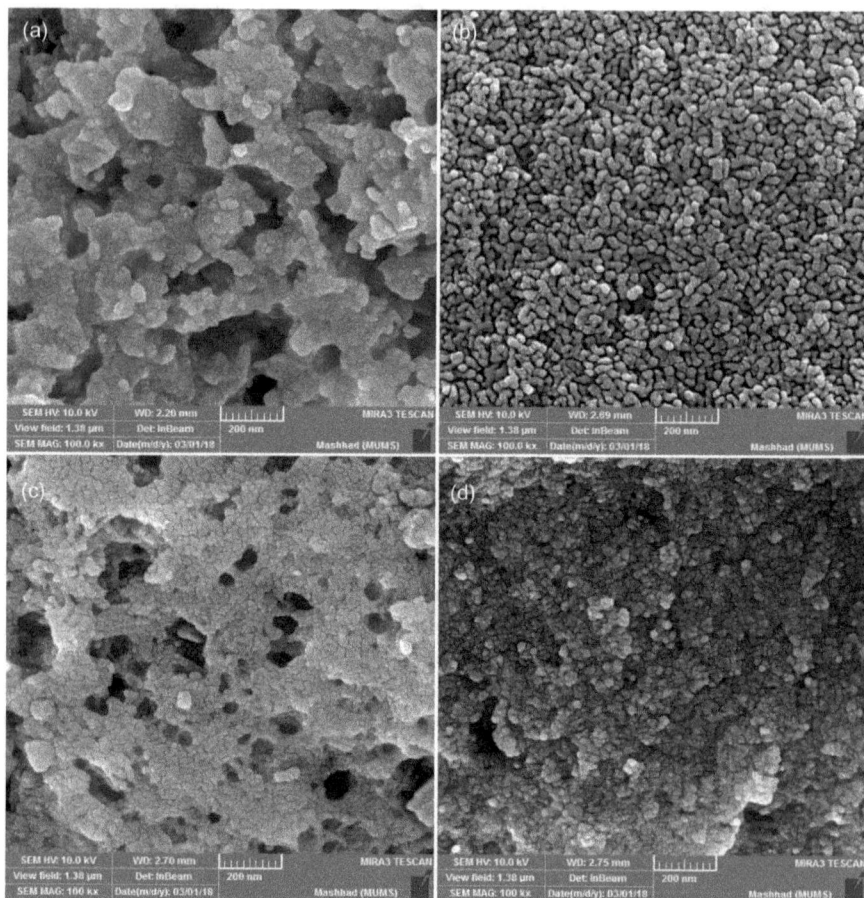

FIGURE 8.5 SEM images of TbFeO$_3$ nanostructures prepared by using (a) sodium dodecyl sulfate, (b) cetyltrimethylammonium bromide, (c) cherry juice, and (d) orange juice as surfactant. Adopted with the permission (Mehdizadeh et al. 2020). Copyright 2020, Elsevier.

the time of synthesis. Synthesis of functionalized magnetic nanoparticle by various strategies is shown in Figure 8.6 and the various methods are as follows:

- Covalent Interaction: Formation of covalent band on the surface of the nanoparticle during the chemical reactions.
- Noncovalent Interaction: Hydrophobic, electrostatic interaction on the surface of the nanoparticle during the chemical reactions.
- Ligand Exchange: Polymers or organic molecules attached to the surface of the nanoparticle to control the nanoparticle size, shape, and enrich their properties.
- In situ surface modification: incorporation of heteroatom and defect formation on the crystal structure of the nanoparticle (Kumar and Sinha Ray 2018; Bhatia et al. 2021; Sneharani and Byrappa 2021). Among the various

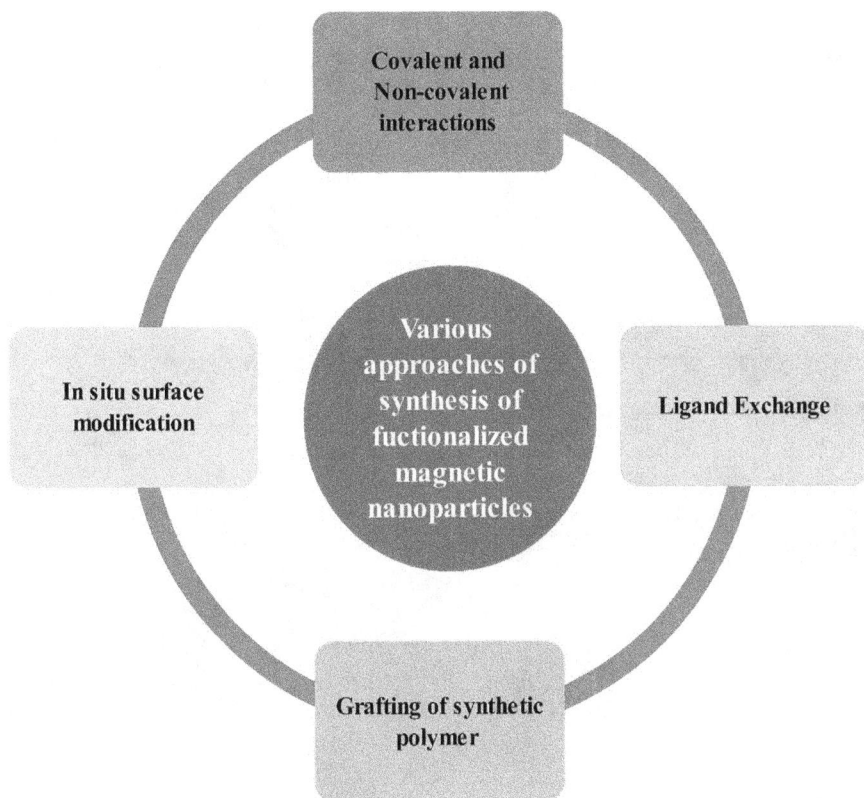

FIGURE 8.6 Various strategies of functionalized magnetic nanoparticles synthesis (Kumar and Sinha Ray 2018; Bhatia et al. 2021; Sneharani and Byrappa 2021)

strategies, in situ surface modifications are widely preferred for preparation of functionalized magnetic nanoparticle by green synthesis method.

8.5.1 METAL NANOCOMPOSITE

Doping and incorporating with various metals ions are important functionalization strategies for enhancing the optical and electrical properties of magnetic nanoparticles. Doping and incorporation of metals such as N, S, P, and O deviate the band length and angle, which help to provide more active sites on the surface of the nanomaterials (Kumar and Sinha Ray 2018). There are several reports on increasing the optical and photocatalytic efficiency of the magnetic nanoparticle photocatalyst by doping and forming nanocomposite with the metal atoms. From Figure 8.7, 1% Y-doped $BiFeO_3$ has smaller particle size, which contributes large surface-active sites. They have noticed that the crystallite size of both undoped and Y-doped (with different concentrations of Y) $BiFeO_3$ nanoparticles is almost similar for all the samples, but the band gap was found to be decreased with increasing the concentration of dopant (Abdul Satar et al. 2019). In the case of Zn-doped $CoFe_2O_4$

FIGURE 8.7 FESEM images of undoped and Y-doped BiFeO3 nanoparticles; (a) undoped, (b) 1%, (c) 2%, (d) 5% of Y-doped BiFeO3. Adopted with the permission (Abdul Satar et al. 2019). Copyright 2019, Elsevier.

prepared with curd as a reducing agent, it was discovered that the lattice parameter and peak shift increased as the concentration of Zn was increased (Zn = 0.0, 0.2, 0.4, and 0.6). Figure 8.8 depicts the XRD pattern of $CoFe_2O_4$ together with Zn-doped $CoFe_2O_4$. The reason for the shift in lattice parameter and peak position is because Zn^{2+} has a greater ionic radius than Co^{2+} (Madhukara Naik, Bhojya Naik, Nagaraju, Vinuth, Vinu, et al. 2019). Surendra et al. were able to synthesize Ag-doped $CuFe_2O_4$ nanoparticles using jatropha oil extract as a starting material. Compared with the XRD pattern of bare $CuFe_2O_4$, the XRD pattern of Ag-doped $CuFe_2O_4$ nanoparticle shows two additional peaks corresponding to crystal planes (311) and (440), which are attributed to doping of Ag in $CuFe_2O_4$. Also, $CuFe_2O_4$ particle exhibited porous flakes with agglomeration and Ag $CuFe_2O_4$ nanoparticle showed spherical-shaped agglomeration with numerous trapped pores. The band gaps, calculated from diffuse reflectance spectra, of prepared $CuFe_2O_4$ and Ag-doped $CuFe_2O_4$ nanoparticles were 2.15 and 1.6 eV, respectively. Also, they have

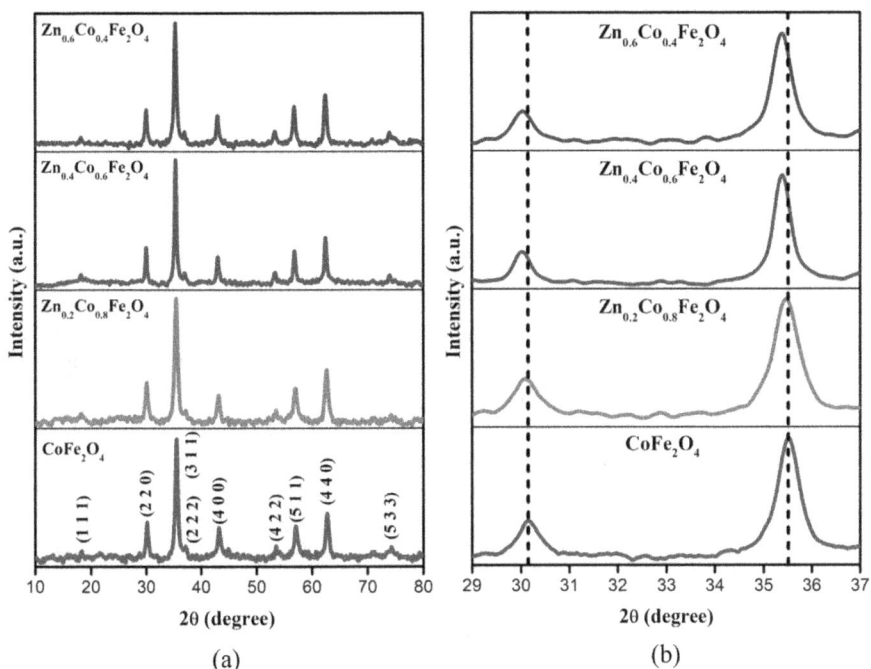

FIGURE 8.8 XRD patterns of Zn-doped CoFe2O4 NPs calcined at 650°C synthesized by combustion method. (b) XRD patterns of the shift in intense peak (2 2 0) and (3 1 1) with an increase in Zn-doping concentration. Adopted with the permission (Madhukara Naik, Bhojya Naik, Nagaraju, Vinuth, Vinu, et al. 2019). Copyright 2019, Elsevier.

concluded that the bandgap of Ag-doped $CuFe_2O_4$ decreases with increase in doping concentration (Surendra 2018).

Atrak et al. have investigated preparation of Al-doped $MgNiFe_2O_4$ magnetic nanoparticles with different dosage concentrations of Al (x= 0.5,1, 1,5) and its photocatalytic degradation ability for the degradation of dye solution using tragacanth gel as natural reducing agent. In their studies, they have found that while increasing the Al doping concentration, the XRD diffraction peaks were broadening due to decrease in crystallite size of prepared nanoparticles. They have also reported that bandgap of $MgNiFe_2O_4$ was found to increase with respect to doping concentration (Atrak, Ramazani, and Taghavi Fardood 2019). Somanathan et al. have incorporated Ce ions on the $NiFe_2O_4$ and they found that the incorporation contributes to more active sites on the surface of the photocatalysts, and also, they have found that the electron–hole recombination rate is reduced. The surface morphology of the magnetic Ce-incorporated $NiFe_2O_4$ photocatalysts exhibits flakes-like structure with agglomeration because of magnetic dipole interaction between magnetic nanoparticles (Somanathan et al. 2019). From the above observation, enhancement of photocatalytic performance of metal nanocomposite because of doping of metal ion on the surface of magnetic nanomaterials can provide more surface-active site, improve the lifetime of charge carriers, and improve the light-gathering capacity due to its bandgap alignment.

8.5.2 Metal Oxide Nanocomposite

The recombination of charge carriers and performance of photocatalytic activity of the magnetic nanomaterials were also improved by forming the nanocomposites with metal oxides. In one of the metal oxide nanocomposites study, $MgFe_2O_4@ \gamma-Al_2O_3$ magnetic nanoparticles were reported without adding any additional agent, using tragacanth gel and metal nitrate as the precursor. $\gamma-Al_2O_3$ has been chosen because of its large surface area and high porous material. Their photocatalytic degradation ability for the degradation of reactive red 195 dye and reactive orange 122 dye was found to be reasonable. In addition to this, they have also studied recovering and reusability of the magnetic nanoparticle and it was found that no significant loss in the photocatalytic activity up to five photocatalytic measurement cycles (Atrak, Ramazani, and Taghavi Fardood 2018). As of now, there are many reports available on green synthesis of magnetic nanoparticles and metal oxides nanoparticle, but green synthesis of magnetic nanocomposite photocatalysts is not yet much explored.

8.5.3 Carbon-Based Nanocomposite

There are very few reports on carbon-based magnetic nanocomposites, which were observed to provide high surface area, enhanced charge transportation, and tunable bandgap. Magnetic nanoparticle with carbon-based nanohybrid was prepared by Krishnan et al. for photodegradation of organic pollutant via hydrothermal method. On the surface of the rGO sheets, $ZnFe_2O_4$ nanoparticles were highly dispersed, which results in reduction of recombination of charge carriers. The reduced recombination rate of photoexcited electron–hole pair was confirmed by PL spectra (Krishnan et al. 2021). Even though there are good number of reports available on carbon-based magnetic nanophotocataysts, studies relevant to green synthesis methods are very much limited and yet to be developed.

8.6 RETRIABLE PHOTOCATALYST FOR DEGRADATION OF ORGANIC POLLUTANT

Photocatalysis is an advanced oxidation process for eliminating organic pollutants present in the industrial effluents and a good photocatalyst must possess large surface area, abundant surface-active sites, tunable bandgap, and better charge separation. Especially, bandgap is the key factor for the photocatalysis, where the absorption of light is of primary importance. Hence, doping with metals is one of the solution, for reducing the charge recombination, increasing the surface-active cites, and adjusting bandgap alignment, to improve the efficiency of a photocatalyst.

Green-synthesized magnetic nanophotocatalysts such as Fe_3O_4, Fe_2O_3, FeO and M Fe_2O_4(M =Mn, Ba, and Ni), and M FeO_3 (M= Bi, Sr and Tb) using various natural sources, their photocatalytic application for the degradation of various organic pollutants, and their degradation efficacies are listed in Table 8.1.

TABLE 8.1

Summary of Reported Green-Synthesized Magnetic Nanomaterials for Photocatalytic Pollutant Degradation

S. No.	Magnetic Material	Preparation Method	Natural Source	Reducing / Capping Agent/ Stabilizer/ Surfactant	Pollutant	Light Source	Degradation Efficiency/Time	Reference
1	Yt/BiFeO$_3$	Biosynthesis	Kappa-Carrageenan seaweed powder	Solvent	MB	Sunlight	97.6%/180 min	Abdul Satar et al. (2019)
2	BiFeO$_3$	Sol–gel	*Abelmoschus esculentus L* leaf	Capping agent	MB	Sunlight	94.04%/120 min	Indriyani et al. (2021)
3	CaFe$_2$O$_4$	Combustion	Lemon juice	Capping agent	Evans blue	Xenon Arc Lamp (340 nm)	98.11%	Adarsha et al. (2021)
4	Zn/CoFe$_2$O$_4$	Combustion	Curd	Reducing agent	Congo red and Evans blue	Sunlight	96%/150 min	Madhukara Naik et al. (2019)
5	Ag/CuFe$_2$O$_4$	Combustion	Jatropha oil	Reducing agent	MG	Sunlight	85%/240 min	Surendra (2018)
6	α-Fe$_2$O$_3$	Chemical Reduction	Papaya leaf	Reducing agent	Remazol yellow RR	Sunlight	77%/150 min	Bhuiyan et al. (2020)
7	Fe$_2$O$_3$ and FeO	Chemical Reduction	Henna leaf	Reducing agent	MB	Visible light	88%/70 min	Abid and Kadhim (2020)
8	Al/MgNiFe$_2$O$_4$	Sol–gel method	Tragacanth gel	Reducing agent	Direct blue 129	Fluorescent lamp (λ > 400 nm, 90 W)	94%	Atrak, Ramazani, and Taghavi Fardood (2019)
9	Fe$_2$O$_4$	Combustion	Kappaphycus alvarezii	Stabilizer	MB	High pressure Hg lamp (420 nm, 300 W)	75%/<180 min	Arularasu et al. (2018)

(Continued)

TABLE 8.1 (Continued)

Summary of Reported Green-Synthesized Magnetic Nanomaterials for Photocatalytic Pollutant Degradation

S. No.	Magnetic Material	Preparation Method	Natural Source	Reducing / Capping Agent/ Stabilizer/ Surfactant	Pollutant	Light Source	Degradation Efficiency/Time	Reference
10	Fe_3O_4	Biosynthesis	Calliandra haematocephala leaf	Reducing agent	MG	Sunlight	100%/55 min	Sirdeshpande et al. (2018)
11	FeO	Biosynthesis	Avocado fruit peel	Reducing agent	Congo red, safranin, CV, MG and MO	Sunlight	~ 90%/300 min	Kamaraj et al. (2019)
12	FeO	Biosynthesis	Composite of Psidium guavaja-Moringa oleifera leaf	Reducing and capping agent	MB	Sunlight	60 min	Madubuonu et al. (2019)
13	FeO	Biosynthesis	Daphne mezereum leaf	Reducing and stabilizing agent	MO	Sunlight	81%/360 min	Beheshtkhoo et al. (2018)
14	FeO	Chemical Reduction	Withania coagulans fruit	Reducing and capping agent	Safranin dye	Sunlight	180 min	Qasim et al. (2020)
15	$\gamma-Al_2O_3/MgFe_2O_4$	Sol–gel	Tragacanth gel	Reducing agent	reactive red 195 and reactive orange 122	Fluorescent lamp ($\lambda > 400$ nm, 80 W)	~95%/60 min	Atrak, Ramazani, and Taghavi Fardood (2018)
16	$Ce/NiFe_2O_4$	Microwave-assisted combustion	Calotropis gigantean leaf	Reducing and capping agent	MB	Halogen lamp ($\lambda > 420$ nm, 150 W)	99%/150 min	Somanathan et al. (2019)
17	$ZnFe_2O_4$	Hydrothermal	Orange peel	Reducing agent	MB	Xenon lamp (300 W)	92.4% in 140 min	Krishnan et al. (2021)

(Continued)

TABLE 8.1 (Continued)

Summary of Reported Green-Synthesized Magnetic Nanomaterials for Photocatalytic Pollutant Degradation

S. No.	Magnetic Material	Preparation Method	Natural Source	Reducing / Capping Agent/ Stabilizer/ Surfactant	Pollutant	Light Source	Degradation Efficiency/Time	Reference
18	$ZnFe_2O_4$	Combustion	Jatropha oil	Reducing agent	MG	–	98%	Surendra et al. (2020)
19	$ZnFe_2O_4$	Microwave synthesis	*Limonia acidissima* juice	Reducing agent	Evans blue and MG	Tungsten lamp (300W)	EB 89%/90 min MB 99.66%/90 min.	Madhukara Naik, et al. (2019)
20	$SrFe_{12}O_{19}$– $SrTiO_3$	Sol–gel	Lemon juice	Capping and chelating agent	Acid black, acid brown, MR and MO	–	~95%/~40 min	Karahroudi, Hedayati, and Goodarzi (2020)
21	$TbFeO_3$	Sonication	Cherry and Orange juice	Surfactant	MO, acid blue 92, acid black 1 and acid Brown 214	–	MO-98%/120 min	Mehdizadeh et al. (2020)

Abdul Satar et al. (2019), Surendra (2018), and Somanathan et al. (2019) have reported doping of metal ion on the surface of the magnetic nanoparticle, which was found to provide more surface-active sites on the surface of the catalyst. Also, they have observed extended lifetime of the excited charge carriers and attributed them as the reason for the enhanced photocatalytic activity compared to nondoped catalysts.

Sirdeshpande et al. (2018) and Surendra et al. (2020) have prepared Fe_3O_4 mesoporous magnetic nanoparticle and $ZnFe_2O_4$ magnetic nanoparticles and studied their photocatalytic dye degradation performances. In their studies, small crystallite size, large surface area, and morphology were attributed to the enhanced activity.

Madhukara Naik et al. reported the preparation of Zn-doped cobalt ferrites ($Zn_xCO_{1-x} Fe_2O_4$) using curd as a fuel by combustion method. Lactic acids and carbohydrates present in the curd acts as a powerful reducing agent and they are responsible for the combustion process. In their studies, they have compared the photocatalytic activity of $Zn_xCO_{1-x} Fe_2O_4$ catalysts, with various doping ratio of Zn+ ions, for the degradation of Congo red and Evans blue dyes under visible light irradiation. They have proposed a photocatalytic mechanism for $Zn_xCO_{1-x} Fe_2O_4$ catalyst and it is shown in Figure 8.9 (Madhukara Naik, Bhojya Naik, Nagaraju, Vinuth, Vinu, et al. 2019). They have concluded that the presence of Zn+ ion doping at the Cobalt site prevents the recombination of electron–hole pairs and also defects formation in the energy levels, was responsible for the improved photocatalytic activity and better light sensitivity of the catalyst. Surendra et al. have reported photocatalytic decomposition of Malachite green (MG) dye by using Ag-doped $CuFe_2O_4$ and $CuFe_2O_4$ and its photocatalytic activity result, as shown in Figure 8.10. Compared to dye decomposition ability of undoped $CuFe_2O_4$ nanoparticle, almost complete decomposition was found in Ag-doped $CuFe_2O_4$ under UV light due to the presence of more surface-active sites, smaller particle size, and decreased bandgap (Surendra 2018).

There are few reports in which the effect of the light source, catalyst dosage, initial concentration, and irradiation time on the photodegradation ability for magnetic nanomaterials such as $Mg_{0.5}Ni_{0.5}Al_xFe_{2-x}O_4$ (x=0.5, 1, 1.5), $TbFeO_3$, and $MgFe_2O_4@$ $\gamma-Al_2O_3$ has been investigated (Atrak, Ramazani, and Taghavi Fardood 2019; Atrak, Ramazani, and Taghavi Fardood 2018; Mehdizadeh et al. 2020). Among all the

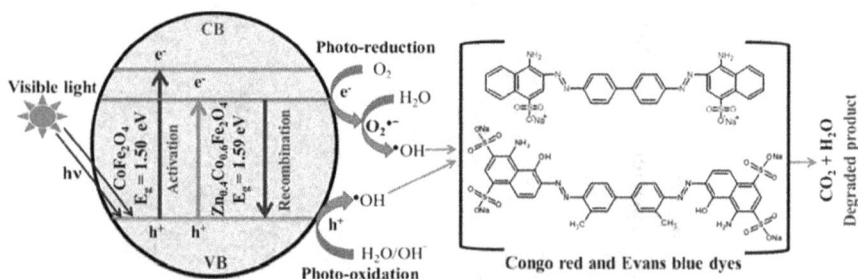

FIGURE 8.9 Photocatalytic mechanism of Zn-doped $CoFe_2O_4$ for degradation of Congo red and Evans blue dyes. Adopted with the permission (Madhukara Naik, Bhojya Naik, Nagaraju, Vinuth, Vinu, et al. 2019). Copyright 2019, Elsevier.

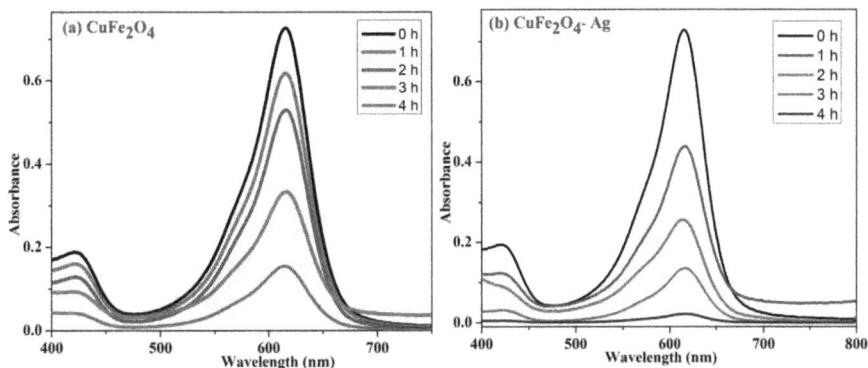

FIGURE 8.10 (a) % decomposition of MG under UV light for the synthesized host CuFe2O4 nanoparticles; (b) % decomposition of MG under UV light for the synthesized Ag-doped $CuFe_2O_4$ nanoparticles. Adopted under the terms of the Creative Commons CC-BY license (Surendra 2018). Copyright 2018, The Authors. Published by Elsevier-VNU.

studies, the concentration of Al_x (x = 1.5) was better efficient for the degradation of Direct Blue 129 dye under visible light (Atrak, Ramazani, and Taghavi Fardood 2019). On increasing the concentration of dye solution, the photocatalytic performance of the $TbFeO_3$ catalyst was found to be reduced. Compared with synthetic surfactant SDS used $TbFeO_3$ nanoparticle, orange juice assisted $TbFeO_3$ nanoparticle showed higher photocatalytic activity due to the smaller particle size, high purity, and high crystalline structure (Mehdizadeh et al. 2020).

Decoloration of methyl orange with and without H_2O_2 in the dye solution was reported by Beheshtkhoo et al. In their studies, the addition of hydrogen peroxide to the FeO nanoparticle provides additional hydroxyl radical for degradation process and it exhibited efficient degradation compared with other solutions (Beheshtkhoo et al. 2018). The photocatalytic activity of green and chemical synthesis of iron oxide nanorods was investigated by Qasim et al. for the degradation of safranin dye under the sunlight irradiation. In comparison with chemically prepared iron oxide, green-synthesized iron oxide nanorods show enhanced rate of photocatalytic degradation because of the phytochemicals present in the plant extract, which has a tendency to increase the oxidation and reduction potential (Qasim et al. 2020).

Formation of nanohybrid with rGO sheet on the $ZnFe_2O_4$ nanoparticle showed significant improvement in lifetime of the excited charge carriers and their photocatalytic degradation ability is found to be noticeable (Krishnan et al). Figure 8.11 shows the photocatalytic mechanism of $ZnFe_2O_4$/ rGO nanohybrids under visible light irradiation. Krishnan et al. have observed that when $ZnFe_2O_4$/rGO was exposed to sunlight irradiation, photogenerated electrons in the $ZnFe_2O_4$ valance band got excited into the conduction band of $ZnFe_2O_4$ and photogenerated electrons were transferred to the rGO, which reduced charge recombination. Photogenerated electrons in the rGO undergo the reduction reaction, and a photogenerated hole in the $ZnFe_2O_4$ undergoes the oxidation reaction, degrading the MB dye (Krishnan et al. 2021).

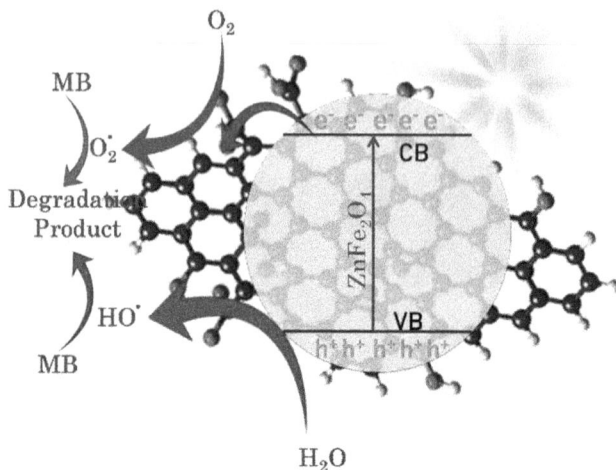

FIGURE 8.11 Proposed degradation mechanism of MB dye molecules over $ZnFe_2O_4/rGO$ nanohybrids. Adopted with the permission (Krishnan et al. 2021). Copyright 2021 Elsevier.

Somanathan et al. and Abdul Satar et al. have investigated the stability and reusability of the magnetic photocatalysts $Ce-NiFe_2O_4$ and Y-doped $BiFeO_3$. They have found that after the fifth cycle, photocatalytic activity reduced slightly due to the decrement in active sites, absorption of contaminant, and loss of catalyst during the washing process (Somanathan et al. 2019; Abdul Satar et al. 2019).

From the above reported observations, it is possible to tune properties of magnetic photocatalysts such as particle size, synthesis time and temperature, morphology, bandgap, surface-active site, and catalyst dosage for improving their photocatalytic activity by green synthesis approach.

8.7 FUTURE SCOPE

The photocatalytic performance and stability of green-synthesized and functionalized magnetic nanomaterials are found to be reasonable and comparable to those of conventionally synthesized magnetic nanomaterials. As of now, many reports are available on the green synthesis of functionalized magnetic nanomaterials using plant extracts for dye degradation applications. However, there are a few gaps in the research area of green synthesis of magnetic nanophotocatalysts, and some of the most important gaps identified are listed below:

- Preparation of green synthesis of functionalized magnetic nanoparticles using bio-organism and other photocatalytic applications are not yet explored much.
- In most of the green synthesis processes, plant extracts are used as reducing agent, capping agent, and surfactant and not used as a metal source for preparing magnetic nanoparticles.

- Also, in some of the reports chemical reducing agents, capping agent and surfactant are additionally used along with plant extracts during the synthesis process.
- Green-synthesized composite of carbon-based nanomaterial and metal oxide with magnetic nanoparticle for photocatalytic applications is not much studied yet and further investigations are needed on this above-mentioned research gaps.

8.8   CONCLUSION

This chapter discusses various green synthesis methods for functionalized magnetic nanoparticles that use natural sources as reducing, capping, stabilizing, and surfactant agents, as well as their photocatalytic activity. Compared with the conventional synthesis route, green-synthesized magnetic nanoparticles exhibit improved photocatalytic performance and are easily recoverable and reused for several cycles. The pH, synthesis temperature, calcination temperature, and incubation time affect the properties of the green-synthesized magnetic nanoparticles, and they must be carefully noted at the time of synthesis for better performance. Greenly synthesized and functionalized magnetic nanoparticles are stable and simple to recover and reuse for photocatalytic dye degradation applications. There are many more materials that need to be investigated for their suitability in green synthesis methods. Hence, further research is needed in the field of green synthesis as well as on various functionalization strategies.

REFERENCES

Abdul Satar, Nurul Syamimi, Rohana Adnan, Hooi Ling Lee, Simon R. Hall, T. Kobayashi, Mohamad Haafiz Mohamad Kassim, and Noor Haida Mohd Kaus. 2019. "Facile Green Synthesis of Ytrium-Doped BiFeO3 with Highly Efficient Photocatalytic Degradation towards Methylene Blue." *Ceramics International* 45 (13): 15964–73. https://doi.org/10.1016/j.ceramint.2019.05.105.

Abid, Muslim A., and Duha A. Kadhim. 2020. "Novel Comparison of Iron Oxide Nanoparticle Preparation by Mixing Iron Chloride with Henna Leaf Extract with and without Applied Pulsed Laser Ablation for Methylene Blue Degradation." *Journal of Environmental Chemical Engineering* 8 (5): 104138. https://doi.org/10.1016/j.jece.2020.104138.

Adarsha, J. R., T. N. Ravishankar, C. R. Manjunatha, and T. Ramakrishnappa. 2021. "Green Synthesis of Nanostructured Calcium Ferrite Particles and Its Application to Photocatalytic Degradation of Evans Blue Dye." *Materials Today: Proceedings* 49 (xxxx): 777–88. https://doi.org/10.1016/j.matpr.2021.05.293.

A.H. Sneharani, and K. Byrappa. 2021. *Facile Chemical Fabrication of Designer Biofunctionalized Nanomaterials. Functionalized Nanomaterials I.* Vol. 1.

Alizadeh, Sajad, Narges Fallah, and Manochehr Nikazar. 2019. "An Ultrasonic Method for the Synthesis, Control and Optimization of CdS/TiO2 Core–Shell Nanocomposites." *RSC Advances* 9 (8): 4314–24. https://doi.org/10.1039/C8RA10155H.

Arularasu, M. v., J. Devakumar, and T. v. Rajendran. 2018. "An Innovative Approach for Green Synthesis of Iron Oxide Nanoparticles: Characterization and Its Photocatalytic Activity." *Polyhedron* 156: 279–90. https://doi.org/10.1016/j.poly.2018.09.036.

Atrak, Kobra, Ali Ramazani, and Saeid Taghavi Fardood. 2018. "A Novel Sol–Gel Synthesis and Characterization of MgFe2O4@γ–Al$_2$O$_3$ Magnetic Nanoparticles Using Tragacanth Gel and Its Application as a Magnetically Separable Photocatalyst for Degradation of Organic Dyes under Visible Light." *Journal of Materials Science: Materials in Electronics* 29 (8): 6702–10. https://doi.org/10.1007/s10854-018-86565.

Atrak, Kobra, Ali Ramazani, and Saeid Taghavi Fardood. 2019. "Eco-Friendly Synthesis of Mg0.5Ni0.5AlxFe2-XO4 Magnetic Nanoparticles and Study of Their Photocatalytic Activity for Degradation of Direct Blue 129 Dye." *Journal of Photochemistry and Photobiology A: Chemistry* 382 (June): 111942. https://doi.org/10.1016/j.jphotochem.2019.111942.

Beheshtkhoo, Nasrin, Mohammad Amin Jadidi Kouhbanani, Amir Savardashtaki, Ali Mohammad Amani, and Saeed Taghizadeh. 2018. "Green Synthesis of Iron Oxide Nanoparticles by Aqueous Leaf Extract of Daphne Mezereum as a Novel Dye Removing Material." *Applied Physics A: Materials Science and Processing* 124 (5): 0. https://doi.org/10.1007/s00339-018-1782-3.

Bhatia, Amarpreet K., Shippi Dewangan, Ajaya K. Singh, and Sónia. A.C. Carabineiro. 2021. "Functionalized Nanomaterial (FNM)–Based Catalytic Materials for Energy Industry." *Functionalized Nanomaterials for Catalytic Application*, 53–88. https://doi.org/10.1002/9781119809036.ch2.

Bhuiyan, Md Shakhawat Hossen, Muhammed Yusuf Miah, Shujit Chandra Paul, Tutun das Aka, Otun Saha, Md Mizanur Rahaman, Md Jahidul Islam Sharif, Ommay Habiba, and Md Ashaduzzaman. 2020. "Green Synthesis of Iron Oxide Nanoparticle Using Carica Papaya Leaf Extract: Application for Photocatalytic Degradation of Remazol Yellow RR Dye and Antibacterial Activity." *Heliyon* 6 (8): e04603. https://doi.org/10.1016/j.heliyon.2020.e04603.

Bishnoi, Shahana, Aarti Kumar, and Raja Selvaraj. 2018. "Facile Synthesis of Magnetic Iron Oxide Nanoparticles Using Inedible Cynometra Ramiflora Fruit Extract Waste and Their Photocatalytic Degradation of Methylene Blue Dye." *Materials Research Bulletin* 97 (January): 121–27. https://doi.org/10.1016/J.MATERRESBULL.2017.08.040.

Chislova, I. v., A. A. Matveeva, A. v. Volkova, and I. A. Zvereva. 2011. "Sol-Gel Synthesis of Nanostructured Perovskite-like Gadolinium Ferrites." In *Glass Physics and Chemistry* 37: 653–60 https://doi.org/10.1134/S1087659611060071.

Cristiani, Cinzia, Giovanni Dotelli, Mario Mariani, Renato Pelosato, and Luca Zampori. 2013. "Synthesis of Nanostructured Perovskite Powders via Simple Carbonate Co-Precipitation." *Chemical Papers* 67 (5): 526–31. https://doi.org/10.2478/s11696-013-0306-z.

Das, N., and S. Kandimalla. 2017. "Application of Perovskites towards Remediation of Environmental Pollutants: An Overview: A Review on Remediation of Environmental Pollutants Using Perovskites." *International Journal of Environmental Science and Technology* 14: 1559–72. https://doi.org/10.1007/s13762-016-1233-7.

Faraji, M., Y. Yamini, and M. Rezaee. 2010. "Magnetic Nanoparticles: Synthesis, Stabilization, Functionalization, Characterization, and Applications." *Journal of the Iranian Chemical Society* 7 (1): 1–37. https://doi.org/10.1007/BF03245856/METRICS.

Gu, Jianmin, Siheng Li, Enbo Wang, Qiuyu Li, Guoying Sun, Rui Xu, and Hong Zhang. 2009. "Single-Crystalline α-Fe2O3 with Hierarchical Structures: Controllable Synthesis, Formation Mechanism and Photocatalytic Properties." *Journal of Solid State Chemistry* 182 (5): 1265–72. https://doi.org/10.1016/J.JSSC.2009.01.041.

Hayashi, Hiromichi, and Yukiya Hakuta. 2010. "Hydrothermal Synthesis of Metal Oxide Nanoparticles in Supercritical Water." *Materials (Basel)* 3 (7): 3794–3817. https://doi.org/10.3390/ma3073794.

Hedayati, Kambiz, Marjan Joulaei, and Davood Ghanbari. 2020. "Auto Combustion Synthesis Using Grapefruit Extract: Photocatalyst and Magnetic Mgfe2o4-Pbs Nanocomposites." *Journal of Nanostructures* 10 (1): 83–91. https://doi.org/10.22052/JNS.2020.01.010.

Helmiyati, Helmiyati, Nurani Fitriana, Metha Listia Chaerani, and Fitriyah Wulan Dini. 2022. "Green Hybrid Photocatalyst Containing Cellulose and γ–Fe2O3–ZrO2 Heterojunction for Improved Visible-Light Driven Degradation of Congo Red." *Optical Materials* 124 (September 2021): 111982. https://doi.org/10.1016/j.optmat.2022.111982.

Indriyani, Asih, Yoki Yulizar, Rika Tri Yunarti, Dewangga Oky Bagus Apriandanu, and Rizki Marcony Surya. 2021. "One-Pot Green Fabrication of BiFeO3 Nanoparticles via Abelmoschus Esculentus L. Leaves Extracts for Photocatalytic Dye Degradation." *Applied Surface Science* 563 (January): 150113. https://doi.org/10.1016/j.apsusc.2021.150113.

Irshad, Muneeb, Quar tul Ain, Muhammad Zaman, Muhammad Zeeshan Aslam, Naila Kousar, Muhammad Asim, Muhammad Rafique, et al. 2022. "Photocatalysis and Perovskite Oxide-Based Materials: A Remedy for a Clean and Sustainable Future." *RSC Advances* 12 (12): 7009–39. https://doi.org/10.1039/d1ra08185c.

John, Varsha Lisa, Yamuna Nair, and T. P. Vinod. 2021. "Doping and Surface Modification of Carbon Quantum Dots for Enhanced Functionalities and Related Applications." *Particle & Particle Systems Characterization* 38 (11): 2100170. https://doi.org/10.1002/PPSC.202100170.

Kamaraj, M., T. Kidane, K. U. Muluken, and J. Aravind. 2019. "Biofabrication of Iron Oxide Nanoparticles as a Potential Photocatalyst for Dye Degradation with Antimicrobial Activity." *International Journal of Environmental Science and Technology* 16 (12): 8305–14. https://doi.org/10.1007/s13762-019-02402-7.

Karahroudi, Zahra Hajian, Kambiz Hedayati, and Mojtaba Goodarzi. 2020. "Green Synthesis and Characterization of Hexaferrite Strontium-Perovskite Strontium Photocatalyst Nanocomposites." *Main Group Metal Chemistry* 43 (1): 26–42. https://doi.org/10.1515/mgmc-2020-0004.

Khedr, M. H., K. S. Abdel Halim, and N. K. Soliman. 2009. "Synthesis and Photocatalytic Activity of Nano-Sized Iron Oxides." *Materials Letters* 63 (6–7): 598–601. https://doi.org/10.1016/J.MATLET.2008.11.050.

Krishnan, S., S. Murugesan, V. Vasanthakumar, A. Priyadharsan, Murad Alsawalha, Thamer Alomayri, and Baoling Yuan. 2021. "Facile Green Synthesis of ZnFe2O4/ RGO Nanohybrids and Evaluation of Its Photocatalytic Degradation of Organic Pollutant, Photo Antibacterial and Cytotoxicity Activities." *Colloids and Surfaces A: Physicochemical and Engineering Aspects* 611 (October): 125835. https://doi.org/10.1016/j.colsurfa.2020.125835.

Kumar, Aniket, Lipeeka Rout, L. Satish K. Achary, Sangram Keshari Mohanty, and Priyabrat Dash. 2017. "A Combustion Synthesis Route for Magnetically Separable Graphene Oxide–CuFe2O4–ZnO Nanocomposites with Enhanced Solar Light-Mediated Photocatalytic Activity." *New Journal of Chemistry* 41 (19): 10568–83. https://doi.org/10.1039/C7NJ02070H.

Kumar, Neeraj, and Suprakas Sinha Ray. 2018. *Synthesis and Functionalization of Nanomaterials. Springer Series in Materials Science.* Vol. 277. Springer International Publishing. https://doi.org/10.1007/978-3-319-97779-9_2.

Liaskovska, Mariia, Tetiana Tatarchuk, Mohamed Bououdina, and Ivan Mironyuk. 2019. "Green Synthesis of Magnetic Spinel Nanoparticles." *Springer Proceedings in Physics* 222: 389–98. https://doi.org/10.1007/978-3-030-17755-3_25.

Lohrasbi, Sajedeh, Mohammad Amin Jadidi Kouhbanani, Nasrin Beheshtkhoo, Younes Ghasemi, Ali Mohammad Amani, and Saeed Taghizadeh. 2019. "Green Synthesis of Iron Nanoparticles Using Plantago Major Leaf Extract and Their Application as a Catalyst for the Decolorization of Azo Dye." *BioNanoScience* 9 (2): 317–22. https://doi.org/10.1007/s12668-019-0596-x.

Madhukara Naik, M., H. S. Bhojya Naik, G. Nagaraju, M. Vinuth, H. Raja Naika, and K. Vinu. 2019. "Green Synthesis of Zinc Ferrite Nanoparticles in Limonia Acidissima Juice: Characterization and Their Application as Photocatalytic and Antibacterial Activities." *Microchemical Journal* 146 (February): 1227–35. https://doi.org/10.1016/j.microc.2019.02.059.

Madhukara Naik, M., H. S. Bhojya Naik, G. Nagaraju, M. Vinuth, K. Vinu, and R. Viswanath. 2019. "Green Synthesis of Zinc Doped Cobalt Ferrite Nanoparticles: Structural, Optical, Photocatalytic and Antibacterial Studies." *Nano-Structures and Nano-Objects* 19: 100322. https://doi.org/10.1016/j.nanoso.2019.100322.

Madubuonu, N., Samson O. Aisida, A. Ali, I. Ahmad, Ting kai Zhao, S. Botha, M. Maaza, and Fabian I. Ezema. 2019. "Biosynthesis of Iron Oxide Nanoparticles via a Composite of Psidium Guavaja-Moringa Oleifera and Their Antibacterial and Photocatalytic Study." *Journal of Photochemistry and Photobiology B: Biology* 199 (June): 111601. https://doi.org/10.1016/j.jphotobiol.2019.111601.

Malakootian, Mohammad, Alireza Nasiri, Ali Asadipour, and Elham Kargar. 2019. "Facile and Green Synthesis of ZnFe2O4@CMC as a New Magnetic Nanophotocatalyst for Ciprofloxacin Degradation from Aqueous Media." *Process Safety and Environmental Protection* 129: 138–51. https://doi.org/10.1016/j.psep.2019.06.022.

Mamba, Gcina, and Ajay Mishra. 2016. "Advances in Magnetically Separable Photocatalysts: Smart, Recyclable Materials for Water Pollution Mitigation." *Catalysts 2016, Vol. 6, Page 79* 6 (6): 79. https://doi.org/10.3390/CATAL6060079.

Mehdizadeh, Pourya, Yasin Orooji, Omid Amiri, Masoud Salavati-Niasari, and Hossein Moayedi. 2020. "Green Synthesis Using Cherry and Orange Juice and Characterization of TbFeO3 Ceramic Nanostructures and Their Application as Photocatalysts under UV Light for Removal of Organic Dyes in Water." *Journal of Cleaner Production* 252: 119765. https://doi.org/10.1016/j.jclepro.2019.119765.

Mishra, Priti, Sulagna Patnaik, and Kulamani Parida. 2019. "An Overview of Recent Progress on Noble Metal Modified Magnetic Fe3O4 for Photocatalytic Pollutant Degradation and H2 Evolution." *Catalysis Science & Technology* 9 (4): 916–41. https://doi.org/10.1039/C8CY02462F.

Noroozifar, M., M. Yousefi, and S. Jahani. 2013. "Chemical Synthesis and Characterization of Perovskite NdfeO 3 Nanocrystals via a Co-Precipitation Method." *International Journal of Nanoscience and Nanotechology* 9 (1): 7–14.

Pal, Gaurav, Priya Rai, and Anjana Pandey. 2019. *Green Synthesis of Nanoparticles: A Greener Approach for a Cleaner Future. Green Synthesis, Characterization and Applications of Nanoparticles.* Elsevier Inc. https://doi.org/10.1016/b978-0-08-102579-6.00001-0.

Qasim, Shaheen, Ayesha Zafar, Muhammad Saqib Saif, Zeeshan Ali, Maryem Nazar, Muhammad Waqas, Ain Ul Haq, et al. 2020. "Green Synthesis of Iron Oxide Nanorods Using Withania Coagulans Extract Improved Photocatalytic Degradation and Antimicrobial Activity." *Journal of Photochemistry and Photobiology B: Biology* 204 (January): 111784. https://doi.org/10.1016/j.jphotobiol.2020.111784.

Roy, Saikatendu Deb, Krishna Chandra Das, and Siddhartha Sankar Dhar. 2021. "Conventional to Green Synthesis of Magnetic Iron Oxide Nanoparticles; Its Application as Catalyst, Photocatalyst and Toxicity: A Short Review." *Inorganic Chemistry Communications.* Elsevier B.V. https://doi.org/10.1016/j.inoche.2021.109050.

Saeid Taghavi Fardood, Farzaneh Moradnia, Miad Mostafaei, Zolfa Afshari, Vahid Faramarzi, Sara Ganjkhanlu. 2019. "Biosynthesis of MgFe 2 O 4 Magnetic Nanoparticles and Their Application in Photodegradation of Malachite Green Dye and Kinetic Study." *Nanochem Res* 4 (1): 86–93. https://doi.org/10.22036/ncr.2019.01.010.

Sarkar, Shubhrajit, Santanu Sarkar, and Chiranjib Bhattacharjee. 2019. "Green Synthesis of Novel Photocatalysts," 241–61. https://doi.org/10.1007/978-3-030-10609-6_9.

Shaker Ardakani, Leili, Vahid Alimardani, Ali Mohammad Tamaddon, Ali Mohammad Amani, and Saeed Taghizadeh. 2021. "Green Synthesis of Iron-Based Nanoparticles Using Chlorophytum Comosum Leaf Extract: Methyl Orange Dye Degradation and Antimicrobial Properties." *Heliyon* 7 (2): e06159. https://doi.org/10.1016/j.heliyon.2021. e06159.

Sirdeshpande, Karthikey Devadatta, Anushka Sridhar, Kedar Mohan Cholkar, and Raja Selvaraj. 2018. "Structural Characterization of Mesoporous Magnetite Nanoparticles Synthesized Using the Leaf Extract of Calliandra Haematocephala and Their Photocatalytic Degradation of Malachite Green Dye." *Applied Nanoscience (Switzerland)* 8 (4): 675–83. https://doi.org/10.1007/s13204-018-0698-8.

Somanathan, T., A. Abilarasu, B. Rabindran Jermy, Vijaya Ravinayagam, and D. Suresh. 2019. "Microwave Assisted Green Synthesis Ce0.2Ni0.8Fe2O4 Nanoflakes Using Calotropis Gigantean Plant Extract and Its Photocatalytic Activity." *Ceramics International* 45 (14): 18091–98. https://doi.org/10.1016/j.ceramint.2019.06.031.

Surendra, B. S. 2018. "Green Engineered Synthesis of Ag-Doped CuFe2O4: Characterization, Cyclic Voltammetry and Photocatalytic Studies." *Journal of Science: Advanced Materials and Devices* 3 (1): 44–50. https://doi.org/10.1016/j.jsamd.2018.01.005.

Surendra, B. S., H. P. Nagaswarupa, M. U. Hemashree, and Javeria Khanum. 2020. "Jatropha Extract Mediated Synthesis of ZnFe2O4 Nanopowder: Excellent Performance as an Electrochemical Sensor, UV Photocatalyst and an Antibacterial Activity." *Chemical Physics Letters* 739: 136980. https://doi.org/10.1016/j.cplett.2019.136980.

Vasantharaj, Seerangaraj, Selvam Sathiyavimal, Palanisamy Senthilkumar, Felix LewisOscar, and Arivalagan Pugazhendhi. 2019. "Biosynthesis of Iron Oxide Nanoparticles Using Leaf Extract of Ruellia Tuberosa: Antimicrobial Properties and Their Applications in Photocatalytic Degradation." *Journal of Photochemistry and Photobiology B: Biology* 192 (November 2018): 74–82. https://doi.org/10.1016/j.jphotobiol.2018.12.025.

Zhong, Yuanhong, Lin Yu, Zhi Feng Chen, Hongping He, Fei Ye, Gao Cheng, and Qianxin Zhang. 2017. "Microwave-Assisted Synthesis of Fe3O4 Nanocrystals with Predominantly Exposed Facets and Their Heterogeneous UVA/Fenton Catalytic Activity." *ACS Applied Materials and Interfaces* 9 (34): 29203–12. https://doi.org/10.1021/ACSAMI.7B06925/SUPPL_FILE/AM7B06925_SI_001.PDF.

9 Phytogenic Magnetic Nanoparticles (PMNPs)

Synthesis, Properties, Characterization, and Its Potential Application in Waste Water Treatment

J. Jenifer Annis Christy, S. Sham Sait,
M. Karthikeyan, and RM. Murugappan
Thiagarajar College

S. Benazir Begum
V.O. Chidambaram College

CONTENTS

DOI: 10.1201/9781003335580-9

9.1 INTRODUCTION

Modern lifestyle and booming industrialization provides us the luxury of using various products to make our life more comfortable and easy, but it comes at a price namely the mass generation and erratic release of wastewater (Xiao, Liu, and Ge 2021). Wastewater is "utilized water from any mix of residential, mechanical, business or farming exercises, surface overflow or storm water, and any sewer inflow or sewer penetration" (Almuktar, Abed, and Scholz 2018). Fortunately, reuse of wastewater after treatment is one of the common strategies to control water pollution and conservation. The treatment process depended upon the type of wastewater and its innate characteristics.

Although wastewater treatment strategies are advantageous, implementation of sophisticated discharge standards could potentially result in the pollution burden shifting from aquatic environments to other ecosystems due to extensive resource consumption and environmental emissions (Parra-Saldivar, Muhammad, and Iqbal 2020). Therefore, effective decision-supported tools are being utilized to identify eco-friendly wastewater treatment strategies.

9.2 WASTE WATER TREATMENT STRATEGIES

Conventional wastewater treatment strategies include physical, chemical, biological, and sludge treatment. Utilization of a novel approach for sustainable wastewater treatment still remains a major challenge. Significant efforts were made to develop novel techniques in the treatment of wastewater that have been contaminated with carcinogenic dyes, heavy metals, pesticides, antibiotics, microbes, etc., Due to technological development, extensive research was carried out in wastewater treatment process. In this backdrop, utilization of nanomaterials in wastewater treatment is catalyzing rapid development in recent years. In this scenario, utilization of nanoparticles (NPs) has emerged as an efficient tool in wastewater treatment. The intrinsic properties like small size, large surface area and quantum effect of MNPs have paved way for diversified application in wastewater treatment.

Nanoparticles have been applied in various sectors like, electronics, textiles, energy, food, healthcare, optics, and agriculture. Recent advances in nanoremediation pave the ways in treating wastewaters in small and large-scale setups. Due to large surface area, nanoparticles facilitate grater absorption and high degree of interaction with pollutants present in the wastewater. Metal nanoparticles (MtNPs), nanopolymers (NP), nanosorbents (NS), and nanocatalysts (NC) have been designed and developed to treat different quality and quantity of wastewater.

9.3 NANOTECHNOLOGY IN WASTEWATER TREATMENT

In the 21st century developed and developing countries face a lot of problems due to the discharge of wastewater from the industries. Release of hazardous wastes into the environment without proper treatment leads to the degradation of the ecosystem. Food, pharmaceutical, cosmetic, leather, and textile industries discharge hazardous, nondegradable waste into water bodies and damage aquatic life.

Hence, treatment of hazardous waste before their release is an inevitable task. Various physical, chemical, and biological technologies have been utilized in the treatment of industrial effluents (Weng et al. 2013). However, the main impediments for the implementation of these technologies are high installation and operational costs, production of substantial amount of toxic biomass, and low treatment performance (Ali et al. 2017).

Effective treatment of industrial effluents is still an open challenge (Cheera et al. 2016). Activated carbon is utilized more frequently in conventional adsorption process. However, the adsorption capacity is inversely proportional to the concentration of organic content in wastewater and reported to be substantially decreased at high concentration.

To fulfill the demand, nanoparticles are increasingly utilized in wastewater treatment (WWT) process. NPs are nothing but submicron moieties made of inorganic and organic materials, and the size of the NPs ranges from 1 to 100 nm. Because of the so-called "quantum size effects", greater permeability and high stability NPs are getting much research attention. Size, morphology, and different phases had an influence on the properties and application potential (Figure 9.1). Further, due to their dynamic morphology, desired size, and high saturation potential, they can be efficiently employed in WWT (Sivashankar et al. 2014).

Physical and chemical properties of nanoparticles are under extensive research. Particularly, the magnetic property of nanomaterial is greater than the massive (bulk) material. Several earlier reports illustrate that nanoparticles show differences in the Curie (T_C) or Neel (T_N) temperatures to that of bulk materials. Similarly, properties like magnetization (per atom) and the magnetic anisotropy were reported to be higher in nanoparticles. Magnetic nanoparticles (MNPs) have gained attention in various fields with the advent of nanobiotechnology. MNPs are the group of engineered materials that of size less than 100 nm that can be manipulated under the effect of an external magnetic field (Indira and Lakshmi 2010). In furtherance,

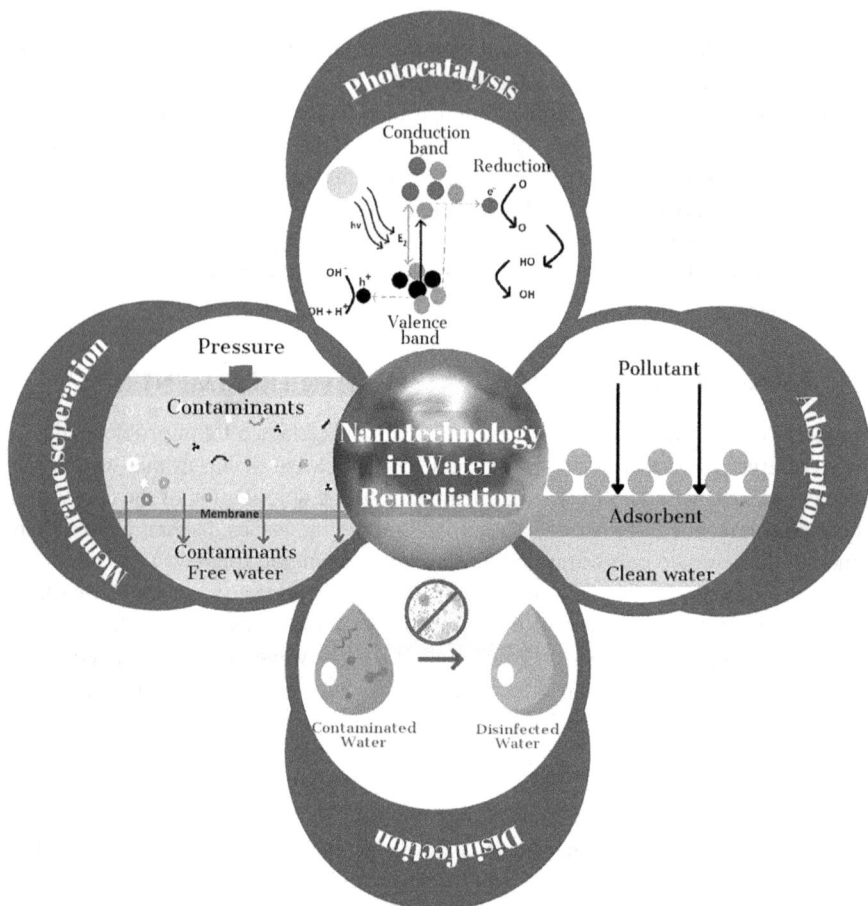

FIGURE 9.1 Nanomaterial-based wastewater remediation process.

magnetic nanomaterials are found to possess high magnetoresistance, magnetocaloric potential, etc. Due to the fast development and application of magnetic materials in various fields, utilization of magnetic nanoparticles in wastewater treatment seems to be promising. The magnetic nature facilitates their easy separation and reuse after purification process. However, more research is requested on optimizing mass production, surface functionalization, biocompatibility, and stability of nanoparticles.

9.4 TYPES OF NANOMATERIALS IN WASTEWATER TREATMENT

Nanotechnology offers leapfrogging opportunities in next-generation water remediation process. Further, nanomaterials possess multifunctional pollution remediation strategies. Unique properties like high reactivity, distinct quantum, surface properties, deep penetration, strong sorption, high separation efficiency, good recyclability, environment friendliness, and easy fabrication (Lu and Yuan 2018) facilitate the vivid application of nanomaterials in wastewater treatment. At present, zero-valent metal nanoparticles, metal oxide nanoparticles, carbon nanotubes (CNTs), and nanocomposites (Table 9.1) are extensively used in wastewater treatment.

9.5 ZERO-VALENT METAL NANOPARTICLES

Ag NPs have been extensively utilized in the treatment process as it (silver) is said to be highly toxic to a wide range of microorganisms inhabiting wastewater. Antimicrobial efficiency of Ag NPs is not clearly illustrated. However, several theories have been put forward; for example, Ag NPs have been reported to adhere to the microbial cell wall and cause structural changes by increasing its permeability (Sondi and Salopek-Sondi 2004). Punnoose and Mathew (2018) reported the degradation of various organic pollutants by photosynthesized silver metal nano photocatalyst.

9.6 METAL OXIDES

Metal oxide nanoparticles are used as photocatalysts, and adsorbents are used in the removal of heavy metal from wastewater. The most common and significant metal oxides are iron oxides, manganese oxides, aluminum oxides, titanium oxides, and silver and zinc oxides (Ahmad et al. 2020). Metal oxide nanoparticles act as a catalyst in different oxidation reactions and change pollutant molecules into environmentally suitable products. As metal nanoparticles are shape controlled, stable, monodispersed, and notable for their antifungal, antibacterial, and antiviral activity, they are used as disinfectants (Shao et al. 2017).

Titanium dioxide (TiO_2) is one of excellent photocatalysts due to its low cost, toxic free property, chemical stability, easy availability, super hydrophilic, and superoleophobic (Tan, Qiu, and Ting 2015). TiO_2 is reported to exist in three states, anatase, rutile, and brookite. Till date, anatase is considered as an efficient nanophotocatalyst material. TiO_2 nanomaterial generates photocatalytic reactive oxygen species and reported to be less toxic to human beings. Nanotubes, nanowires and nanosheets can be prepared with the help of TiO_2 (Ibhadon and Fitzpatrick 2013). Synthesis of titanium dioxide (TiO_2) is facilitated by

- Sol-gel process.
- Metal organic chemical vapor deposition (MOCVD).
- Wet chemical synthesis by precipitation of hydroxides from salts (Lazar, Varghese, and Nair 2012).

In addition to filtration, TiO_2 nanowire membrane is used in photocatalytic oxidation of organic contaminants. Excitation and charge separation within TiO_2 nanoparticles were induced ultraviolet radiations. Since TiO_2 possesses low selectivity, it has the capability to degrade all kinds of contaminants (Serra et al. 2017). For improving the photocatalytic activity of TiO_2, metal doping has been demonstrated and silver has attracted the attention for doping (Ahmed et al. 2018). For

TABLE 9.1

Application of Nanomaterials in Wastewater Treatment Process (Qu, Alvarez, and Li 2013; Jain et al. 2021)

Applications	Examples of Nanomaterials	Some of Novel Properties	Enabled Technologies
Adsorption	Carbon nanotubes	High specific surface area, excellent adsorption capacity, high porosity, diverse pollutant-CNT interactions, tunable surface chemistry, reusable.	Contaminant preconcentration/detection, adsorption of recalcitrant contaminants (e.g., Mn^{7+} ion, metronidazole and levofloxacin).
	Nanostructured metal oxides	Large specific surface area, Small intraparticle diffusion distance, more adsorption sites, compressible without significant surface area alteration, recyclability, some are superparamagnetic.	Adsorptive media filters, slurry reactors.
	Nanofibers with core–shell structure	Tailored shell surface chemistry for selective pollutant elimination, high surface area for fast decontamination, reactive core for degradation, Short internal diffusion distance.	Reactive nano-adsorbents (e.g., Malachite Green and Pb^{2+} removal).
Disinfection and microbial control	Nanocarbon titanium	Photocatalytic ROS generation, high chemical stability, low human toxicity, and cost.	POU to full scale disinfection and decontamination.
	CNT	Antimicrobial activity, fiber shape, and conductivity.	Point-of-use water disinfection, anti-biofouling surface.
	Nano-Ag	Strong and wide-spectrum antimicrobial activity, low human toxicity, ease of use	POU water disinfection, anti-biofouling surface.
	Zeolites	High surface area, Controlled release of nanosilver, bactericidal	Disinfection
Photocatalysis	Nano-TiO_2	Photocatalytic activity in UV and possibly visible light range, breakdown ample range of organic materials, low human toxicity, high stability, long life time, and low cost.	Photocatalytic reactors, solar disinfection systems.
	Nano-magnetite	Tunable surface chemistry, superparamagnetic	Forward osmosis

(Continued)

TABLE 9.1 (Continued)

Application of Nanomaterials in Wastewater Treatment Process (Qu, Alvarez, and Li 2013; Jain et al. 2021)

Applications	Examples of Nanomaterials	Some of Novel Properties	Enabled Technologies
Membranes and membrane processes	Nanofibers	High porosity, tailor-made, higher permeate efficiency.	Filter cartridge, ultrafiltration, prefiltration, water treatment, stand-alone filtration device.
	Nano-zeolites	Molecular sieve, hydrophilicity	High-permeability thin-film nanocomposite membranes
	Nano-Ag	Strong and wide-spectrum antimicrobial activity, low toxicity to humans	Anti-biofouling membranes
	Carbon nanotubes	Antimicrobial activity (unaligned carbon nanotubes)	Anti-biofouling membranes
		Small diameter, atomic smoothness of inner surface, tunable opening chemistry, high mechanical and chemical stability	
	Aquaporin	High-permeability and ionic selectivity	Low pressure desalination
	Nano-TiO_2	Photocatalytic activity, hydrophilicity, high chemical stability	Reactive membranes, high performance thin-film nanocomposite membranes
	Nano-magnetite	Tunable surface chemistry, superparamagnetic	Forward osmosis
Sensing and monitoring	Quantum dots	Broad absorption spectrum, narrow, bright, and stable emission which scales with the particle size and chemical component	Diazinon pesticide and antibiotic-resistant pathogen detection
	Noble metal nanoparticles	Proficient recognition of trace contaminants, Enhanced localized surface plasmon resonances, high conductivity, fast analysis	Optical and electrochemical detection (e.g, Caffeine)
	Dye-doped silica nanoparticles	High sensitivity and stability, rich silica chemistry for easy conjugation	Optical detection (e.g., triclosan)
	Carbon nanotubes	Large surface area, high mechanical strength and chemical stability, excellent electronic properties	Electrochemical detection, sample Preconcentration (e.g., Cr).
	Magnetic nanoparticles	Tunable surface chemistry, Superparamagnetism, simple separation	Heavy metal organic pollutant removal. Sample preconcentration and purification

industrial applications, doping with anions is considered as suitable and much cost effective. Recently, doped TiO_2 magnetic nanoparticles have been synthesized in a spinning disk reactor to achieve a feasible recovery of the nanoparticles by a magnetic trap (Sacco et al. 2012).

9.7   CARBON NANOMATERIALS

Graphene (allotrope of carbon) and carbon nanotubes (CNTs) are used as efficient adsorbents in wastewater treatment process. Discovery of the first fullerene in 1985 has opened a new horizon for the synthesis of CNTs and it was fabricated in the year 1991 (Iijima 1991). Synthesis of CNTs from graphite is facilitated by laser ablation, arc discharge, or by chemical vapor deposition from carbon-containing gas. CNTs have gained considerable interest in wastewater treatment process because of hydrophobic and electrostatic interactions, hydrogen bonding, and ion exchange with the pollutants. Compared to granular or powder activated carbons, CNTs have high sorption capability. Along with removal, CNTs even have the potential to inactivate pathogenic bacteria. Parham et al. (2012) reported that the CNT filter shown yeast filtration.

Carbon nanotubes are open ended or capped cylindrical sheet of graphene with diameter less than or equal to 1 nm. CNTs possess high adsorption efficiency toward dichlorobenzene, ethyl benzene, Zn^{2+}, Pb^{2+}, Cu^{2+}, and Cd^{2+}, and other dyes due to the high surface area to volume ratio and abundant porous structures. Graphene oxides are demonstrated to have high adsorbent capacity due to their large specific area, electron rich environment, and their strong functional groups (e.g., carboxyl, hydroxyl, and phenol) (Machado et al. 2019). Surwade et al. (2015) demonstrated that nanoporous graphene function as a selective membrane for water desalination. Functionalized graphene oxide (GO) with hydrophilicity and negative charge density has desalination potential (Zahid et al. 2018). To improve the adsorption efficiency, a composite was prepared by incorporating a metal to that of a CNT.

9.8   HYBRID NANOMATERIALS

Hybrid nanomaterials have a wide range of applications in health and environment. Integration of different nanomaterials results in a new multifunctional entity (Barsan et al. 2016). Compared to an individual single entity nanoparticle, additional nanoparticles adhering to a scaffold improve the stability of the material and enhance the activity. Khaydarov, Khaydarov, and Gapurova (2013) incorporated photo catalytic functional groups with nanocarbon titanium. Under sunlight irradiation, the nanomaterial has shown time dependent antimicrobial activity against *E.coli*.

Ahmed et al. (2015) synthesized ZnO/multi-walled CNT hybrid by dispersing MWCNT into a Zn-based solution by sonication. A high concentration of the hybrid nanomaterial (4 mg mL^{-1}) or reported to possess better antibacterial property compared to low concentration (0.125 mg mL^{-1}). Although the above-mentioned nanoparticles are effective for pollutants removal, disadvantage like removal of NPs from effluent shifts the focus toward the development of MNPs for WWT.

9.9 MAGNETIC NANOPARTICLE COMPOSITION

Magnetic elements present in the nanomaterial are manganese, iron, chromium, nickel, cobalt, gadolinium, and other chemical compounds. The most widely explored and used magnetic nanoparticles are magnetite and/or maghemite, the two different forms of iron oxide. Superparamagnetic metal particles (nanoscale size) have wide applications either as a single entity or along with different functional group. In addition, the functional group prevents the aggregation and minimizes the interaction of the particles. The resultant surface coated MNPs are used in different fields for example in medical field for drug delivery, sensors, and disease diagnosis, further they also used in industrial and environmental applications.

9.10 MAGNETISM

Magnetic effects are induced by movement of particles like electrons, protons, and charged ions. A spinning electric-charged particle creates a magnetic dipole called magneton (Akbarzadeh, Samiei, and Davaran 2012). A magnetic domain refers to a volume of material in which all magnetons pointing in the same direction and acting cooperatively are separated by domain walls.

Materials are classified as diamagnetic, paramagnetic, ferromagnetic, anti-ferromagnetic, and ferrimagnetic based on the magnetic moment orientation in a material/response to an externally applied magnetic field (Table 9.2) (Faraji, Yamini, and Rezaee 2010).

9.11 SUPERPARAMAGNETISM

Ferro and ferrimagnetism properties are useful in developing permanent or hard magnets. But, they remain magnetized after the removal of magnetic field and forms reversible clumps and aggregates. An ideal NP for biological application should have high magnetic susceptibility (material's degree of magnetization in response to applied magnetic field x=M/B), zero coercivity, no hysteresis, and no magnetic memory. By reducing the size of ferro- or ferromagnetic material (multidomain materials) into superparamagnetic diameter (single domain particle) by modern nanotechnology techniques, the MNPs are able to return quickly to nonmagnetized state when the external magnetic field is removed. (Faraji, Yamini, and Rezaee 2010) Superparamagnetism occurs when the thermal fluctuations demagnetize a previously saturated assembly. Magnetic domain theory illustrate that the critical size of the domain is influenced by the factors like the strength of the crystal anisotropy, exchange forces, domain wall energy, and the value of magnetic saturation.

9.12 APPROACHES AND TECHNIQUES IN FABRICATION OF MAGNETIC NANOPARTICLES

Novel and effective methods have been adopted to synthesize MNPs *viz.,* chemical coprecipitation, microemulsion, thermal decomposition, electrochemical, and sol-gel precipitation. MNPs are synthesized in different shapes, such as nanorods, nanotubes, and hierarchical superstructures by the above methods. However, these methods are

TABLE 9.2

Magnetism and Magnetic Moment Behavior

Type of Magnetism	Observed In	Magnetic Moment
Diamagnetism	Material with filled electric subshells like water	Magnetic moments are paired and overall cancel each other (zero net magnetic moment). Small negative magnetic response. Weakly repel external magnetic field (e.g., Ag, Au, Co)
Paramagnetism	Material partially attributed to unpaired electrons in its atomic shells like oxygen	Uncoupled atomic magnetic movements (e.g., magnesium, lithium) Slight positive magnetic response when apply an external magnetic field.
Ferromagnetism	Material partially attributed to unpaired electrons in its atomic shells	In the absence of applied magnetic field, aligned atomic magnetic moment with equal magnitude confers spontaneous magnetization which is known as hard magnets which will form irreversible clumps and aggregates (magnetic domain also called Weiss domain). They show direct coupling interaction due to crystalline structure which enhances the flux density (e.g., Fe, Ni)
Antiferromagnetism	Material partially attributed to unpaired electrons in its atomic shells	Equal magnitude atomic magnetic moment with materials arranged in an antiparallel fashion leaving zero net magnetization (e.g., MnO, NiO)
Ferrimagnetism	Atoms or ions in ordered but nonparallel fashion in zero applied magnetic field below Neel temperature (T_N).	

facing serious challenges because of the application of high pressure and temperature, utilization of toxic and hazardous chemicals in the fabrication of these metal oxides NPs (Ali et al. 2017). Because of these drawbacks, green fabrication of MNPs serves as an Elsevier in wastewater treatment process. In phytofabrication of MNPs, plant metabolites are being utilized as reducing and capping agent due to their reductive capacities (Shamaila et al. 2016). Phytofabricated MNPs are reported to be efficient in terms of adsorptive removal and resources recovery than conventional nanoparticles (Cheera et al. 2016). Phytofabricated MNPs are employed in the removal of aqueous pollutants and the recovery of metallic ions due to the presence of organic functional groups (e.g., polyphenols, amino acids, sugars, alkaloids, terpenoids, proteins, carbonyl, carboxyl, and polysaccharides), unique morphology, desired size, super paramagnetic behavior and high saturation magnetization valve from wastewater (Machado et al. 2015). Plant-based metabolites present in PMNP acts like ion-exchange resins and facilitates the removal of pollutants via electrostatic attractions.

9.13 COPRECIPITATION

In addition to phytofabrication, coprecipitation is a widely used method for the synthesis of MNPs with desired sizes and magnetic properties (Sandeep Kumar 2013).

Because of the ease of application, synthetic procedures, and less harmful materials, MNPs are extensively used for wastewater treatment. In coprecipitation process, MNPs are prepared from aqueous salt solutions, by the addition of a base under an inert atmosphere at room temperatures or at high temperature (Faraji, Yamini, and Rezaee 2010).

9.14 MICROEMULSION

Microemulsion involves isotropic dispersal of oil, water, and surfactant, in combination with cosurfactant and it is said to be thermodynamically stable. By mixing two identical water-in-oil microemulsions consisting of the chosen reactants, the microdroplets will continuously collide, coalesce, and break again, and finally a precipitate form in the micelles (Faraji, Yamini, and Rezaee 2010).

The surfactants in water endure rapid coalescence facilitating rapid mixing, followed by precipitation and aggregation of nanodroplets in the synthesis of MNPs. Compared to other methods, fabrication of MNPs using microemulsion is advantageous as they have control over particle size, composition, and crystalline nature. Feltin and Pileni (1997) synthesized MNPs with average sizes ranging from 4 to 12 nm, and standard deviation ranging from 0.2 to 0.3, using microemulsions. Micellar solution of ferrous dodecyl sulfate ($Fe(DS)_2$) is used for the synthesis of MNP and their size is controlled by surfactant concentration and temperature.

9.15 POLYOL METHOD

NP synthesis by polyol technique is a most promising approach. It could be efficiently applied in biomedical applications such as magnetic resonance imaging. Fine metallic particles can be produced by reducing dissolved metallic salts and directly precipitating metals from a solution including a polyol (Nejati-Koshki et al. 2014). Nanocrystalline alloys and bimetallic clusters can be synthesized by polyol method.

In polyol method, the liquid act as a solvent for reduce the metallic precursor also act as a complexing agent for the metallic cations. The solution is stirred and heated to the boiling point of the polyol. A better control of the mean size of the metal particles can be achieved by seeding the reactive medium with foreign particles (heterogeneous nucleation) (Majidi et al. 2014).

9.16 HYDROTHERMAL METHOD

Hydrothermal method also called as the solvothermal method, is one of the most successful ways to grow crystals of many different materials (Faraji, Yamini, and Rezaee 2010). This process crystallizes material in a sealed container, from aqueous solution at the high temperature range of 130°Δ C to 250°Δ C, and at high vapor pressure, generally in the range of 0.3 to 4 MPa (Wu, He, and Jiang 2008). However, despite several studies, hydrothermal approaches still fail to obtain quality hydrophilic nanocrystals smaller than 10 nm (Stojanovic et al. 2013).

9.17   CHEMICAL VAPOR DEPOSITION (CVD)

Here in the CVD process, the substrate is exposed to a hot-wall reactor that contains a volatile precursor induces nucleation and facilitates the deposition of the substrate in order to obtain the desired thin film. The immense flexibility of CVD process facilitates the utilization of chemical materials from the database chemistries. The precursors can be solid, liquid, or gas at ambient conditions but are delivered to the reactor as a vapor (from a bubbler or sublimation source, as necessary) (Majidi et al. 2014).

9.18   SPRAY PYROLYSIS AND SONOCHEMICAL METHOD

In spray pyrolysis process a solution is sprayed into a series of reactors where in the aerosol droplets endure evaporation, with the subsequent condensation of solute within the droplet, followed by vaporization and precipitation of substrate. Finally, nucleation and development of thin metal film occur. Spray pyrolysis utilizes simple equipment and said to be cost effective and very efficient. Produced thin films have large surface area of substrate coverage. Due to concerns like poor quality of thin film, thermal decomposition and vapor convection spray pyrolysis seems not useful (Majidi et al. 2014).

Sonochemical method is one of the simplest methods utilizes sonochemistry for obtaining magnetic nanomaterials. High intensity ultrasound is utilized in sonochemical method for the production of novel magnetic nanomaterials. Acoustic cavitation is implosive collapse of gas bubbles within the liquid phase that generates localized hot spots. Hot spots produced and cavitation process concomitantly triggers oxidant radicals production and promotes the formation of final stable phases.

Ultrasonic irradiation facilitates solute diffusion; that subsequently results in increased nucleation rate, crystallization process followed by reduction in induction time and zone width of the metabolite. The method was found to reduce the chemical compound's toxic environmental impact and reduces the infrastructure requirements and incubation time (Majidi et al. 2014).

Although the above-mentioned nanoparticles are effective for pollutants removal, shortcomings like separation of NPs from the effluent and adverse ecotoxicological effects by chemical and physical synthesis methods, hinders the application of NPs in wastewater treatment (WWT) process. In this regard, phytofabrication of MNP using plant has received major attention in WWT.

9.19   PHYTOFABRICATION AND
CHARACTERIZATION OF PMNPs

There is no single optimal method for the synthesis of MNPs, which vary slightly from one another in choice of research findings and application purpose include type of MNP, starting material, selection of metal salt, presence or absence of NaOH, variable reaction condition, and the method of collecting synthesized NPs (Ali et al. 2021). In current era, due to the great variety of applications involving magnetic nanoparticles and toxic impacts of the conventional physical and chemical methods, research focus has shifted toward the development of clean and earth-friendly promising future protocols for MNPs synthesis. Greenly orchestrated magnetic nanoparticles are more effective, fiscally smart, effortlessly scaled up, nontoxic, and easy to characterize. In addition, they can be easily separated from the final reaction mixture by applying external magnetic field and reused in consecutive treatment process (Priya et al. 2021). Utilization of plants and their phytochemicals are extremely practical as they are nontoxic. The plant extract with compounds like amino acids, flavonoids, saponins, terpenoids, aldehydes, carboxylic acid, proteins, vitamins, enzymes, nitrogenous bases, reducing sugars, polyphenols and pigments acts as reducer, stabilizer, redox mediators, and capping agents in PMNPs synthesis (Gour and Jain 2019). Commonly used reducing and capping chemical structure of plant metabolites and phytofabrication process is shown in Figure 9.2. As the phytochemicals in the plant product reduce metal ions compared to other biological method in shorter time and

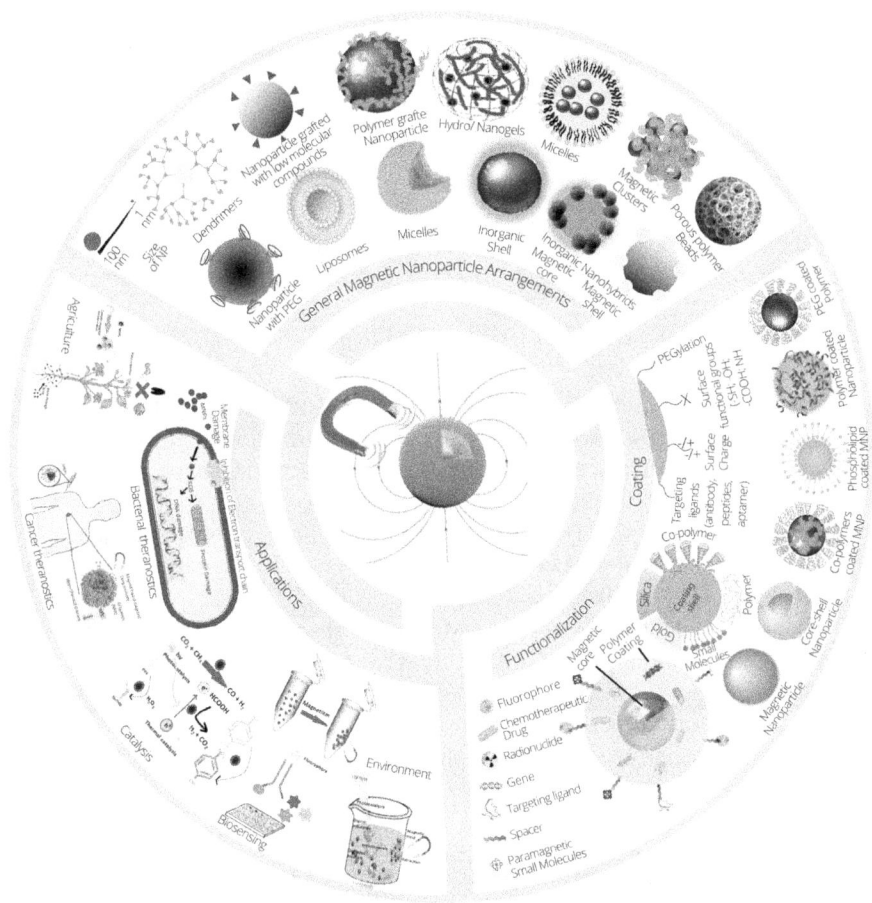

FIGURE 9.2   Schematic representation on the fabrication of phytogenic magnetic nanoparticles (PMNPs) and the stabilization process.

produce stable NPs, phytogenic MNP synthesis method is superior than other methods (Zhang et al. 2020).

In plant product-mediated "green" synthesis of magnetic nanoparticles, the plant metabolites act as reducing and capping agent. Reduction of metal ion happens on the addition of sodium acetate (electrostatic stabilizing agent) to the ion precursor and plant extract mixer. Oxidation of plant residues induce the free radicals (R: –OH– group) formation that interact with the metal ions and forms reduced metal atoms. Depending upon the fabrication conditions, metal ions form green nano zero-valent metal or metal oxides. In furtherance, organic molecules stabilize the nanoparticles to attain desired shape and size. The final black precipitation indicates the fabrication of PMNPs. For further application, by applying magnetic field or by simple filtration method, the precipitate is to be separated, cleansed with solvent, and oven dried. To functionalize the PMNPs for different applications, various ligands were used.

Equations (9.1)–(9.3) explain the fabrication mechanism of PMNPs (Ali et al. 2017).

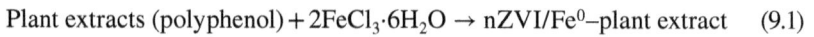

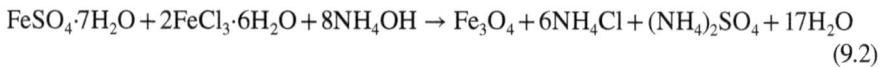

$$\text{Plant extracts (polyphenol)} + 2FeCl_3 \cdot 6H_2O \rightarrow nZVI/Fe^0\text{–plant extract} \quad (9.1)$$

$$FeSO_4 \cdot 7H_2O + 2FeCl_3 \cdot 6H_2O + 8NH_4OH \rightarrow Fe_3O_4 + 6NH_4Cl + (NH_4)_2SO_4 + 17H_2O \quad (9.2)$$

$$Fe_3O_4 + \text{plant waste} + 17H_2O \rightarrow \text{plant waste–}Fe_3O_4 \quad (9.3)$$

To ponder the morphology and other conformational subtle elements of PMNPs, various methods are most widely utilized (Faraji, Yamini, and Rezaee 2010; Ali et al. 2021).

For determining the size and surface morphology,

1. Dynamic light scattering (DLS) method and Brunauer Emmet Teller—to determine the size, surface area, and particle distribution.
2. Atomic force microscopy—to illustrate the surface roughness, position of the particle, and step height.
3. X-ray diffraction (XRD)—to define the crystallinity of NPs that includes crystal size, crystal perfection, structural parameters, and degree of substitution of metal by other trivalent cations.
4. Photon correlation spectroscopy, Mossbauer spectroscopy, high-resolution TEM, and field emission SEM are used to determine the particle size and distribution.

Elemental composition and surface morphology can be determined by wide array of methods:

1. Scanning electron microscopy (SEM) provides data on surface topography of the sample and composition.
2. Transmission electron microscopy (TEM) determines the composition, morphology, aggregation state of NPs, electron space shift, lattice spacing, and size of NPs.

3. AAS is used determine the elemental composition of liquid NPs whereas for solid NPs an acid or base should be used as solvent.
4. XPS provides information on composition and chemical state of NPs.
5. UV–Vis absorption spectroscopy confirms elemental composition like metal and metal oxides.
6. Energy dispersive X- ray diffraction (EDAX) determines the surface morphology, size, and shape of the nanoparticles.

Structure and bonding of MNPs are determined by the following methods:

1. Fourier transmission infrared (FTIR) spectroscopy confirms the presence of various reducing and stabilizing functional groups. Against potassium bromide background at 400–4,000 cm^{-1} using the spectrophotometer, the spectrum were attained.
2. FT-IR illustrates the bonding between pattern of organic and inorganic particles.
3. XPS determines the mechanism of reaction that takes place on the surface of MNPs.
4. Raman spectroscopy used to find out the structure and spinal lattice of the compound.
5. TGA technique estimates binding efficiency of the MNPs by coating formation between surfactants and polymers.
6. XAS determines the oxidation states and elements required for electronic configuration.

Changes in the above characters lead to changes in physicochemical properties of PMNP.

9.20 FACTORS INFLUENCE THE PHYTOFABRICATION OF MNPs

Several factors such as pH, temperature, pressure, light intensity, concentration of plant extract, concentration of metal solution, and incubation/reaction time affect the synthesis, morphology, magnetic property, and application of nanoparticles (Ahmad et al. 2020). However, only a few reports are available on the factors that influence the formation of PMNPs. Major factors that influence phytofabrication of nanoparticle are discussed below.

9.21 INFLUENCE OF pH

pH of a solution serves a crucial role in green synthesis of magnetic nanoparticles. Different plant extracts and extracts from different parts of the same plant have different pH, which influence the rate of bio reduction concomitantly affects the size and shape of NPs (Mittal et al. 2014). Mechanism behind the effect of pH is it affects the electrical loads of biomolecules in a medium which consequently affect biomolecules stability, growth as well as their capping ability. Development of nucleation center (protonation–deprotonation) increases as pH increases, and expansion of nucleation center enhances the formation of metal nanoparticles by the reduction of metal ions. Moreover, high pH favors reduction and stabilizing capacity of the NPs. At low pH, especially under acidic condition (pH <7), the reaction time is longer and the NPs synthesized are usually larger (Ahmad et al. 2020).

9.22   REACTANT CONCENTRATION

The concentration of plant extracts influences the shape, size, and rate of reduction in metal nanoparticles formation. Several earlier reports illustrate that the concentration of plant extract and size of the nanoparticles are inversely proportional.

9.23   INCUBATION TIME

Incubation time is the time interval between the addition of the reactant to the mixture and completion of all the steps. Incubation period significantly affects the size morphology, stability, and degree of phytofabrication of NPs. The stirring time or incubation period enables the proper interaction of metal salt and complex compounds of plant for reduction. The reaction time depends on the reaction mixture pH of the solution, light intensity, temperature, and enzyme mediated reducing power of plant extract. The reaction mixture containing more secondary metabolites reduces the metal salt in less time. If the plant contains less secondary metabolites, the phytofabrication of NP takes place quickly (Ahmad et al. 2020). Also, insufficient biomolecules bind with metal precursor, which finally results in the formation of instable NPs (Fazlzadeh et al. 2016). Increase in the reaction time resulted in an increase in the nanoparticle size by aggregation or shrinkage of particles (Saif, Tahir, and Chen 2016).

9.24   EFFECT OF METAL ION CONCENTRATION

Concentration of precursor influences the synthesis time of iron oxide NPs. Increase in concentration of precursor increases the rate of synthesis and vice versa. This phenomenon is attributed to inadequate proportion of biomolecules in the plant extract to the precursor for growth and nucleation of nanoparticles in the solution (Zhu et al. 2012). By increasing the metal ion concentration, the absorbance peak and particle size were increased in metal nanoparticle. (Dubey, Lahtinen, and Sillanpaa 2010). Dehsari et al. (2017) reported that the decrease in precursor concentration was found to yield iron oxide NPs with decreased size and vice versa.

9.25   REACTION TEMPERATURE

Temperature is one among the factors that influence the size, shape, and reaction rate of nanoparticle synthesis. Increasing temperature to an optimum level enhances metal nanoparticle synthesis and stability. Iravani and Zolfaghari (2013) observed that increase in temperature (25°C, 50°C, 100°C, and 150°C) leads to increase in sharpness of adsorption peak and decrease in size of AgNPs. Earlier studies have

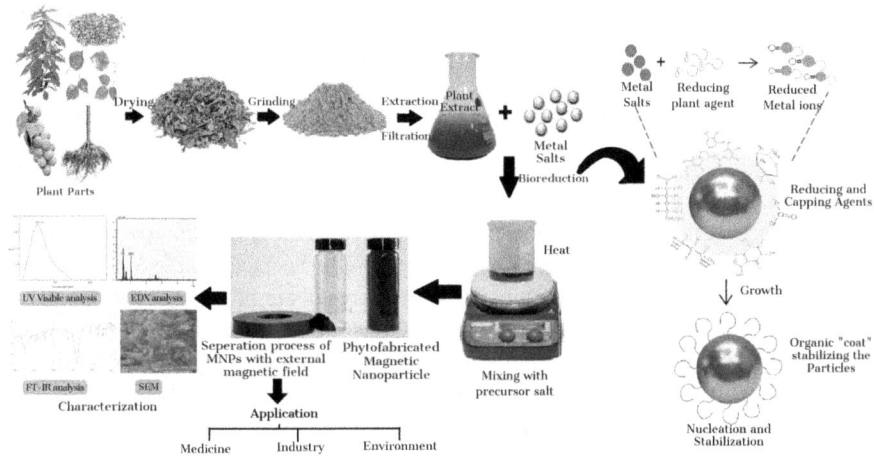

FIGURE 9.3 Magnetic nanoparticle organization, arrangement, and applications.

reported that magnetic nanoparticles with controlled size, shape, and magnetization can be efficiently exploited in various biomedical and biotechnological applications.

9.26 MAGNETIC NANOPARTICLE STABILIZATION

Synthesis method and chemical structure are said to influence the properties of MNPs. The smaller size and large surface area of MNPs are very sensitive to agglomeration or precipitation. A very simple method to protect its pure metal core is known as surface passivation by mild oxidation. The straight forward method is to improve the chemical stability of the surface of the MNP that protect the magnetic core against the environment. Carbon, silica, organic polymers, surfactants, precious metals, and metal oxides (Figure 9.3) have been successfully employed in encapsulation of MNPs. Stabilization of magnetic nanoparticle is achieved by equilibrium between electrostatic or steric repulsion.

9.26.1 ORGANIC COATINGS

Surfactants and polymers are either chemically anchored or physically adsorbed as single or double layer over the surface of MNPs during synthesis or after synthesis. Surfactants bind covalently to the native hydroxyl groups present on the surface of metal oxide. Phosphates and sulfates bind to the surface of magnetic particle. Polymers receive more attention than surfactants due to its high repulsive capacity. To coat polymers on the surface of MNPs two approaches have been developed. In the first approach, macromolecular chains are irreversibly attached by chemisorptions. In another approach, to give brush like hybrid particle, higher number of polymer chains attached directly to the particle surface through polymerization method. The drawback behind organic coated MNPs is they are easily leached by acidic solution and their low intrinsic stability at higher temperature. Surface-modified PMNPs with biocompatible polymers are intensively studied for magnetic field-directed biological applications.

9.26.2 Inorganic Coating

Titanium oxide, zirconium oxide, and aluminum oxide are coated on the surface of metal core by controlled oxidation process. The above-mentioned metal oxide surfaces are readily modified with phosphorylated molecules. Due to its specific surface derivative property, gold is widely used to coat MNPs. Surface of gold is well known for self-assembling of thiolated molecules, which results in monolayer formation.

The most common method to deposit precious metal on MNPs surface is redox transmetalation, reactions in microemulsion, and sonolysis (produces air-stable magnetic iron NPs). Magnetic nanoparticle coated with noble metal shells results in enhanced chemical stability and high electrical conductivity.

Silica coated magnetic microspheres show stable negative charge ranging from pH 6 to 7. The magnetic core covered by silica has unique magnetic responsivity, low cytotoxicity, and easy control of interparticle interactions. These magnetic properties can be fine-tuned by varying chemical composition and the shell's thickness. Two approaches called stober process and microemulsion synthesis were employed to generate silica coating. Silica was formed through *in situ* hydrolysis and condensation of a sol-gel precursor in stober process. In microemulsion synthesis, micelles or inverted micelles control the silica coating on the core–shell surface.

Carbon-encapsulated MNPs have received much attention for many biological applications. In addition, carbon is stable both chemically and thermally, further, cheap and biocompatible. Owing to the above advantages various methods, chemical vapor depositions, combustion, pulse laser decomposition, arc discharge techniques, and pyrolysis of metal complexes were developed actively.

9.26.3 Organic and Inorganic Combinations

In order to increase the chemical stability of magnetic nanoparticles, organic shell components are cross-linked with polymeric chains, which also protect the magnetic core from chemical and physical decomposition.

9.27 STRATEGIES TO FUNCTIONALIZATION

Ideal MNPs should be chemically stable and possess liquid medium dispersion ability. Surface functional group influences the particle interactions with environment. Therefore, in post synthetic modification of MNPs, a suitable functional group is to be added, which render MNPs to be chemically stable. The electrostatic chimioadsorption and covalent conjugation are the two strategies used for surface functionalization includes either ligand addition or surface ligand exchange. In ligand addition to the original ligands, incoming ligands were added and form double-layer-like structure. In ligand exchange, a new bifunctional ligand replaces the original ligand. In pollutant extraction and removal, surface-modified MNPs have been proven to be efficient.

9.28 APPLICATION OF MAGNETIC NANOPARTICLE

Magnetic nanoparticle with different size and shape is tailored to explore their magnetic behavior and possibility of application in new areas. Due to their extraordinary magnetic behavior, magnetic nanomaterials in imaging, magnetic hyperthermia, biosensors, pollutant remediation, and other applications are possible.

In magnetic hyperthermia, MNP converts magnetic energy to heat, which is used to detect, capture, and eradicate the cancerous cell in free form or at the targeted region.

In magnetic separation, specific target analytes like cells, proteins, DNA, and pathogenic substances are selectively removed (Zhang et al. 2013). Owing to the characters like biocompatibility, durability, safety, and magnetic signaling, the MNPs-based biosensors have shown remarkable applications in clinical diagnosis, food industry, and environment monitoring. Recently, MNPs-based strategies have developed to kill microbes through disrupting plasma membrane, releasing toxic metals and generation of reactive oxygen species (ROS), e.g., Fe_3O_4 magnetic microbots (Dong et al. 2021).

9.29 ROLE OF PMNPS IN THE REMOVAL OF POLLUTANTS

Due to ever increasing population, severe droughts, and strict water quality standards, there is an increasing demand for clean water. In recent years, application of nanotechnologies in water treatment process has become popular. Requirements of the future generation on freshwater demand can be met by the application of MNPs in wastewater treatment. Phytofabricated magnetic nanoparticles possess unique physical, chemical, and biological properties, thereby able to sense water pollution and offer quantum scale wastewater treatment and prevention at low yield cost. The high surface area to volume ratio of MNPs enhances the reactivation rate with contaminants and provides everlasting wastewater treatment strategies. Use of PMNP membrane enables water reuse and desalination that provides water quality and quantity in the long run.

PMNPS has the potential to provide both water quality and quantity in the long run through the use of membranes enabling water reuse and desalination. The unique properties of MNPs *viz.*, high absorption potential, strong interaction, and efficient reaction capabilities with pollutants on the aqueous suspensions display quantum size effects (Amin, Alazba, and Manzoor 2014).

A wide verity of metallic ions in pollutant water is efficiently adsorbed by magnetic nanomaterials. Resin-based magnetic chelating materials and magnetic hydrogels are also used in the removal of Cu(II), Co(II), and Ni(II) ions. Acrylate-based polymer composites exhibit a selective and orderly manner (Cu>Cr>Zn>Ni) in the removal of heavy metals. Magnetic zeolite composites are used in the removal of heavy metal contaminants.

Dyes in the wastewater had a lethal impact on human health. In addition to heavy metals, magnetic nanoparticles are efficient adsorbents in the removal of dyes released from industries. Therefore, magnetic nanoparticles are fabricated for the efficient adsorption of dyes released from industries and chemicals from pharmaceutical industries. Earlier literature illustrate that magnetic nanoparticle removes nearly 95% of pharmaceutical wastes. Gallic acid coated magnetic nanoparticles (GA-MNP) are used as a photocatalyst in the degradation of meloxicam, a nonsteroidal anti-inflammatory drug (Sharma et al. 2018).

9.30 DISINFECTION

In addition to removal of pollutants, disinfection process is also facilitated by magnetic nanomaterials. Magnetic nanomaterials possess reasonably good removal efficiencies of biological contaminants. The contaminants are categorized into three types, namely, natural organic matter (NOM), biological toxins, and microorganisms. Disinfection of waste sources using chlorine was already reported to be carcinogenic and lead to the formation of harmful by-products. UV radiation mediated disinfection requires longer exposure time. Utilization of magnetic nanoparticles in disinfection technologies is in limelight due to their potential in the elimination of pathogens, stability, large-scale adoption, low toxicity, and microbial inactivation in water (Amin, Alazba, and Manzoor 2014).

In PMNPs, plant metabolites act as the reducing and capping agent facilitate the removal of toxic heavy metals. The −OH groups in the plant extract facilitate the removal of metal ions by electrostatic attraction process. Compared to other NPs, PMNPs are reported to be superior in the removal of toxic dyes like methylene blue, methyl orange, and its degradation. Anionic and cationic dyes are removed by adsorption behavior of PMNPs. In addition to dyes, organic pollutants are also efficiently removed from wastewater. Adsorptive removal of nitrate from wastewater was facilitated due to chemisorptions mechanism. Various studies have been carried out on the utilization of PMNPS in the removal of organic pollutants; however, more research efforts are required to investigate PMNPs potential in the removal of persistent or refractory organic pollutants (Ali et al. 2017).

9.31 CONCLUSION

To conclude in the backdrop of conventional and advanced wastewater treatment technique, all the techniques to be adopted should be scrutinized thoroughly on the cost, feasibility, output, pollution removal index, and eco-friendly nature. Magnetic nanoparticles (MNPs) mediated environmental remediation offers great promise and said to be fast and effective. Like other new and remediation technologies, there are concerns regarding safety and effectiveness on the utilization of metal nanoparticles. For example, there is no widely accepted ecotoxicological standard to compare particle toxicity. Furthermore, any alterations made to the nanoparticles, either a new coating or addition of catalyst in order to enhance the remediation is said to alter the particle behavior. Therefore, to conclude, the above characteristics restrict the usage of phytofabricated magnetic metal nanoparticles in large scale in WWT process. In situ pilot-scale tests have to be carried out to highlight the positive outcomes on the usage of magnetic metal nanoparticles in WWT.

REFERENCES

Ahmad, N. A., P. S. Goh, L. T. Yogarathinam, A. K. Zulhairun and A. F. Ismail. 2020. Current advances in membrane technologies for produced water desalination. *Desalination*, 493, p. 114643. doi:10.1016/j.desal.2020.114643.

Ahmed, D., A. Abed, A. Bohan, and J. Rbat. 2015. Effect of (ZnO/MWCNTs) hybrid concentrations on microbial pathogens removal. *Engineering and Technology Journal*, 33(8), pp. 1402–1411.

Ahmed, S. N. and W. Haider. 2018. Heterogeneous photocatalysis and its potential applications in aater and wastewater treatment: A review. *Nanotechnology*, 29(34), p. 342001. doi:10.1088/1361-6528/aac6ea.

Akbarzadeh, A., M. Samiei and S. Davaran. 2012. Magnetic nanoparticles: Preparation, physical properties, and applications in biomedicine. *Nanoscale Research Letter*, 7, p. 144. doi:10.1186/1556-276X-7-144.

Ali, I., C. Peng, I. Naz, Z. M. Khan, M. Sultan, T. Islam and I. A. Abbasi. 2017. Phytogenic magnetic nanoparticles for wastewater treatment: A review. *RSC Advances*, 7, pp. 40158–40178. doi:10.1039/C7RA04738J.

Ali, A., T. Shah, R. Ullah, P. Zhou, M. Guo, M. Ovais, Z. Tan and Y. Rui. 2021. Review on recent progress in magnetic nanoparticles: Synthesis, characterization, and diverse applications. *Frontiers in Chemistry*, 9, p. 629054. doi:10.3389/fchem.2021.629054.

Almuktar, S. A. A. A. N., N. S. Abed, and M. Scholz. 2018. Wetlands for wastewater treatment and subsequent recycling of treated effluent: A review. *Environmental Science and Pollution Research*, 25(24), pp. 23595–23623. doi:10.1007/s11356-018-2629-3.

Amin, M. T., A. A. Alazba, and U. Manzoor. 2014. A review of removal of pollutants from water/wastewater using different types of nanomaterials. *Advances in Materials Science and Engineering*, 825910, pp. 1–24. doi:10.1155/2014/825910.

Barsan, M. M., V. Pifferi, L. Falciola and C. M. A. Brett. 2016. New CNT/poly (brilliant green) and CNT/poly(3,4-ethylenedioxythiophene) based electrochemical enzyme biosensors. *Analytica Chimica Acta*, 927, pp. 35–45. doi:10.1016/j.aca.2016.04.049.

Cheera, P., S. Karlapudi, G. Sellola, and V. Ponneri. 2016. A facile green synthesis of spherical Fe_3O_4 magnetic nanoparticles and their effect on degradation of methylene blue in aqueous solution. *Journal of Molecular Liquids*, 221, pp. 993–998. doi:10.1016/j.molliq.2016.06.00.

Dehsari, H. A., A. H. Ribeiro, B. Ersoz, W. Tremel, G. Jakob, and K. Asadi. 2017. Effect of precursor concentration on size evolution of iron oxide nanoparticles. *CrystEngComm*, 19, pp. 6694–6702. doi:10.1039/C7CE01406F.

Dong, Y., L. Wang, K. Yuan, F. Ji, J. Gao, Z. Zhang, D. Xingzhou, Y. Tian, Q. Wang, and L. Zhang. 2021. Magnetic microswarm composed of porous nanocatalysts for targeted elimination of biofilm occlusion. *ACS Nano*, 15(11), pp. 5056–5067. doi:10.1021/acsnano.0c10010.

Dubey, S., M. Lahtinen, and M. Sillanpaa. 2010. Tansy fruit mediated greener synthesis of silver and gold nanoparticles. *Process Biochemistry*, 45(7), pp. 1065–1071. doi:10.1016/j.procbio.2010.03.024.

Faraji, M., Y. Yamini, and M. Rezaee. 2010. Magnetic nanoparticles: Synthesis, stabilization, functionalization, characterization, and applications. *Journal of Iranian Chemical Society*, 7, pp. 1–37. doi:10.1039/b815548h.

Fazlzadeh, M., K. Rahmani, A. Zarei, H. Abdoallahzadeh, F. Nasiri and R. Khosravi. 2016. A novel green synthesis of zero valent iron nanoparticles (NZVI) using three plant extracts and their efficient application for removal of Cr(VI) from aqueous solutions. *Advanced Powder Technology*, 28(1), pp. 122–130. doi:10.1016/j.apt.2016.09.003.

Feltin, N. and M. P. Pileni. 1997. New technique for synthesizing iron ferrite magnetic nanosized particles. *Langmuir*, 13(15), pp. 3927–3933. doi:10.1021/la960854q.

Gour, A. and N. K. Jain. 2019. Advances in green synthesis of nanoparticles. *Artificial Cells, Nanomedicine, and Biotechnology*, 47(1), pp. 844–851. doi:10.108 0/21691401.2019.1577878.

Ibhadon, A. and P. Fitzpatrick. 2013. Heterogeneous photocatalysis: Recent advances and applications. *Catalysts*, 3(1), pp. 189–218. doi:10.3390/catal3010189.

Iijima, S. 1991. Helical microtubules of graphitic carbon. *Nature*, 354(6348), pp. 56–58. doi:10.1038/354056a0.

Indira, T. K. and P. K. Lakshmi. 2010. Magnetic nanoparticles - A review. *International Journal of Pharmaceutical Sciences and Nanotechnology (IJPSN)*, 3(3), pp. 1035–1042. doi:10.37285/ijpsn.2010.3.3.1.

Iravani, S. and Zolfaghari, B. 2013. Green synthesis of silver nanoparticles using *Pinus eldarica* bark extract. *Biomed Research International*, p. 639725. doi:10.1155/2013/639725.

Jain, K., Patel, A. S., Pardhi, V. P. and S. J. S. Flora. 2021. Nanotechnology in wastewater management: A new paradigm towards wastewater treatment. *Molecules*, 26(1797): pp. 1–26. doi:10.3390/molecules26061797.

Khaydarov, R. A., R. R. Khaydarov and O. Gapurova. 2013. Nano-photocatalysts for the destruction of chloro-organic compounds and bacteria in water. *Journal of Colloid and Interface Science*, 406, pp. 105–110. doi:10.1016/j.jcis.2013.05.067.

Lazar, M., S. Varghese, and S. Nair. 2012. Photocatalytic water treatment by titaniumdioxide: Recent updates. *Catalysts*, 2(4), pp. 572–601. doi:10.3390/catal2040572.

Lu, Y. and W. Yuan. 2018. Superhydrophobic three-dimensional porous ethyl cellulose absorbent with micro/nano-scale hierarchical structures for highly efficient removal of oily contaminants from water. *Carbohydrate Polymers*, 191, pp. 86–94. doi:10.1016/j.carbpol.2018.03.018.

Machado, A. B., G. Z. P. Rodrigues, L. R. Feksa, D. B. Berlese and J. G. Tundisi. 2019. Applications of nanotechnology in water treatment. *Revista Conhecimento Online*, 1, pp. 3–15. doi:10.25112/rco.v1i0.1706.

Machado, S., J. P. Grosso, H. P. A. Nouws, J. T. Albergaria, and C. Delerue-Matos. 2015. Characterization of green zero-valent iron nanoparticles produced with tree leaf extracts. *Science of the Total Environment*, 533, pp. 76–81. doi:10.1016/j.scitote nv.2015.06.091.

Majidi, S., F. Z. Sehrig, S. M. Farkhani, M. S. Goloujeh and A. Akbarzadeh. 2014. Current methods for synthesis of magnetic nanoparticles. *Artificial Cells, Nanomedicine, and Biotechnology*, 44(2), pp. 722–734. doi:10.3109/21691401.2014.982802.

Mittal, J., A. Batra, A. Singh, and M. M. Sharma. 2014. Phytofabrication of nanoparticles through plant as nanofactories. *Advances in Natural Sciences: Nanoscience and Nanotechnology*, 5, p. 043002. doi:10.1088/2043-6262/5/4/043002.

Nejati-Koshki, K., M. Mesgari, E. Ebrahimi, A. Abhari, S. F. Aval, A. A. Khandaghi, and A. Akbarzadeh. 2014. Synthesis and in-vitro study of cisplatin-loaded Fe_3O_4 nanoparticles modified with PLGA-PEG6000 copolymers in treatment of lung cancer. *Journal of Microencapsulation*, 31(8), pp. 815–823. doi:10.3109/02652048.2014.940011.

Parham, H., S. Bates, Y. Xia, and Y. Zhu. 2012. A highly efficient and versatile carbon nanotube/ceramic composite filter. *Carbon*, 54, pp. 215–223. doi:10.1016/j.carbon.2012.11.032.

Parra-Saldivar, R., B. Muhammad and H. M. N. Iqbal. 2020. Life cycle assessment in wastewater treatment technology. *Current Opinion in Environmental Science & Health*, 13(4), pp. 80–84. doi:10.1016/j.coesh.2019.12.003.

Priya, N., K. Kaur and A. K. Sidhu. 2021. Green synthesis: An ecofriendly route for the synthesis of iron oxide nanoparticles. *Frontiers in Nanotechnoogy*, 3, p. 655062. doi:10.3389/fnano.2021.655062.

Punnoose, M. S. and B. Mathew. 2018. Treatment of water effluents using silver nanoparticles. *Material Science and Engineering*, 2(5), pp. 159–166. doi:10.15406/mseij.2018.02.00050.

Qu, X., Alvarez, P. J. J., and Q. Li. 2013. Applications of nanotechnology in water and wastewater treatment. *Water Research*, 47, pp. 3931–3946. doi:10.1016/j.watres.2012.09.058.

Sacco, O., M. Stoller, V. Vaiano, P. Ciambelli, A. Chianese and D. Sannino. 2012. Photocatalytic degradation of organic dyes under visible light on N-doped photocatalysts. *International Journal of Photoenergy*, 626759, pp. 1–8. doi:10.1155/2012/626759.

Saif, S., A. Tahir and Y. Chen. 2016. Green synthesis of iron nanoparticles and their environmental applications and implications. *Nanomaterials*, 6(11), p. 209. doi:10.3390/nano6110209.

Sandeep Kumar, V. 2013. Magnetic nanoparticles-based biomedical and bioanalytical applications. *Jounal of Nanomedicine and Nanotechology*, 4, e130. doi:10.4172/2157-7439.1000e130.

Serra, A., S. Grau, C. Gimbert-Suriñach, J. Sort, J. Nogués and E. Valles. 2017. Magnetically-actuated mesoporous nanowires for enhanced heterogeneous catalysis. *Applied Catalysis B: Environmental*, 217, pp. 81–91. doi:10.1016/j.apcatb.2017.05.071.

Shamaila, S., A. K. L. Sajjad, S. A. Farooqi, N. Jabeen, S. Majeed and I. Farooq. 2016. Advancements in nanoparticle fabrication by hazard free eco-friendly green routes. *Appllied Materials Today*, 5, pp. 150–199. doi:10.1016/j.apmt.2016.09.009.

Shao, F., C. Xu, W. Ji, H. Dong, Q. Sun, L. Yu, and L. Dong. 2017. Layer-by-layer self-assembly TiO_2 and graphene oxide on polyamide reverse osmosis membranes with improved membrane durability. *Desalination*, 423, pp. 21–29. doi:10.1016/j.desal.2017.09.007.

Sharma, M., P. Kalita, K. K. Senapati and A. Garg. 2018. Study on magnetic materials for removal of water pollutants. In *Emerging Pollutants - Some Strategies for the Quality Preservation of Our Environment*, edited by Sonia Soloneski and Marcelo L. Larramendy, pp. 61–78. IntechOpen, London, UK. doi:10.5772/intechopen.75700.

Sivashankar, R., A. B. Sathya, K. Vasantharaj, and V. Sivasubramanian. 2014. Magnetic composite an environmental super adsorbent for dye sequestration - A review. *Environmental Nanotechnology, Monitoring and Management*, 1, pp. 36–49. doi:10.1016/j.enmm.2014.06.001.

Sondi, I. and B. Salopek-Sondi. 2004. Silver nanoparticles as antimicrobial agent: A case study on *E. coli* as a model for gram-negative bacteria. *Journal of Colloid and Interface Science*, 275, pp. 177–182. doi:10.1016/j.jcis.2004.02.012.

Stojanovic, Z., M. Otonicar, J. Lee, M. M. Stevanovic, M. P. Hwang, K. H. Lee, J. Choi, D. Uskokovic. 2013. The solvothermal synthesis of magnetic iron oxide nanocrystals and the preparation of hybrid poly (1-lactide)-polyethyleneimine magnetic particles. *Colloids and Surfaces B: Biointerfaces*, 109, pp. 236–243. doi:10.1016/j.colsurfb.2013.03.053.

Surwade, S., Smirnov, S., Vlassiouk, I. R. Unocic, G. Veith, S. Dai and S. Mahurin. 2015. Water desalination using nanoporous single-layer graphene. *Nature Nanotechnology*, 10, pp. 459–464. doi:10.1038/NNANO.2015.37.

Tan, M., G. Qiu, and Y. P. Ting. 2015. Effects of ZnO nanoparticles on wastewater treatment and their removal behavior in a membrane Bioreactor. *Bioresource Technology*, 185, pp. 125–133. doi:10.1016/j.biortech.2015.02.09.

Weng, X., L. Huang, Z. Chen, M. Megharaj, and R. Naidu. 2013. Synthesis of iron-based nanoparticles by green tea extract and their degradation of malachite. *Industrial Crops and Products*, 51, pp. 342–347. doi:10.1016/j.indcrop.2013.09.024.

Wu, W., Q. He, and C. Jiang. 2008. Magnetic iron oxide nanoparticles: Synthesis and surface functionalization strategies. *Nanoscale Research Letters*, 3, p. 397. doi:10.1007/s11671-008-9174-9.

Xiao, L., L. Liu, and J. Ge. 2021. Dynamic game in agriculture and industry cross sectoral water pollution governance in developing countries. *Agriculture Water Management*, 243, p. 10641. doi:10.1016/j.agwat.2020.106417.

Zahid, M., A. Rashid, S. Akram, Z. A.Rehan, and W. Razzaq. 2018. A comprehensive review on polymeric nano-composite membranes for water treatment. *Journal of Membrane Science & Technology*, 8(1). doi:10.4172/2155-9589.1000179.

Zhang, D., X-l. Ma, Y, Gu, H. Huang, and G-w. Zhang. 2020. Green synthesis of metallic nanoparticles and their potential applications to treat cancer. *Frontiers in Chemistry*, 8, p. 799. doi:10.3389/fchem.2020.00799.

Zhang, X., J. Wang, R. Li, Q. Dai, R. Gao, Q. Liu, and M. Zhang. 2013. Preparation of Fe_3O_4@C@ layered double hydroxide composite for magnetic separation of uranium. *Industrial and Engineering Chemistry Research*, 52(30), pp. 10152–10159. doi:10.1021/ie3024438.

Zhu, M., Y. Wang, D. Meng, X. Qin, and G. Diao. 2012. Hydrothermal synthesis of hematite nanoparticles and their electrochemical properties. *Journal of Physical Chemistry C*, 116, pp. 16276–16285. doi:10.1021/jp304041m.

10 Magnetic Nanomaterials for Solar Energy Conversion Applications

R. Dhivya
Sri Sarada Niketan College of Science for Women

N. Karthikeyan
Ramco Institute of Technology

S. Kulandai Tererse
Nirmala College for Women

R. Thenmozhi
Sakthi College of Arts and Science for Women

A. Judith Jayarani
Bishop Heber College (Autonomous)

M. Swetha
MVJ College of Engineering

F. Caballero-Briones
Instituto Politecnico Nacional, CICATA Altamira

G. Sahaya Dennish Babu
Chettinad College of Engineering and Technology

CONTENTS

DOI: 10.1201/9781003335580-10

10.1   FUNCTIONAL NANOMATERIALS AND THEIR IMPORTANCE

People have been interested in finding a suitable advanced nanomaterial for upcoming energy storage and conservation applications for the past two decades. In this regard, the creation of nanomaterials or nanostructures has been recognized as crucial for future generations' efforts [1,2]. Functional nanomaterials have the potential to be used in numerous technical applications, such as solar cells, batteries, catalysis, drug delivery, and wastewater treatment, due to their exceptional physical, chemical, electrical, and optical capabilities. Therefore, research into novel, highly functional nanomaterials is crucial for applications involving renewable energy in the future [3].

Advanced nanotechnology materials with potential features at the nanoscale are called functional nanomaterials. It can also be described as "a chemical that is well controlled and ordered by the molecular body that is available at the nanometer level size" [2,3]. It has undergone postprocessing, both chemically and physically, to acquire certain attributes that the original material lacked, such as absorbance, fluorescence, strength, ductility, solubility, and wettability. They can be employed extensively in the medical field to penetrate biological systems for purposes such as drug administration and cancer treatment because of their extreme nano size [4]. These substances are highly reactive in catalytic and adsorption applications, which aid in separating molecules with greater sensitivity for improved catalytic effectiveness. So, by tuning their structural and chemical properties, they can be transformed into new multifunctional nanomaterials which are essential for energy device applications.

Functional nanomaterials and devices hold great promise for use in various industries, including catalysis, waveguide, chemical and biosensors, solar cells, nanoscale electronics, light-emitting nanodevices, laser technology, and waveguide. The scientific community has identified a large number of functional materials that are frequently employed to create incredibly innovative and cutting-edge technologies for a variety of uses [5,6]. Particularly praiseworthy is the use of functional nanomaterials in the fields of medicine, optoelectronics, and catalysis. Functional nanomaterials have a crucial role in all cutting-edge scientific and technological sectors, as illustrated in Figure 10.1.

At the moment, nanoparticles with very tiny sizes are highly helpful in the medical disciplines for a variety of applications. Additionally, these nanomaterials have expanded their range of uses to include the development of molecular machinery and pharmaceuticals. Nanomaterials are widely used in the semiconductor and electrical industries in addition to the medical sector. The better electrical and optical properties of nanoparticles and quantum dots, all of which are smaller than 10nm, make them more appropriate for use in the creation of high-resolution electronic devices

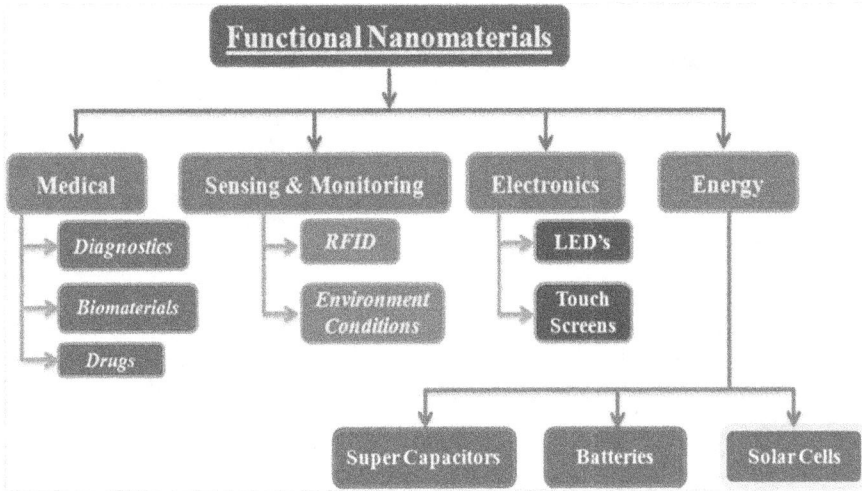

FIGURE 10.1 Applications of functional materials in various fields.

[6]. In addition to electrical devices, nanomaterials are frequently utilized in the creation of extremely sensitive sensors, RFID sensors, light-emitting diodes (LEDs), and touch displays in contemporary smartphones. Advanced functional nanomaterials technology has effectively brought about this transition from large-scale to nanoscale devices.

10.2 FUNCTIONAL NANOMATERIALS IN SOLAR CELLS

Due to their exceptional optical and electrical capabilities, these functional nanomaterials are increasingly important in solar cells and other renewable energy technology. The significant functional nanomaterials currently employed in solar cells are depicted in Figure 10.2. Additionally, the following are some advantages of these useful nanomaterials in the field of solar cells:

- Mainly, it helps in reducing the manufacturing cost and facilitates to get high efficiency in new-generation solar cells [7].
- It widely imparts to modify the traditional and conventional wafer solar cells to semiconductor solar cells with low-cost fabrication and installation [8].
- Solar cells can be fabricated even at room temperature instead of making at high vacuum deposition processes.
- High conversion efficiency can be achieved by using quantum dots and nanocomposites to gather more sunlight over a wide spectrum [9].
- Plasmonic nanomaterials also offer operative ways to diminish the thickness of absorber coatings [10].
- Spectral tuning in quantum-confined materials, sensitizing dyes, and polymers provides a superior effect on optical band gap, which helps to produce photon-generated charge carriers [11].

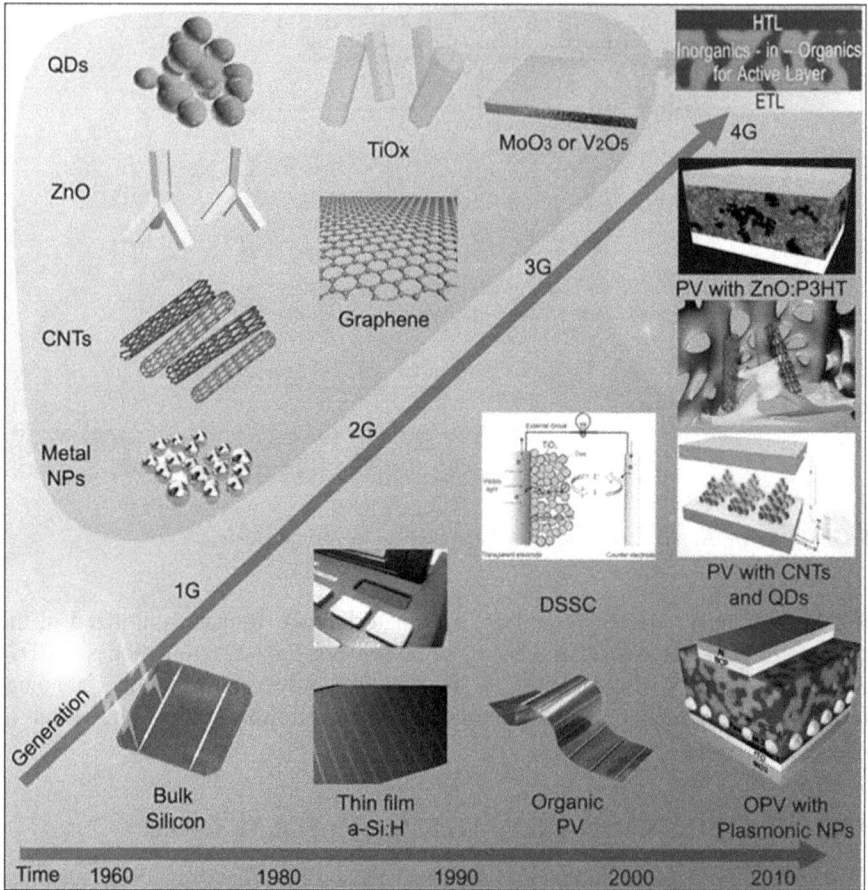

FIGURE 10.2 Various functional nanomaterials used in solar cells. *Reprinted from Reference 11 with permission.*

Considering the energy demand of the future world, these functional nanomaterials may provide new nanotechnological mechanisms and devices to get more efficient utilization of sunlight energy. Thus, these functional semiconducting nanomaterials have the potential to produce more and more new next-generation nanodevices for a variety of applications in the near future.

10.3 SOLAR CELLS

The solar cell is a device that converts sunlight into electricity. Solar energy is only the answer to future energy crisis problems. This can be understood from the quote "the one and only solution for the future energy crisis daily appears at east in the morning and disappears at west in evening." Solar energy is abundant in the world. The simplest way to convert solar energy into useful electricity is achieved by the use

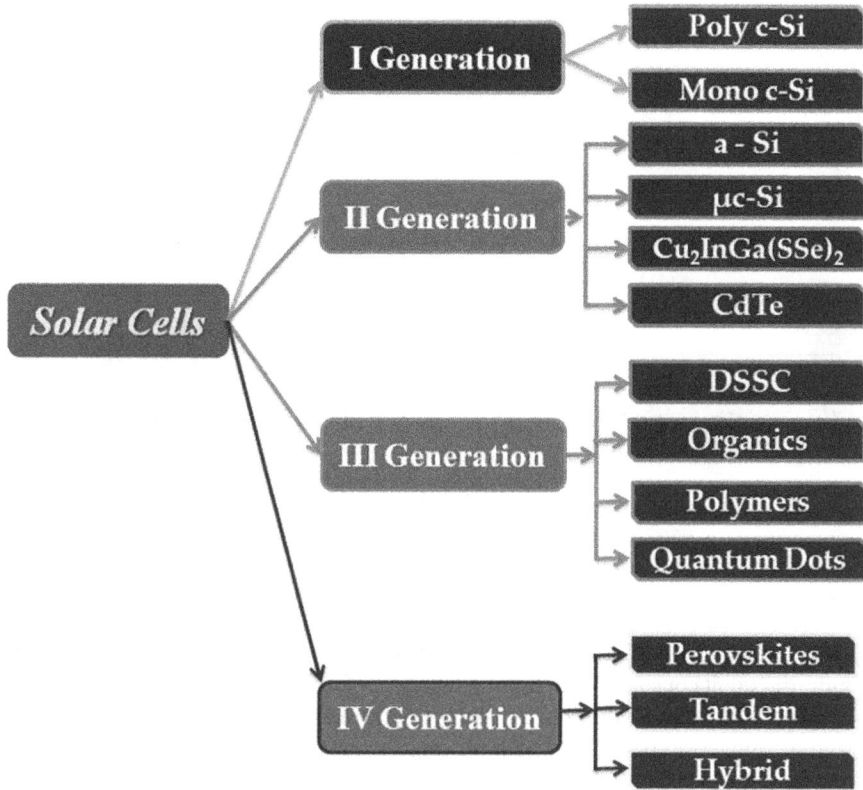

FIGURE 10.3   Types of solar cells and their materials.

of photovoltaic solar cell technology [12,13]. It is a clean, renewable energy technology with carbon-free and environmentally clean. Among the other renewable energy technologies, solar cell technology has the ability to resolve the future terawatt energy demand of the world. Solar cells can be classified into four main categories based on the materials which are used for the absorption of sunlight and their typical device architecture [14]. Types of solar cells and the typical materials used were mentioned in Figure 10.3.

10.3.1   Silicon Solar Cells

Solar cells are mainly made of silicon (Si) materials that are mono- and poly-types of silicon wafers, which are made from large-scale Si single crystals. The silicon solar cells are fabricated by producing single crystals of silicon into silicon wafers, which act light harvesting part of the solar cell. However, these Si solar cells are classified into three categories as wafer-based crystalline silicon, gallium arsenide (GaAs), and multi-junction solar cells. Si solar cell technology is a well-established one that holds a maximum of 27.6% power conversion efficiency (PCE) in multi-junction

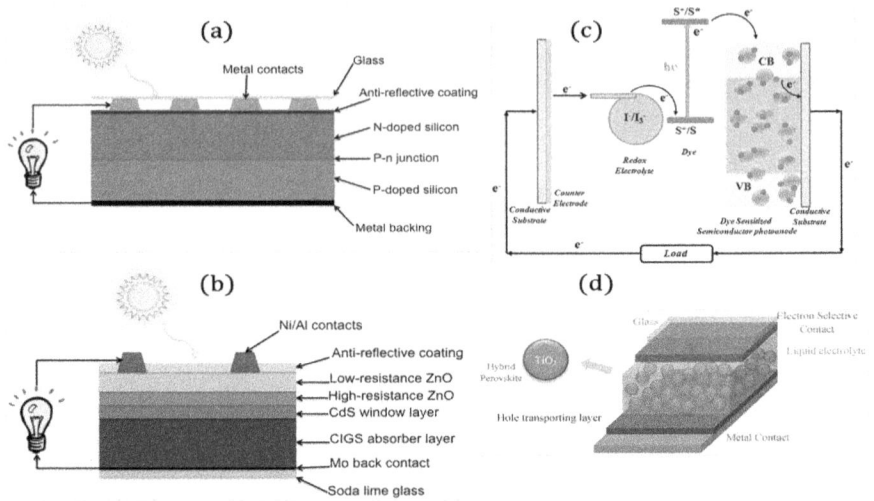

FIGURE 10.4 Schematic structures of (a) Si, (b) CIGS, (c) DSSC, and (d) perovskite solar cells. *Reprinted from References 42 with permission.*

Si solar cells [15]. A typical schematic structure of Si solar cells is represented in Figure 10.4(a). It consists of an aluminum backing for energy transfer, a p-type Si and an n-type Si for electron production and recombination, and an antireflective coating (preferably SiN_x or TiO_2), which prevents air moisture and is followed by metal contacts. However, due to its indirect band gap value of Si materials (1.1 eV) at room temperature leads to fabricating a high-thickness absorber layer in the order of millimeter size [16]. Si solar cells have a very thick absorber, high manufacturing costs, expensive precursors, a high vacuum manufacturing method, and the hard and brittle character of Si materials, which has encouraged researchers to develop a substitute material for it.

10.3.2 THIN FILM SOLAR CELLS

The second-generation solar cells are mainly divided into two categories such as copper indium gallium sulfide/selenide (CIGSSe) and cadmium telluride (CdTe) solar cells, where these two p-type semiconductor materials are used as an absorber material [17]. From Figure 10.4(b), a simple difference between the Si solar cells and CIGS solar cells is that the semiconductor material used in the solar cell has a direct optical band gap instead of an indirect band gap Si and therefore it relies on p–n junction design. CdTe, $Cu_2InGa(SSe)_4$, and $CuIn(SSe)_2$ are the most common absorber materials, which were used in the thin film solar cells [18]. Here, Al-ZnO and MgF_2 are two common anti-reflection coating materials used in thin film solar cells. Ni/Al contacts were added to the structure for electrical energy flow. These solar cells have reached a high PCE of 23.6%, which is very close to crystalline Si solar cells. It also has some significant drawbacks like high-cost and less-abundant precursor materials, and most important is the presence of environmentally toxic materials [19].

Recently, researchers have drawn their interest to identifying and synthesizing new functional nanomaterials to replace these toxic and high-cost materials deprived of negotiating on the power conversion efficiency.

10.3.3 THIRD-GENERATION SOLAR CELLS

Dye-sensitized solar cells (DSSCs) and polymer and organic solar cells are types of third-generation solar cells, which have several advantages such as flexibility, low-cost processability, and ease of fabrication at room temperature. Among those advantages, the flexibility gives an additional feature to the polymer solar cells, which allows the solar cells to be incorporated into all kinds of applications, most notably where flexibility is needed. Compared with previously discussed, these solar cells can be made by simple solution processing method at room temperature environment [20,21]. This feature allows for scaling up these solar cells in a large manner and thus reducing the production cost. DSSCs are not considered as organic solar cells because they use organometallic dyes and inorganic semiconducting material as photoanode. A typical schematic of DSSC is shown in Figure 10.3(c). The only disadvantage of these materials is their long-term stability [22]. Researchers are working to solve this stability problem and make this material a potential one.

10.3.4 FOURTH-GENERATION SOLAR CELLS

Emerging solar cells comprise a more sophisticated form of solar cell that frequently employ massive cost-conducting polymers behaving both n-type semiconductors with organic-inorganic hybrid materials working for p-type semiconductors. The standard structure of a perovskite solar cell is shown in Figure 10.4(d). Perovskites including those composed of methylammonium lead halide ($CH_3NH_3PbX_3$) were commonly implemented active light harvesters, whereas various polymeric polymers are being used as materials that transport holes within perovskite solar cells. [23]. Due to their high efficiency attained in a relatively short amount of time as well as their superior material characteristics, which enable to absorb a greater number of light photons from sunlight, these solar cells have recently received a great deal of attention from researchers. The main disadvantages of these materials, however, were lead toxicity, reproducibility, and stability [24,25]. These limitations are now being overcome, and these materials may conceivably be used for solar cell applications in the future.

10.4 MAGNETIC NANOMATERIALS

Iron, nickel, cobalt, chromium, manganese, gadolinium, and other magnetic elements and their chemical compounds make up magnetic nanoparticles, which are nanomaterials [26]. Due to their nanoscale size, magnetic nanoparticles are superparamagnetic and have significant promise in a wide range of applications, whether they are uncoated or coated with functional groups chosen for particular applications. The most researched magnetic nanoparticles, in particular ferrite nanoparticles, can be significantly boosted by grouping many individual superparamagnetic nanoparticles into clusters to form magnetic beads [27]. An electromagnet or permanent magnet can generate an external magnetic field that can transport magnetic nanoparticles to a specific location after already being selectively linked to a functional molecule [28]. A surface layer may be necessary to stop aggregation and reduce the contact of the particles with the system environment. To boost their stability in solution, spinel ferrites are frequently surface-modified with surfactants, silica, silicones, or phosphoric acid derivatives.

Due to their potential use in sophisticated optoelectronic devices and nano-medicine, functional magnetic nanomaterials, whose properties are fundamentally different from those of their bulk counterparts, have garnered interest on a global scale. Materials that have at least one three-dimensional dimension that is either in the nanometer size range (1–100 nm) or is made up entirely of them are referred to as nanomaterials. Solar energy conversion equipment primarily uses inorganic magnetic nanoparticles like Fe_3O_4. Magnetite (Fe_3O_4), maghemite (-Fe_2O_3), and hematite (-Fe_2O_3) are the three types of iron oxides that occur most frequently in nature [28,29]. Currently, the primary techniques used to prepare magnetic nanopar-ticles include coprecipitation, high-temperature decomposition, microemulsion, and reverse micelle. Magnetic nanoparticles are easily modified, have a large specific surface area, and require little preparation. However, when solely Fe_3O_4 and -Fe_2O_3 are utilized as adsorbents, particles have a tendency to clump together, which reduces adsorption selectivity and the enrichment effect. We must conduct more studies in the future to develop a more straightforward and workable preparation method that will swiftly produce magnetic nanoparticles with a narrower particle size dispersion, high purity, and good magnetic fineness [29]. In the meantime, new modification techniques must be developed to cover the magnetic particles, increase chemical stability, stop oxidation, and give them a unique function.

10.5 PROPERTIES OF MAGNETIC NANOMATERIALS

Iron nanoparticles react violently with oxidizing substances, especially air. Each nanoparticle is covered with a thin coating to prevent it from oxidizing completely and permanently. For this purpose, silica and gold were being used for coatings; how-ever, these diminish the magnetic properties of the nanoparticles. By creating appro-priate techniques, significant advancements have been made in the two major factors that influence the significance of iron oxide magnetic nanoparticles, namely, control over the size and form. Additionally, a magnesium coating is being used, which has little effect on the magnetic properties of the iron particles [27]. Iron nanoparticles are mixed with smaller-than-microscopic magnesium particles in the substance that is created, making it complex. Iron carbide coating is the most practical way for creating virtually totally magnetic iron particles that are shielded from oxidation; nevertheless, the resulting particles are larger (20–100 nm), polydisperse, and fer-romagnetic, therefore, they are not so ideal [28]. Nevertheless, this is a significant advancement. Due to their magnetic characteristics and numerous uses, iron oxide nanoparticles are extremely important. Colloidal iron and iron oxide nanoparticles have significant size-dependent structural and optical features that are related to electrical structure and quantum size effects [29]. Their size and crystal structure might also vary depending on the synthesis process used.

10.6   MAGNETIC NANOMATERIALS FOR SOLAR ENERGY CONVERSION

10.6.1 Fe_3O_4 IN BULK HETEROJUNCTION (BHJ) SOLAR CELLS

Because of their advantages in regard to cost, flexibility, and light weight, polymer solar cells (PSCs) featuring bulk heterojunction (BHJ) have garnered a lot of interest. The photoactive layer within BHJ structure, which has been described as the most effective architecture of PSCs so far with, has been constituted of a conjugated polymer benefactor, including such poly(3-hexylthiophene-2,5-diyl) (P3HT), and a soluble fullerene acceptor, commonly [6,6]-phenyl-C61-butyric acid methyl ester (PCBM) [30,31]. While there has been notable development research on novel donor and acceptor materials and additives used in new types of BHJ-PSCs, attaining the PCE has remained an obstacle. The emphasis of most recent research on P3HT:PCBM BHJ-PSCs has focused on postthermal processing of photoactive layers or strengthening microphase separation structure by selecting appropriate solvents for the treatment of the photoactive layers [32].

Importantly, it has been demonstrated that the solvent that dissolves the P3HT:PCBM mix has a massive effect on PCE for P3HT:PCBM BHJ-PSCs. This is mostly because charge carrier mobility is reliant just on solvents used to process the photoactive layers and is sensitive to the nanoscale morphology of the thin film made up of the paired photoactive materials [33]. The open circuit voltage (V_{oc}), fill factor (FF), and short-circuit current (J_{sc}) of PCE are now all strongly linked with the difference between both the HOMO level of the donor and the LUMO level of the acceptor. The FF is formed by the charge carriers that emerge at the electrodes when the constructed field is reduced near the open circuit voltage. The J_{sc} density and the charge carrier mobility inside the photoactive material are determined by the amplification of the photo-generated charge carrier. It is generally known that in organic compounds like P3HT, the principal charge carriers generated by light are so-called excitons, which can exist in singlet states or triplet states with varying lifetimes. Dissociated excitons or electron-hole pairs bound via Coulomb were the only charge carriers that could contribute to the photocurrent [34]. For the efficiency of this procedure, the fact that the singlet lifespan in P3HT is approximately 300 ps and that the succeeding singlet exciton diffusion length is reported to be in the range of 3–6 nm, which is a significant bottleneck. The triplet exciton diffusion length is in the region of 100 nm, in contrast, with both the P3HT triplet lifespan being within the range of 10 ms. This enhances the photovoltaic process in organic solar cells by enabling the triplet exciton's diffusion process toward donor-acceptor junctions smoother [35]. The increased triplet concentration results in a higher probability of interfacial exciton dissociation and charge separation. Although triplet excitons have a longer lifespan and, as a result, longer diffusion than singlet excitons, they have been utilized as a remedy for BHJ cells [36].

Fe_3O_4 magnetic nanoparticles are doped into P3HT: PCBM BHJ PSC by Zhang et al. They discover that PCE is being enhanced [33]. The factor (V_{oc}, J_{sc}, FF) most responsible for the improved PCE is identified, and the impact of Fe_3O_4 magnetic

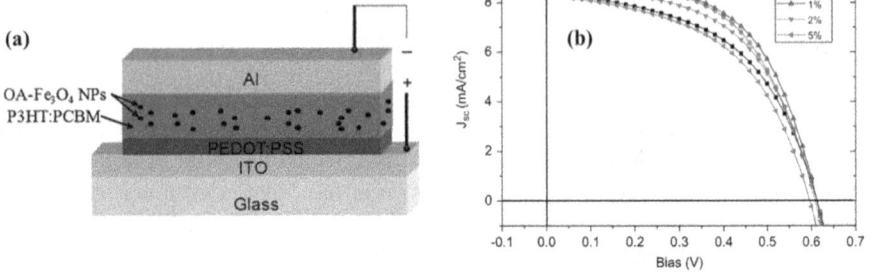

FIGURE 10.5 (a) Schematic structure of OA-Fe$_3$O$_4$ in PSCs device and (b) J–V curves of with and without OA-Fe$_3$O$_4$ PSCs. *Reprinted from Reference 33 with permission.*

nanoparticles is considered. When exposed to simulated AM 1.5 radiation, PCE of the device was measured in an air atmosphere (Figure 10.5).

The current–voltage (J–V) curves were obtained at various nanoparticle doping ratios. The P3HT:PCBM gadget in its purest form displays a PCE of 2.62% utilizing chloroform as the solvent. This is mostly due to the fact that fabrication processes are carried out in an air atmosphere in vacuum level (10^{-5} torr) is insufficient for the top Al electrode's deposition. Each PCE-determining parameter, including J_{sc}, V_{oc}, and FF, is compared in order to identify the factor responsible for the PCE's improvement with the doping of OA-Fe$_3$O$_4$ nanoparticles from 0.5% to 2%. In contrast to the pure device, the V_{oc} of OA-Fe$_3$O$_4$ devices remains constant 0.60-0.61 V [33]. Nevertheless, the J_{sc} of OA-Fe$_3$O$_4$ devices increase significantly from 8.41 to 9.55 mA/cm^2.

They also found that the PCE of the device rose by 18% when Fe$_3$O$_4$ was subsequently inserted. This improvement is mostly owing to a 14% rise in J_{sc}, which is attributable to the magnetic field influence of the superparamagnetic Fe$_3$O$_4$ nanoparticles, which elevates the population of triplet excitons [33,34]. This investigation confirmed that the observed enhancement of PCE were caused solely either by magnetic property of Fe$_3$O$_4$ nanoparticles. It is also determined if Fe$_3$O$_4$ nanoparticles have an effect on the improvement of PCE of the device by contrasting various doping techniques. Additionally, they discovered that even in the air atmosphere, the magnetic field effect depends on by Fe$_3$O$_4$ nanoparticles is efficiently achieved, making efficiency improvement relatively simple [36,37]. Because applying an external magnetic field for large-area PSCs, which is often required for commercial applications, is unrealistic, our discovery provides a simple and practical method to achieve the magnetic field effect produced by magnetic nanoparticles in PSCs.

10.6.2 ROLE OF MAGNETIC NANOPARTICLES IN DSSC

Recently, numerous initiatives have been made to increase the DSSCs effectiveness by customizing its various components. The photoanode's morphological, structural, and chemical alterations have received the most attention among them. Regardless of the existence of numerous metal oxide semiconductors (MOS), titanium oxide (TiO$_2$)

has been widely used as a photoanode material in DSSCs due to the rapid injection rate of electrons from excited dye into the TiO_2 conduction band (CB). The PCE of a device is decreased by TiO_2's poor electron mobility, which leads in exceptionally high recombination reaction rates [38]. In DSSC and perovskites cells, the optoelectronic properties of various other oxides have also been thoroughly investigated [37–39].

The investigation from several potential materials to make cheap counter electrodes is the other component in the efficient production of DSSC. Because CE is typically far more expensive than a photoanode due to its fabrication from precious metals like platinum (Pt), it incurs a higher capital cost [39]. An ideal CE should have the following characteristics: relatively high adhesion, moderate cost, increased catalytic activity, substantial reflectivity, large surface area, porosity, stability, and energy level that is properly matched with electrolyte potential. 2D graphene has lately received a lot of attention because of its numerous beneficial characteristics. Reduced graphene oxide or rGO has excellent dispersibility and solution processability. Since magnetite (Fe_3O_4) nanoparticles have a vast surface area and good dispersion, they have been widely used in many fields. Compared to other iron oxides, Fe_3O_4 nanoparticles display higher optoelectronic properties and improved optical absorption in solar cells. To show an enhancement in PCE, Zhou et al. coated Fe_3O_4 NPs on the shallow of rGO [40]. Therefore, graphene oxide and metal oxides can be used together as CE in DSSC to replace limited Pt and provide better optoelectronic properties. Akbhar Ali Qureshi et al. made the two functioning electrodes of a DSSC to investigate the effects of change on photovoltaic properties [38]. A simple solvothermal method was used to create hybrid Fe_3O_4@rGO CE and SnO_2-TiO_2 nanocomposite-based photoanodes.

Figure 10.6 represents the Fe_3O_4@rGO DSSCs cyclic voltammetry (CV) curves at a scan rate of 100 mV. (a) Based on the two pairs of oxidation/reduction peaks on the CV voltammogram, the hybrid Fe3O4@rGO nanocomposite has a respectable electrochemical capacity for iodide/tri-iodide redox processes. A reasonable catalytic activity for tri-iodide reduction too is revealed either by hybrid Fe3O4@rGO nanocomposite's narrow Epp value. Based on the CV results, a counter electrode composed of a Fe3O4@rGO nanocomposite is used, which significantly increased the surface area of the FTO substrate, optimizing the electrochemical attributes of the DSSC [38].

The electrochemical impedance spectroscopy (EIS) spectra are shown in Figure 10.6(c) for various CEs. To compare the Fe_3O_4@rGO CE's charge transfer properties to those of traditional (Pt) CE in the DSSC, the high-frequency semicircular area in the two semicircles can be studied to assess charge transfer resistance (R_{ct}) in order to understand the enhancement in catalytic performance due to the large number of active sites present. The hybrid counter electrode has multiple active sites for tri-iodide reduction for enhanced catalytic performance. It is understood from this study that R_{ct} of the device is greater than Pt and it is depicted in Figure 10.6(d). The slant was used to measure the exchange current density (J_{sc}) [38]. The findings show that hybrid CE, when compared to platinum, has a good diffusion velocity for the reduction of tri-iodide, which is due to the wrapping of Fe_3O_4@rGO nanosheets.

From Figure 10.7(a), it is understood that PCE of 3.28% with standard Pt CE was shown by the dye-sensitized solar cells which has SnO_2-TiO_2 nanocomposite-based

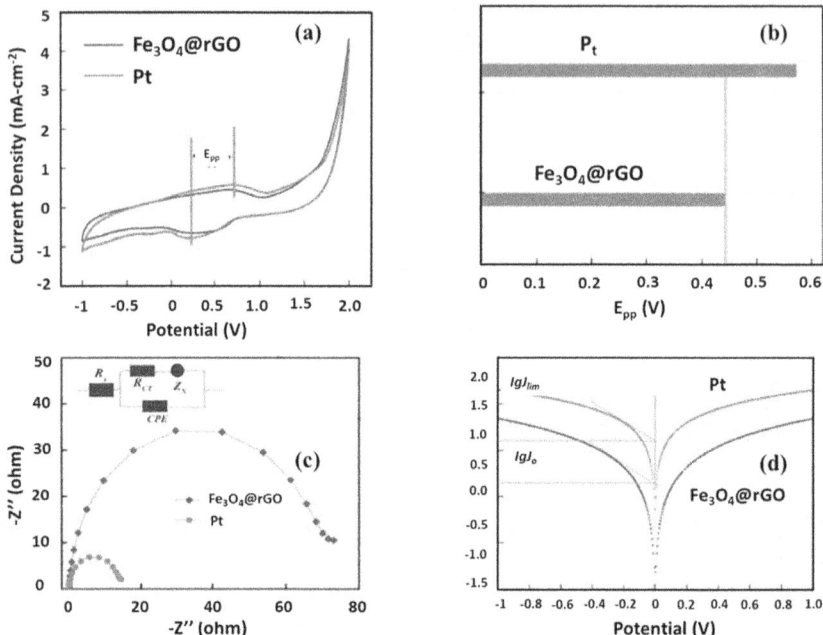

FIGURE 10.6 (a)–(d) Electrochemical analysis of Fe_3O_4@rGO CE based DSSC. *Reprinted from Reference 38 with permission.*

photoanode. In order to improve the photovoltaic characteristics of DSSC, multiple benefits, particularly band gap reduction and enhanced light scattering effects, played a key role. By giving tri-iodide additional oxidation sites, Fe_3O_4 nanoparticles with superior optoelectronic properties and rGO with higher electronic conductivity significantly improved DSSC performance [37,38]. It is conceivable that the making of SnO_2-TiO_2 nanostructure through Fe_3O_4@rGO-based CE in DSSC is what produced the noteworthy upsurge in the V_{OC} rate. Since TiO_2 has an isoelectric level that is significantly greater than that of SnO_2, it pushes and increases V_{oc}. As a result, the DSSCs hybrid Fe_3O_4@rGO CE and SnO_2-TiO_2 photoanode displayed a decent PCE that was relatively equivalent to Pt. As a result, conventional pricey Pt CE can be replaced with low-cost hybrids.

The best spectrum response when compared to Pt is depicted in Figure 10.7(b) [38]. The DSSCs exhibit dominance values of J_{sc} because of the interaction between the altered functional electrodes, which increased charge transfer and enabled for maximal dye loading [39,40]. Additionally, they agreed that Pt-free DSSCs may well be produced that used a hybrid Fe_3O_4@rGO CE and a photoanode by integrating those elements in a cooperative fashion.

10.6.3 Fe_2O_3 Nanoparticles in Perovskite Solar Cells

The perovskite solar cells (PerSCs) considered as one of the significant solar cell technology in recent year. The PCE of PerSCs solar cell improved from 3.8% to more

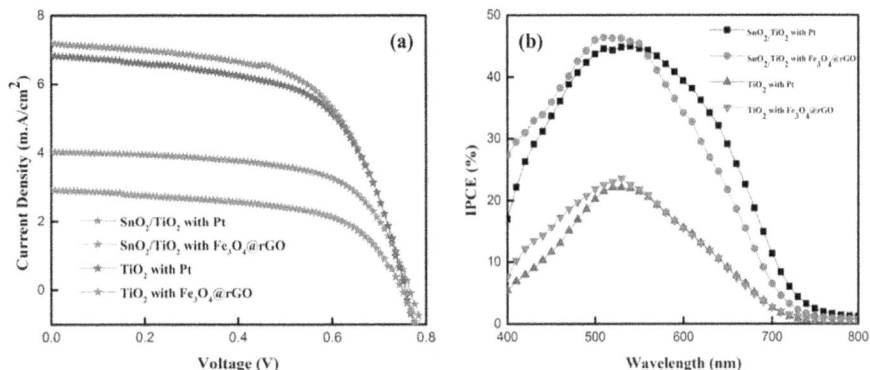

FIGURE 10.7   (a) PCE performance of DSSCs on various CE including Fe_3O_4 in rGO. *Reprinted from Reference 38 with permission.*

than 25% from around 10 years [41,42]. Perovskite solar cells have a high PCE, but their stability is still insufficient for practical applications. The primary cause is water and oxygen in the air corroding solar cells, which can be prevented by adequately packing the solar cells or by preparing the solar cells' outer surfaces to repel water [43]. To achieve stable device functioning in this scenario, we can substitute less thermally stable perovskite materials with better thermally stable perovskite materials or 2D perovskite materials [44]. Because the layer deteriorates due to the UV portion of sunlight reacting with ETLs like TiO_2, research into the use of an ultraviolet protective coating to increase stability is significantly required.

Visible light is absorbed by iron oxide (Fe_2O_3), an n-type semiconductor with an appropriate energy band location, excellent chemical stability, and a reasonable cost [43]. In perovskite solar cells, Fe_2O_3 has been recurrently labored as stable layer. Wang et al. created a thin Fe_2O_3 coating that remained stable over the course of 30 days when exposed to outside air. Fe_2O_3-based devices perform well in terms of stability, although their PCE is still modest [44]. In order to efficiently increase conductivity in perovskite solar cells, Guo et al. used a Ni-Fe_2O_3 thin film as charge transport layers. Compared to devices without Fe_2O_3 doped devices, the PCE of devices influences 14.2%, an increase of 150% [45]. High-quality Fe_2O_3 thin films can create stable perovskite solar cells with high PCE, although doing so is challenging due to Fe_2O_3's poor electrical conductivity and crystallinity.

Figure 10.8 displays the schematics of several Fe_2O_3 films produced employing nanoparticles and $FeCl_3$ solution (a). Due to the exceptional hydrophilicity of water-dispersed nanoparticles, the films may self-assemble on the ITO surface, and high-quality iron oxide films may be produced as a result of the tiny crystal structure change that occurs during the annealing process. The spin-coated $FeCl_3$ film for films made with $FeCl_3$ solution has an excellent density, but during the postannealing process, the iron oxide crystallization process will result in the production of numerous holes in the initially dense film [42]. Figure 10.8 demonstrates the architecture for the charge transport mechanism of several Fe_2O_3 layers (b). Fe_3O_4 nanoparticles were employed to form a thick Fe_2O_3 layer that delivers improved carrier separation.

(a)

Fe₃O₄ Nanoparticles → Compact Fe₂O₃

ITO

Spin-coated → Annealed

FeCl₃ Film → Network Fe₂O₃

(b)

← Perovskite →
← Fe₂O₃ →
← ITO →

10 nm Fe₃O₄ Sample   FeCl₃ Sample

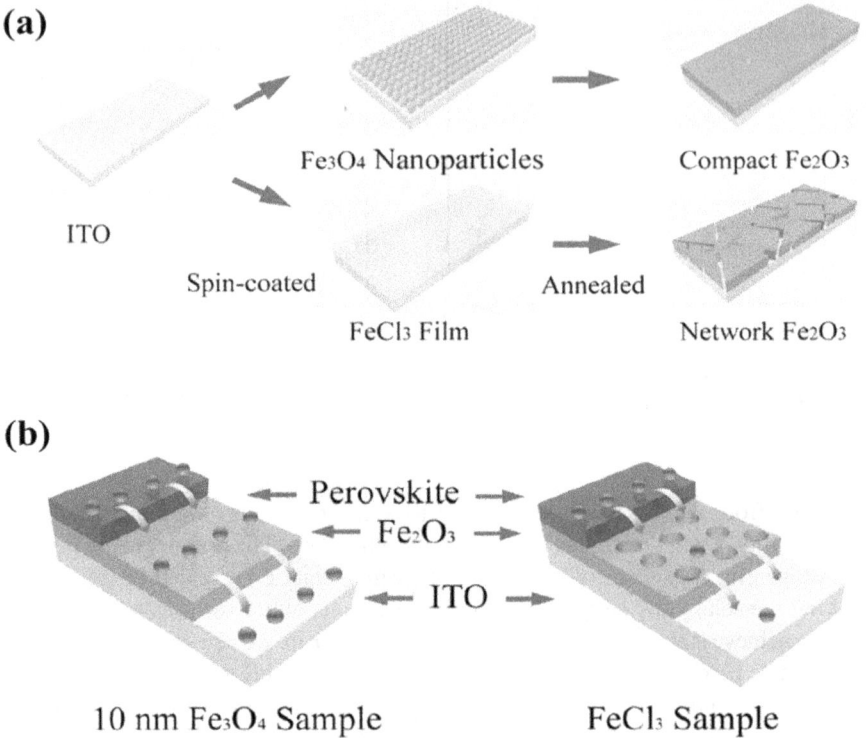

FIGURE 10.8 (a) Schematic of Fe₂O₃ films and (b) Schematic of the charge transport model of Fe₂O₃ devices. *Reprinted from Reference 42 with permission.*

In contrary, the amount of holes in $FeCl_3$-produced Fe_2O_3 layers can enable the perovskite layer to be in direct contact with the ITO layer, making it simple for carriers to recombine during the transmission process.

As shown in Figure 10.9(a), the sample made with $FeCl_3$ has the lowest PV parameters. In this research, they have identified that the Fe_2O_3 layer is not fulfill the role of charge transport layers, which is notified from lowering the V_{oc}. Different V_{oc} may also result from probable differences in the multidimensional Fermi levels of numerous charge layers [42–45]. They stated that the preferable size of the nanoparticles was between 10 and 15 nm. Electrochemical impedance spectroscopy (EIS) Nyquist plots of a number of ETLs are given in Figure 10.9(b), where the tested results are represented by symbol curves and the fitted results are presented with solid line curves [42,45]. They assume that now the semicircle at high frequency depicts the gets transferred (R_{ct}) at the transport layer/perovskite interface while the transmission line at low frequency represents its recombination impedance (R_{rec}) of solar cells.

Figure 10.9(c) demonstrates the J–V curves, which are measurements of the leakage performance of several Fe_2O_3 layers in the dark model (c). Since the higher voltage corresponds to the nanoparticle samples' lowest point on the leaked graph, this indicates that they have higher leakage properties than the $FeCl_3$ sample. Additionally,

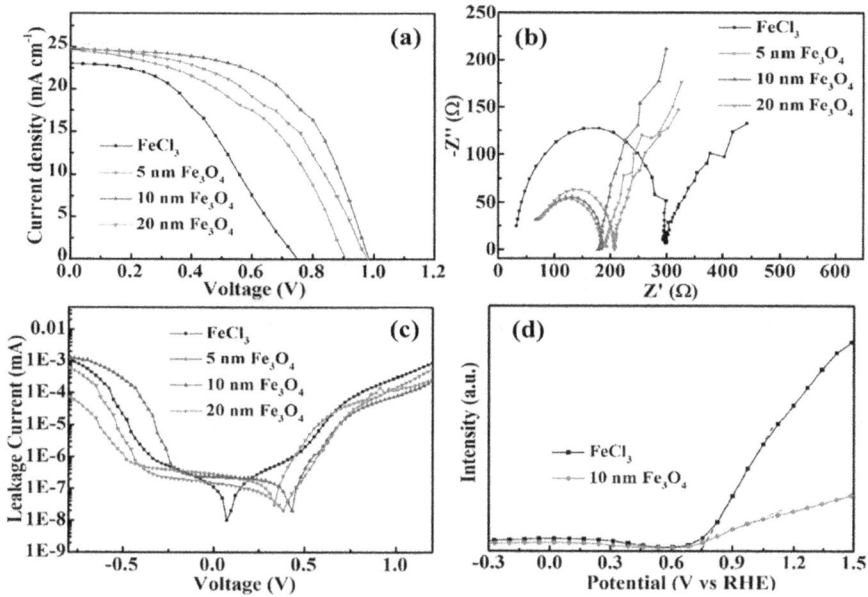

FIGURE 10.9 EIS analysis of perovskite solar cells with diverse Fe_2O_3. *Reprinted from Reference 42 with permission.*

the sample containing high-potential 10 nm nanoparticles exhibits a lower leakage value than the others, suggesting improved anti-leakage capabilities. Figure 10.9(d) illustrates the band location of the Fe_2O_3 layer [46,47]. The test material was an n-type semiconductor since the slope of the curve has an undesirable charge. The probable worth of the produced $FeCl_3$ model is 0.76 V. A higher band position for both the nanoparticle sample than the $FeCl_3$ sample is demonstrated by the lower value, which is in accordance with the V_{oc} tendency as determined by the J–V measurement. The planar Fe_2O_3-based thin film solar cells demonstrated good stability at more than 95% efficiency after irradiation. They also advocated utilizing water-dispersed Fe_3O_4 nanoparticles. [48]. These cells had a 14.3% efficiency achieved by ITO/Fe_2O_3/$(FAPbI_3)0.97(MAPbBr_3)0.03$/Spiro-OMeTAD/Au combination. The superior performance of the solar cells is attributed to the dense Fe2O3 layer's exceptional ultraviolet endurance and effective vacuum distillation of interface flaws [42]. This method has a lot of potential for producing perovskite solar cells that are affordable, UV stable, and effective.

10.7   CONCLUSION AND FUTURE OUTLOOK

Contrary to conventional semiconductor solar cells, ferroelectric solar cells have a developed electric field that extends throughout the bulk region due to the ferroelectric's residue left polarization and an accessible voltage (V_{oc}) that can be four or even more several orders of scale greater than the ferroelectric's bandgap energy. This allows for extremely high-power conversion efficiencies. One of the most significant tasks of the 21st century is to use cheap, effective solar cells to capture the sunlight that strikes the earth. Around the world, efforts are being undertaken to overcome this obstacle by developing solar cells built of nanostructured magnetic nanomaterials (such as nanoparticles and nanowires). Additionally, it is crucial to note that nanostructured materials like nanoparticles and nanowires have three distinctive benefits that can aid in converting solar energy into electricity. Because the two vital phases in converting solar energy into electricity, charge separation and light absorption are made possible by the huge surface and interfacial areas that new magnetic nanoparticles offer per unit volume. It is crucial to understand how charge carrier confinement in nanometer-sized particles enables materials' optical and electrical properties to be tuned in ways that are not feasible with bulk materials. Certain recently discovered quantum mechanical processes, which are seen solely in oxide nanoparticles, can be used to push theoretical efficiency boundaries. Last but not least, it required extensive research on magnetic nanostructured materials such as "magnetic nanoparticle inks," which may help lower the cost of producing solar cells by enabling the mass production of thin films at a low cost using well-established roll-to-roll coating or printing technologies.

REFERENCES

1. Yin, Y. and Talapin, D., 2013. The chemistry of functional nanomaterials. *Chemical Society Reviews, 42*(7), pp.2484–2487.
2. Sattler, K.D., 2010. *Handbook of nanophysics: functional nanomaterials.* CRC Press.
3. Wang, Z.M. ed., 2010. *Toward functional nanomaterials* (Vol. 5). Springer Science & Business Media.
4. Busseron, E., Ruff, Y., Moulin, E. and Giuseppone, N., 2013. Supramolecular self-assemblies as functional nanomaterials. *Nanoscale, 5*(16), pp.7098–7140.
5. Rahimi-Iman, A., 2020. Advances in functional nanomaterials science. *Annalen der Physik, 532*(9), p.2000015.
6. Wu, Y., Wang, D. and Li, Y., 2016. Understanding of the major reactions in solution synthesis of functional nanomaterials. *Science China Materials, 59*(11), pp.938–996.
7. Singh, V.N., 2020. A special section on functional nanomaterials for solar cells. *Journal of Nanoscience and Nanotechnology, 20*(6), pp.3620–3621.
8. Thomas, S., Kalarikkal, N., Oluwafemi, S.O. and Wu, J. eds., 2019. *Nanomaterials for solar cell applications.* Elsevier.
9. Bouziani, I., Essaoudi, I. and Ainane, A., 2022. Two-Dimensional Nanomaterials for Solar Cell Technology. In *Artificial Intelligence of Things for Smart Green Energy Management* (pp. 103–119). Springer, Cham.
10. Xiang, H., Wei, S.H. and Gong, X., 2009. Identifying optimal inorganic nanomaterials for hybrid solar cells. *The Journal of Physical Chemistry C, 113*(43), pp.18968–18972.
11. Yin, Z., Zhu, J., He, Q., Cao, X., Tan, C., Chen, H., Yan, Q. and Zhang, H., 2014. Graphene-based materials for solar cell applications. *Advanced Energy Materials, 4*(1), p.1300574.
12. Nelson, J.A., 2003. *The physics of solar cells.* World Scientific Publishing Company.
13. Fraas, L.M. and Partain, L.D., 2010. *Solar cells and their applications* (Vol. 217). Hoboken, NJ: Wiley.
14. Goetzberger, A., Luther, J. and Willeke, G., 2002. Solar cells: Past, present, future. *Solar Energy Materials and Solar Cells, 74*(1–4), pp.1–11.

15. Wenham, S.R. and Green, M.A., 1996. Silicon solar cells. *Progress in Photovoltaics: Research and Applications*, 4(1), pp.3–33.
16. Smith, D.D., Cousins, P., Westerberg, S., De Jesus-Tabajonda, R., Aniero, G. and Shen, Y.C., 2014. Toward the practical limits of silicon solar cells. *IEEE Journal of Photovoltaics*, 4(6), pp.1465–1469.
17. Chopra, K.L., Paulson, P.D. and Dutta, V., 2004. Thin-film solar cells: An overview. *Progress in Photovoltaics: Research and Applications*, 12(2–3), pp.69–92.
18. Poortmans, J. and Arkhipov, V. eds., 2006. Thin film solar cells: fabrication, characterization and applications (Vol. 18). John Wiley & Sons.
19. Fthenakis, V., 2009. Sustainability of photovoltaics: The case for thin-film solar cells. *Renewable and Sustainable Energy Reviews*, 13(9), pp.2746–2750.
20. Gong, J., Sumathy, K., Qiao, Q. and Zhou, Z., 2017. Review on dye-sensitized solar cells (DSSCs): Advanced techniques and research trends. *Renewable and Sustainable Energy Reviews*, 68, pp.234–246.
21. Gong, J., Liang, J. and Sumathy, K., 2012. Review on dye-sensitized solar cells (DSSCs): Fundamental concepts and novel materials. *Renewable and Sustainable Energy Reviews*, 16(8), pp.5848–5860.
22. Sharma, K., Sharma, V. and Sharma, S.S., 2018. Dye-sensitized solar cells: Fundamentals and current status. *Nanoscale Research Letters*, 13(1), pp.1–46.
23. Correa-Baena, J.P., Saliba, M., Buonassisi, T., Grätzel, M., Abate, A., Tress, W. and Hagfeldt, A., 2017. Promises and challenges of perovskite solar cells. *Science*, 358(6364), pp.739–744.
24. Jung, H.S. and Park, N.G., 2015. Perovskite solar cells: From materials to devices. *small*, 11(1), pp.10–25.
25. Green, M.A., Ho-Baillie, A. and Snaith, H.J., 2014. The emergence of perovskite solar cells. *Nature Photonics*, 8(7), pp.506–514.
26. Kumar, C.S., 2009. Magnetic nanomaterials. John Wiley & Sons.
27. Zhu, K., Ju, Y., Xu, J., Yang, Z., Gao, S. and Hou, Y., 2018. Magnetic nanomaterials: Chemical design, synthesis, and potential applications. *Accounts of Chemical Research*, 51(2), pp.404–413.
28. Hou, Y. and Sellmyer, D.J. eds., 2017. *Magnetic nanomaterials: Fundamentals, synthesis and applications*. John Wiley & Sons.
29. Cantor, B. ed., 2004. *Novel nanocrystalline alloys and magnetic nanomaterials*. CRC Press.
30. Mlinar, V., 2013. Engineered nanomaterials for solar energy conversion. *Nanotechnology*, 24(4), p.042001.
31. Kim, Y.S., Lee, Y., Kim, J.K., Seo, E.O., Lee, E.W., Lee, W., Han, S.H. and Lee, S.H., 2010. Effect of solvents on the performance and morphology of polymer photovoltaic devices. *Current Applied Physics*, 10(4), pp.985–989.
32. Scully, S.R. and McGehee, M.D., 2006. Effects of optical interference and energy transfer on exciton diffusion length measurements in organic semiconductors. *Journal of Applied Physics*, 100(3), p.034907.
33. Zhang, W., Xu, Y., Wang, H., Xu, C. and Yang, S., 2011. Fe_3O_4 nanoparticles induced magnetic field effect on efficiency enhancement of P3HT: PCBM bulk heterojunction polymer solar cells. *Solar Energy Materials and Solar Cells*, 95(10), pp.2880–2885.
34. Xu, J.K., Zhang, F.F., Sun, J.J., Sheng, J., Wang, F. and Sun, M., 2014. Bio and nanomaterials based on Fe3O4. *Molecules*, 19(12), pp.21506–21528.
35. Peng, S. and Sun, S., 2007. Synthesis and characterization of monodisperse hollow Fe3O4 nanoparticles. *Angewandte Chemie*, 119(22), pp.4233–4236.
36. Shakya, P., Desai, P., Kreouzis, T., Gillin, W.P., Tuladhar, S.M., Ballantyne, A.M. and Nelson, J., 2008. The effect of applied magnetic field on photocurrent generation in poly-3-hexylthiophene:[6, 6]-phenyl C61-butyric acid methyl ester photovoltaic devices. *Journal of Physics: Condensed Matter*, 20(45), p.452203.

37. Lei, Y., Song, Q., Zhang, Y., Chen, P., Liu, R., Zhang, Q. and Xiong, Z., 2009. Magnetoconductance of polymer–fullerene bulk heterojunction solar cells. *Organic Electronics*, *10*(7), pp.1288–1292.

38. Qureshi, A.A., Javed, S., Javed, H.M.A., Akram, A., Mustafa, M.S., Ali, U. and Nisar, M.Z., 2021. Facile formation of SnO_2–TiO_2 based photoanode and Fe_3O_4@ rGO based counter electrode for efficient dye-sensitized solar cells. *Materials Science in Semiconductor Processing*, *123*, p.105545.

39. Bagavathi, M., Ramar, A. and Saraswathi, R., 2016. Fe_3O_4–carbon black nanocomposite as a highly efficient counter electrode material for dye-sensitized solar cell. *Ceramics International*, *42*(11), pp.13190–13198.

40. Wang, W., Yao, J. and Li, G., 2018. Dual-functional Fe_3O_4@ N-rGO catalyst as counter electrode with high performance in dye-sensitized solar cells. *Journal of Electroanalytical Chemistry*, *823*, pp.261–268.

41. Yin, J., Zhou, H., Liu, Z., Nie, Z., Li, Y., Qi, X., Chen, B., Zhang, Y. and Zhang, X., 2016. Indium-and platinum-free counter electrode for green mesoscopic photovoltaics through graphene electrode and graphene composite catalysts: Interfacial compatibility. *ACS Applied Materials & Interfaces*, *8*(8), pp.5314–5319.

42. Fang, S., Chen, B., Gu, B., Meng, L., Lu, H. and Li, C.M., 2021. An ultrathin and compact electron transport layer made from novel water-dispersed Fe_3O_4 nanoparticles to accomplish UV-stable perovskite solar cells. *Materials Advances*, *2*(11), pp.3629–3636.

43. Zheng, S., Wang, G., Liu, T., Lou, L., Xiao, S. and Yang, S., 2019. Materials and structures for the electron transport layer of efficient and stable perovskite solar cells. *Science China Chemistry*, *62*(7), pp.800–809.

44. Hu, W., Liu, T., Yin, X., Liu, H., Zhao, X., Luo, S., Guo, Y., Yao, Z., Wang, J., Wang, N. and Lin, H., 2017. Hematite electron-transporting layers for environmentally stable planar perovskite solar cells with enhanced energy conversion and lower hysteresis. *Journal of Materials Chemistry A*, *5*(4), pp.1434–1441.

45. Guo, Y., Liu, T., Wang, N., Luo, Q., Lin, H., Li, J., Jiang, Q., Wu, L. and Guo, Z., 2017. Ni-doped α-Fe_2O_3 as electron transporting material for planar heterojunction perovskite solar cells with improved efficiency, reduced hysteresis and ultraviolet stability. *Nano Energy*, *38*, pp.193–200.

46. Papadas, I.T., Galatopoulos, F., Armatas, G.S., Tessler, N. and Choulis, S.A., 2019. Nanoparticulate metal oxide top electrode interface modification improves the thermal stability of inverted perovskite photovoltaics. *Nanomaterials*, *9*(11), p.1616.

47. Bouhjar, F., Mollar, M., Ullah, S., Mari, B. and Bessaïs, B., 2018. Influence of a compact α-Fe_2O_3 layer on the photovoltaic performance of perovskite-based solar cells. *Journal of The Electrochemical Society*, *165*(2), p.H30.

48. Raj, A., Kumar, M. and Anshul, A., 2021. Recent advancement in inorganic-organic electron transport layers in perovskite solar cell: Current status and future outlook. *Materials Today Chemistry*, *22*, p.100595.

11 Functionalized Magnetic Nanoparticles for Energy Storage Applications

Manas Mandal
Sree Chaitanya College

Krishna Chattopadhyay
University of Calcutta

CONTENTS

DOI: 10.1201/9781003335580-11

11.1 INTRODUCTION

The enormous energy requirement by our modern society triggers the materials scientist to search for sustainable energy storage resources [1]. Recently, electrochemical capacitors, also called supercapacitors or ultracapacitors, are being commercialized due to their high specific power, high specific energy, extraordinary life cycle, and obviously low charging time. A simple supercapacitor contains two active electrodes immersed in electrolytes and a separator that prevent short circuit. Although the electrochemical performance depends on various factors like the type of electrolyte, separator, and the design of the device, the active electrode materials play the main role. Based on the charge storage mechanism, these active electrode materials are primarily categorized into two types such as electrochemical double-layer capacitive (EDLC) material and pseudocapacitive material [2–5]. Carbonaceous materials such as activated carbon, carbon nanotubes, graphene, and graphene oxide provide capacitance by electrostatically storing charge on the electrode/electrolyte interface; therefore, these materials fall under EDLC. On the contrary, transition metal oxides/hydroxides/sulfides and conducting polymers fall under pseudocapacitive material. They store charge by performing fast faradaic redox reactions with electrolyte ions.

　　Magnetic nanoparticles (MNPs) are emerging zero-dimensional (0D) nanomaterials, which have drawn intense attention from researchers as they have already shown remarkable applications in many fields like biomedicine, wastewater treatment, magnetic fluid, electrochemical energy storage material, and magnetic resonance imaging (MRI) [6–10]. Magnetic transition metal oxides such as spinel ferrites and perovskite oxides nanoparticles are a promising class of inorganic materials due to their low cost and ease of large-scale production. These materials have already been established as pseudocapacitive electrodes for high-performance supercapacitors with high specific energy and power and exceptional cyclic stability. Not only that but also MNPs have got attention as interesting capacitive electrode materials owing to their magnetic field-dependent behavior toward specific capacitance [11–13]. Therefore, a correlation between the applied magnetic field and the electrochemical reaction occurring

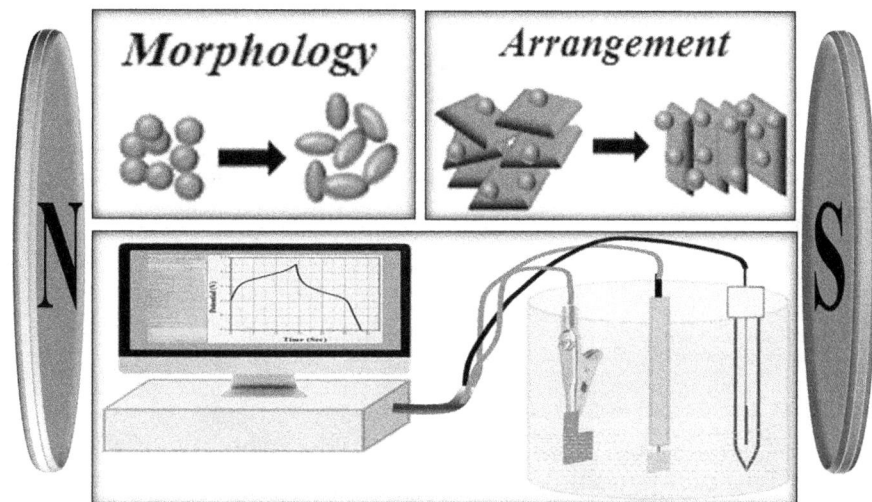

FIGURE 11.1   Schematic diagram depicting the influence of the externally applied magnetic field on active electrode materials preparation and devices.

in the electrochemical capacitor can be drawn. The effect of the external magnetic field is different for different MNPs based on their internal magnetic characteristics. Additionally, the magnetic properties of MNPs depend on their size, which is completely determined by the pathways followed for the synthesis of MNPs [14, 15].

The value of saturation magnetization and coercivity of the MNPs is also equally important as based on these values, MNPs respond toward the externally applied magnetic field. However, the overall magnetic response of MNPs depends on the magnetic domains present in the electrode material. With increasing the magnetic field, these domains start aligning, thus reducing the magnetoresistance in ferromagnetic materials and improving the charge storage capacity. Sometimes, the functionalization of magnetic nanoparticles with other EDLC and/or pseudocapacitive materials can lead to an increase in the electrochemical performance [13, 16]. Not only does the external magnetic field alter the electrochemical performance of MNPs but also various types of porous MNPs having different sizes and shapes can be synthesized in presence of a magnetic field (Figure 11.1).

11.2   SYNTHESIS AND SURFACE MODIFICATION OF MNPs

Different methods can be employed to synthesize various types of MNPs with varying properties. Generally, all the preparation methods can be categorized into three types such as physical methods, chemical methods, and biological methods (Figure 11.2). The physical methods generally follow the top-down strategy, which means the nanoparticles are formed from the bulk materials, whereas the chemical process is based on a bottom-up approach, i.e., nanoparticles develop from the atoms or small molecules.

MNPs Synthesis

Physical Methods	Chemical Methods	Biological Methods
• *Ball-milling method* • *Laser evaporation* • *Electron beam lithography* • *Gas-phase deposition* • *Wire explosion method*	• *Co-precipitation method* • *Thermal decomposition* • *Microemulsion synthesis* • *Hydrothermal method* • *Sol-gel method* • *Electrochemical deposition*	

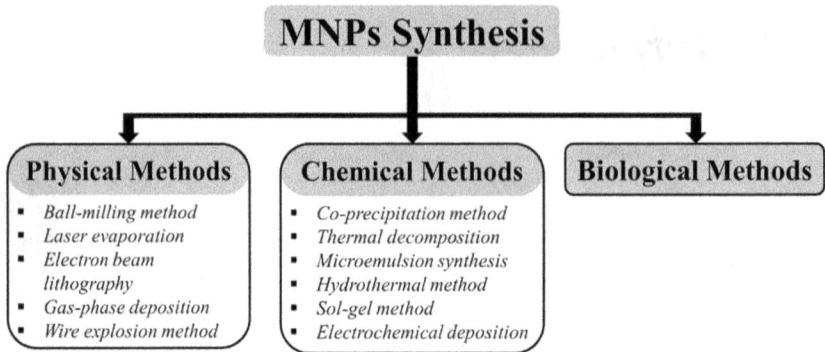

FIGURE 11.2   Synthetic routes for MNPs.

11.2.1 PHYSICAL METHODS

11.2.1.1 Mechanical Ball Milling Method

Ball milling method is a simple and cost-effective "top-down" technique used for the synthesis of MNPs. This method involves crushing of coarse-textured bulk materials into fine-textured MNPs [17]. In 1970, the method was first developed by Benjamin [18]. The process is carried out in a small cylindrical jar with several steel balls. The bulk materials enclosed in the hollow cylinder are ground by the kinetic energy generated due to the collisions between the steel balls and the solid material resulting in nano-sized particles. There are numerous types of milling systems (Figure 11.3), including planetary and shaker mills [19, 20]. The factors that influence the fabrication of the MNPs are the size of the balls, milling time, time of vibration, and ball-to-powder ratio. The main drawback of this synthetic technique is the impurity of the product [21]. The MNPs obtained by this method have a wide range of size distribution when compared to that obtained by chemical synthesis.

11.2.1.2 Laser Evaporation

This is a "bottom-up" technique by which nanoparticles are obtained from gaseous or liquid precursors via condensation [22]. In this method, MNPs are synthesized by fabricating a film and filling the holes in the template using a high-energy laser. This method is highly efficient for the production of iron oxide MNPs [23]. This process involves the evaporation of the coarsed-textured raw materials under the focus of a laser beam. The material is kept in a cell immersed in a liquid while pointing toward the laser beam. The irradiation of the material by the laser beam in solution results in the formation of vapor which is finally cooled down in the gaseous phase. The nanoparticles are formed as a result of fast condensation and nucleation [24]. This synthetic strategy is useful in the large-scale production of improved quality MNPs without using any expensive chemicals or producing any hazardous wastes [25, 26].

11.2.1.3 Electron Beam Lithography

In this synthetic technique, an electron beam is utilized to transform iron elements into iron oxide nanoparticles. Here, the electron beam is passed through the surface

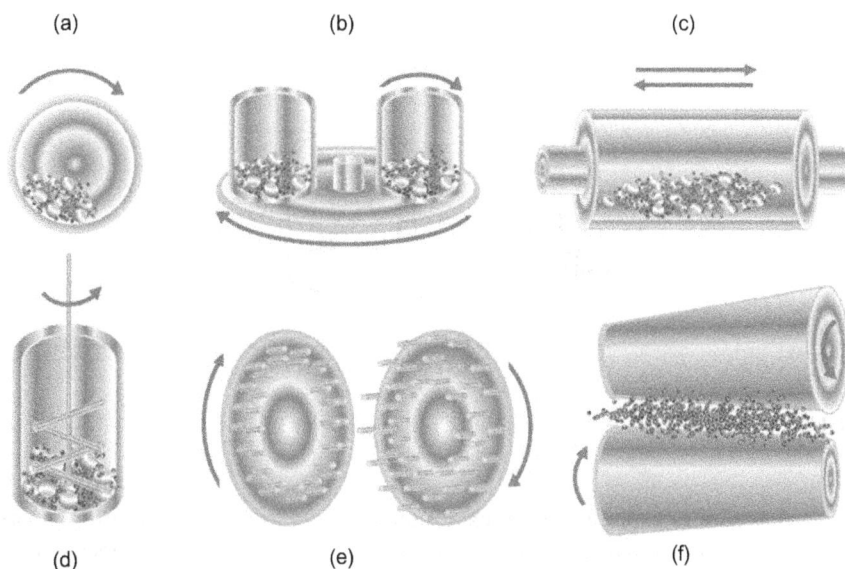

FIGURE 11.3 Various types of ball mill systems: (a) ball mill, (b) planetary mill, (c) vibration mill, (d) attritor (stirring ball mill), (e) pin mill, (f) rolling mill. Adapted with permission from [19]. Copyright (2018) RSC Publishing.

of an iron particle film to produce Fe_3O_4 nanoparticles [27]. Although this method can easily produce nanoparticles, it also has some demerits such as time consuming, expensive, and resolution limitations.

11.2.1.4 Gas-Phase Deposition

This technique can be carried out *via* chemical vapor deposition (CVD) or physical vapor deposition (PVD). Both the routes produce products with different shapes and sizes. Tyurikova et al. reported aerosol CVD for the synthesis of spherical Fe_3O_4 core–carbon shell structures [28]. This method is relatively simple and accomplished with lower consumption of reagents. 1D nanostructure of iron oxide can be synthesized by catalyzed CVD method using $\{Fe(OBut)_3\}_2$ or $Sn(OBut)_4$ precursors on gold-coated alumina substrate [29]. This is a low cost-effective method of synthesis for MNPs. However, PVD techniques need a line of sight between the source and the surface of the substrate.

11.2.1.5 Wire Explosion Method

This is a new single-step physiochemical method for the benign and environmentally safe synthesis of MNPs. This method gives high yields and does not require further steps, including the separation of the nanoparticles and treatment of the side products. The main disadvantage of this method is that the particles obtained by this method are not monodispersed [30]. However, Kurlyandskaya et al. prepared spherical-shaped iron oxide nanoparticles having size of 10 nm by this method (Figure 11.4).

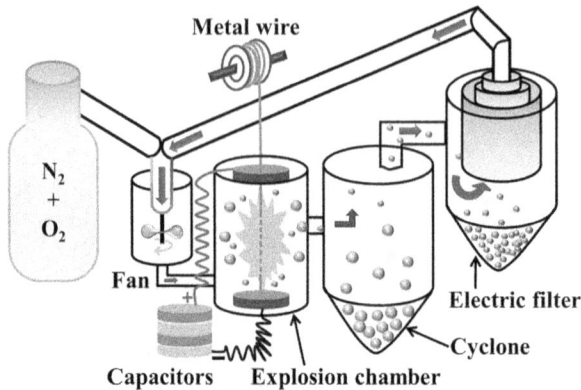

FIGURE 11.4 Schematic diagram of the wire explosion method.

The metal wire is explored in the explosion chamber to produce the iron vapors, which finally formed oxide nanoparticles [31]. The large and small MNPs are accumulated in the cyclone chamber and filter, respectively, by gas flow.

11.2.2 Chemical Methods

The chemical methods of synthesis for the MNPs include various "bottom-up" approaches. A brief depiction of some popular chemical methods that are adopted for the fabrication of MNPs is provided below.

11.2.2.1 Coprecipitation Method

The coprecipitation method of preparation is the most common method for the production of MNPs, specifically when the product is required in large quantities. This technique is very much convenient in the preparation of nanoparticles of precise size and noble magnetic properties [32]. The precursor metal ion salts are dissolved in a suitable solvent and the appropriate precipitating agent is added in order to produce the desired MNPs. During the coprecipitation, the nature and concentration of the metal ions, pH, reaction temperature, etc. play crucial roles in determining the chemical composition, shape, and size of the MNPs [33]. Though the coprecipitation method is chosen because of its simplicity and the uniform size distribution of the MNPs, occasionally, it becomes difficult to regulate the shape of the particles.

11.2.2.2 Thermal Decomposition

Thermal decomposition is highly useful in preparing monodispersed MNPs with high crystallinity, definite shape, and precise size distribution. This method involves the disintegration of organometallic compounds in presence of suitable organic surfactants [34]. This synthetic technique for MNPs utilizes different stabilizing agents like fatty acids, oleic acid, and hexadecylamine. The stabilizers can retard the nucleation process, which regulates the growth of MNPs having spherical shape and desired size. The shape and size of the MNPs are largely controlled by the reaction time, temperature, nature of solvents and surfactants, etc. [35, 36]. The thermal decomposition technique is considered to be one of the best methods for large-scale production of MNPs with identical size and regular shape [37]. The generation of toxic organic compounds limits the adaptation of this synthetic technique in the biomedical field [38]. It is more advantageous than the previous coprecipitation method in the case of the synthesis of MNPs of smaller particle sizes.

11.2.2.3 Microemulsion Synthesis

Microemulsions are transparent isotropic dispersion of water and oil mixed with the help of a surfactant or occasionally with cosurfactants. Three kinds of microemulsions are possible: (a) oil-in-water (O/W) (major part: water, minor part: oil), (b) water-in-oil (W/O) (major part: oil, minor part: water), and (c) microemulsions with comparable amounts of oil and water. In this method, the size and shape of the MNPs are largely dependent on the nature of the microemulsion [39]. Though this method results in uniformly dispersed MNPs, the yield is very low.

11.2.2.4 Hydrothermal Method

In the hydrothermal method, highly pure and crystalline materials with controlled morphology can be achieved from an aqueous salt solution of metal ions under high pressure and high temperature. In this synthetic procedure, MNPs are produced via hydrolysis and oxidation reactions. The main advantage of this method is the uniform size, structural distribution, and high crystallinity of the product [40]. The crystallinity and morphology of the prepared MNPs depend on various factors such as the proper choice of solvent (when the solvent is other than water, the process is called solvothermal), time of reaction, pressure, and temperature. As the hydrothermal synthesis is performed at high temperature and pressure, utmost care should be taken and it requires special equipment. In spite of these few concerns, this process is often preferred compared to other chemical synthetic procedures like microemulsion or sol-gel because of the advantage of fabricating MNPs of desirable size and shape with reliable composition [41].

11.2.2.5 Sol-Gel Method

The sol-gel method of synthesis for MNPs involves hydrolysis and polycondensation of metal alkoxides at room temperature. At first, a sol is prepared by dissolving the metal salt in a suitable solvent. Then, the sol is stirred or heated for enhanced van der Waals interaction between the colloidal particles. After that, the solution is dried by removing the solvent and a gel is formed. The sol-gel method is a room-temperature cost-effective method, which has control over the shape, size, and composition of the MNPs. This method is capable of effectively producing highly pure, crystalline MNPs with good yield. However, the sol-gel method suffers from a few drawbacks like the formation of byproducts, generation of a 3D oxide network, longer reaction time, and use of toxic solvents [6, 42].

11.2.2.6 Electrochemical Deposition

The electrochemical deposition method can be employed for the preparation of highly crystalline and pure nanoparticles [43, 44]. By varying the deposition parameters like potential, current, concentration, and pH of electrolyte, one can control the crystallinity and particle size of the nanoparticles. Karimzadeh et al. prepared magnetic Fe_3O_4 nanoparticles by cathodic electrochemical deposition process [43].

11.2.3 BIOLOGICAL METHOD

Biological method for the synthesis of MNPs involves various living organisms such as plants and microorganisms such as fungi, bacteria, and viruses. Biocompatible MNPs can be developed by this method. This is an efficient, environment-friendly, green process, but it suffers from poor dispersion of the nanoparticles [45]. This synthetic technique is comparatively new and till now, the mechanism of formation of the nanoparticles is not clearly understood properly and it demands further investigations.

A comparative analysis of various synthetic methods is provided in Table 11.1.

11.2.4 FUNCTIONALIZATION OF THE MNPs SURFACE

Stability of the MNPs is of major concern for fruitful application of these materials. Functionalization is the process by which more surface features are added to the surface of MNPs in order to enhance the stability and broaden the scope of applications. Due to the intrinsic hydrophobic nature of the surface MNPs tend to agglomerate resulting in larger size particles. Moreover, pristine MNPs are often nonbiocompatible. Therefore, proper modification of the surface is often required to make these

TABLE 11.1

Various Synthetic Methods for MNPs: Advantages and Disadvantages

Synthetic Procedure	Advantages	Disadvantages
Coprecipitation Method	Simple and proficient	Irregular size distribution, poor crystallinity, and aggregation
Thermal Decomposition	Easy and economical	High-temperature hazards
	High yield, Good regulator of shape and size	
Microemulsion Synthesis	Simple process	Low yield, difficult to remove surfactants, time consuming
	Good control over particle size, homogeneity	
Hydrothermal Method	High-quality crystalline nanoparticles	Requires high temperature and high pressure
		Longer time reaction
Sol-gel Method	Simple, economical, and efficient process	Expensive, long processing time
	Desired size and morphology	
Electrochemical Deposition	Single-step process	Reproducibility
	Stable and highly pure structure	

MNPs application friendly. The various surface functionalization techniques for MNPs are discussed below.

11.2.4.1 Functionalization with Polymer

MNPs have a tendency to form larger clusters under the influence of magnetic force and van der Waal's force. This phenomenon can be minimized via surface modification of the MNPs may be performed with different polymers. This technique enhances the stability of the MNPs by reducing the average particle diameter and polydispersity index. Polymers reported to be used in this functionalization technique are polyaniline (PANI), polypyrrole (PPy), etc. [46, 47]. Besides providing extra stability to the nanoparticles, the conducting polymers greatly improve the electrochemical performance of the materials. PANI coated Fe_3O_4 nanoparticles displayed specific capacitance of 1,669.18 F g^{-1} while bare Fe_3O_4 nanoparticles showed 1,351.13 F g^{-1} specific capacitance at similar conditions. The PANI coating also enhances the cyclic stability of the material from 92% capacity retention for bare Fe_3O_4 to 96.5% retention in case of Fe_3O_4/PANI over 25,000 cycles at a high current density (15 A g^{-1}).

11.2.4.2 Small Molecule Functionalization

Surface functionalization of the MNPs with small molecules can be achieved by suitable chemical reactions. This technique results in improved colloidal stability and reduced particle size [48]. Surface modification of Fe_3O_4 nanoparticles with citrate greatly influences the surface area and morphology. Moreover, the citrate modified Fe_3O_4 showed a specific capacitance of 242 F g^{-1}, which is higher than that of the bare Fe_3O_4 (112 F g^{-1}). The citrate modification enhances the cyclic stability of the material from 35% to 75% over 1,000 cycles.

11.2.4.3 Functionalization with Surfactants

Functionalization of the MNPs surface with surfactants can lead to sufficient repulsive interaction within the system to reduce the agglomeration giving a stable colloidal solution. Surfactant modified MNPs are divided into three categories: oil soluble, water soluble, and amphiphilic. The surfactants modified MNPs exhibit excellent magnetic properties and have various potential applications like energy storage thermal therapy and removal of cationic dyes in sewage. The superparamagnetic Fe_3O_4 MNPs prepared via cetyltrimethylammonium bromide (CTAB) assisted method records improved specific capacitance of 1,192 F g^{-1} at a current density of 1 A g^{-1} [49]. The electrode showed 93% capacity retention over 4,000 cycles with optimum CTAB concentration.

11.2.4.4 Functionalization with Transition Metal Oxides/Hydroxide/Sulfides

For protecting the surface and increasing the stability of the MNPs, the surface of the MNPs is coated with other transition metal oxides/hydroxide/sulfides such as NiO, MnO_2, $Co(OH)_2$, $Ni(OH)_2$, MoS_2, Co_3S_4, and Ni_3S_2. These functionalized materials are not only useful in protecting magnetic characteristics, but also these are reported to have improved electrochemical performance due to presence of more number of electroactive centers.

11.2.4.5 Functionalization with Silicon Dioxide

The nontoxic nature of silica is exploited in functionalization of MNPs to impart extra stability to the nanoparticles. Silica is capable of forming cross-linking and forming an inert coating on the MNPs. This technique is very useful in generating functionalized MNPs having potential application in energy storage, catalysis, adsorption, and magnetic separation [50]. Co_3O_4 magnetic nanoparticles embedded in SiO_2 matrix were produced by citrate-gel method. Electrochemical performance studies confirm the pseudocapacitive nature of the material. A high specific capacitance value of 1,143 F g^{-1} was obtained at a scan rate of 2.5 mV s^{-1} with excellent cyclic stability (> 92%) over 900 cycles.

11.2.4.6 Functionalization with Carbonaceous Materials

Carbonaceous materials possess some unique features like low density, high strength, and exceptional electrical properties. Hence, it is obvious that functionalization of MNPs with carbon will impart excellent electrical properties along with magnetic permeability. The porous carbon materials like activated carbon, graphene, or reduced graphene oxide (rGO), graphene oxide (GO), and carbon nanotubes (CNTs) are the most used for the functionalization of MNPs. Being the double-layer capacitive materials, these greatly improve the specific power of the hybrid electrode materials. The Fe_3O_4 nanoparticles when inserted into the bowl shaped hollow porous carbon nanocapsules (CNB) with high surface area and high conductivity achieved excellent electrochemical performance. The Fe_3O_4@CNB (containing 40.3 wt% Fe_3O_4) electrode displayed a high gravimetric capacitance of 466 F g^{-1} with 92.4% capacity retention over 5,000 cycles [51].

11.3 VARIOUS MNPs IN ELECTROCHEMICAL ENERGY STORAGE

The electrochemical performance of MNPs-based supercapacitive electrodes can be altered by an external magnetic field or/and magnetism possessed by the electrode materials. However, how the internal magnetic property can influence its energy-storing performance has not been completely explored yet. There are mainly following three types of reports describing (i) investigation of only the electrochemical performance of MNPs, (ii) investigation of the magnetic and electrochemical properties, but without drawing any correlation between them, and (iii) influence of external magnetic field on the electrochemical performance of MNPs.

Magnetic metal oxide nanoparticles are an emerging class of inorganic materials due to their low-cost and easy large-volume synthesis. Spinel ferrites of different elements have already been developed as high-performance supercapacitor applications with high specific energy, high specific power, and excellent cyclic stability. Furthermore, inorganic perovskite oxides have also attracted great attention toward high-performance anion-intercalation supercapacitors.

11.3.1 Metal Ferrite Nanoparticles

11.3.1.1 Cobalt Ferrite ($CoFe_2O_4$)

Cobalt ferrite possesses an inverse spinel structure $[M^{III}(M^{II}M^{III})O_4]$ in which Co^{2+} ion is present at the octahedral site and Fe^{3+} ions are present at both octahedral and tetrahedral sites. The ferrite substances have interlocking networks of metal ions and oxides [8]. The nanoparticles of $CoFe_2O_4$ exhibit ferromagnetic behavior. Guerioune et al. reported superparamagnetic $CoFe_2O_4$ nanoparticles fabricated by coprecipitation and hydrothermal methods [52]. They have used various metal ion precursors, i.e., different salts of the metal ions such as chlorides, nitrates, and acetates, and employed different reaction conditions. Detailed morphological and structural studies concluded the formation of mesoporous cobalt ferrite nanoplatelets. The particle size varied within the 11–26 nm range depending on the reaction conditions. The material obtained from the hydrothermal method displayed a specific area of ~34.22 m^2g^{-1}. In a three-electrode cell system, the as-prepared materials showed a high specific capacitance of 429 F g^{-1} at a specific current of 0.5 A g^{-1} with excellent cyclic stability combined with capacitance retention of 98.8% after consecutive galvanostatic charge-discharge (GCD) 6,000 cycles. The $CoFe_2O_4$-containing nanocomposites possess enhanced capacitance and excellent electrochemical behavior. Elseman et al. synthesized $CoFe_2O_4$/carbon sphere nanocomposites via a single-step solvothermal procedure where $CoFe_2O_4$ nanoparticles are combined with glucose, a precursor for the carbon sphere [53]. The electrode displayed appreciable enhancement in the specific capacitance of 600 F g^{-1}, with a 5.9% decrease of its original capacitance over 500 cycles showing a specific energy of 27.08 W h kg^{-1} and a specific power of 750 W kg^{-1}. The improved electrical conductance of the material is ascribed to its hierarchical architecture. The group reported a higher specific capacitance with maximum retention for the composite concluding the potential application of the same in supercapacitors. A simple low-cost coprecipitation method was followed by Vijayalakshmi et al. for the synthesis of ferromagnetic cobalt ferrite [54]. The synthesized $CoFe_2O_4$ nanoparticles were calcined at 350°C and the product was employed for electrochemical studies in aqueous 1 M KOH solution. The synthesized ferrite electrode showed considerably enriched supercapacitive behavior with a specific capacitance of 1,233 F g^{-1} at the scan rate of 5 mVs^{-1} with excellent retention (~90%) up to 5,000 GCD cycles.

11.3.1.2 Nickel Ferrite ($NiFe_2O_4$)

In nickel ferrites ($NiFe_2O_4$), the ferric ions play a crucial role in modulating the morphology, particle size, and magnetic and electrochemical properties. The magnetic interaction arises from the antiparallel arrangement of the spins of Fe^{3+} in the tetrahedral site and that of Ni^{2+} and Fe^{3+} in the octahedral site.

In a report by Arun et al., they have synthesized $NiFe_2O_4$ by chemical oxidation method [55]. They showed that an increase in percentage of Fe^{3+} ion content from 0 to 50 results in thorough phase formation and reduction in particle size from 43 to 29 nm, respectively, with a reduction in magnetic saturation from 45 to 29 emu g^{-1}. The Currie temperature for the material was detected at 584°C. The $NiFe_2O_4$ nanoparticles prepared with 50% Fe^{3+} ion (NF50) displayed a high specific capacitance of 277 F g^{-1}. The asymmetric solid-state cell fabricated with NF50 displayed a maximum specific capacitance of 56 F g^{-1} at a current density of 1 A g^{-1}. The fabricated NF50 electrode showed an improved energy density of 22.5 Wh Kg^{-1}. The GCD study showed 126% capacity retention for the asymmetric cell. Bashir et al. reported a facile and environment-friendly procedure for the synthesis of $NiFe_2O_4$ MNPs using an aqueous extract of Persa Americano seeds [56]. The structural characterizations confirmed the phase purity and spinel structure of $NiFe_2O_4$ MNPs. The band gap energy for bulk $NiFe_2O_4$ and $NiFe_2O_4$ nanoparticles is calculated to be 3.6 and 4.25 eV, respectively. This increment in the band gap can be attributed to the quantum confinement effect (QCE) of the nanoparticles. The presence of lower magnetocrystalline anisotropy of the $NiFe_2O_4$ nanoparticles results in a low value of coercivity. The spin glass behavior of the surface having distorted spins that interact with the ferromagnetically aligned core–shell results in a smaller value of saturation magnetization M_S of the $NiFe_2O_4$ nanoparticles (20.4 emu g^{-1}) compared to that of bulk material (55 emu g^{-1}). The electrochemical characterization of the $NiFe_2O_4$ nanoparticles confirmed that both the charge-transfer and diffusion process control the electrochemical properties. High electronic conductivity and improved electrochemical stability make the particles a suitable candidate for promising electrochemical applications. Sivakumar et al. reported spinel $NiFe_2O_4$/CNT nanocomposite and $NiFe_2O_4$ nanoparticles produced by a chemical method [57]. Electrochemical analysis of the materials confirmed the pseudocapacitive behavior of the $NiFe_2O_4$/CNT nanocomposite exhibiting a faradic energy storage mechanism. The cyclic voltammogram (CV) and GCD analysis of the composite material showed specific capacitance values of 670 F g^{-1} at a scan rate of 10 mV s^{-1} and 343 F g^{-1} at a 1 A g^{-1}, respectively, with 89.16% capacity retention after 5,000 GCD cycles. Moreover, the electrochemical analysis also confirmed the superior capacitive nature of the $NiFe_2O_4$/CNT nanocomposite compared to the $NiFe_2O_4$ nanoparticles. The $NiFe_2O_4$/CNT nanocomposite (as cathode) was then combined with activated carbon (as anode) to fabricate a device which showcased a specific capacitance of 85.94 F g^{-1} at a scan rate of 1 A g^{-1} with 87.27% retention of capacitance after 5,000 cycles. The device displayed a high specific energy of 23.39 W h kg^{-1} and a specific power of 466.66 W kg^{-1}. This study established the potential of $NiFe_2O_4$/CNT nanocomposites as electrode material for energy storage applications.

11.3.1.3   Manganese Ferrite ($MnFe_2O_4$)

Arun et al. reported manganese ferrite ($MnFe_2O_4$) MNPs prepared *via* the chemical oxidation method [58]. They employed Fe^{3+} ions during synthesis to reduce the size of the nanoparticles. Depending on the amount of Fe^{3+} ion used, as-prepared $MnFe_2O_4$ MNPs display saturation magnetization in the range of 45–67 emu g^{-1}. The TEM imaging technique is utilized to investigate the particle size distribution. The small $MnFe_2O_4$ nanoparticles synthesized by using a high concentration of Fe^{3+} ion showed the highest specific capacitance of 415 F g^{-1}. Hence, it becomes obvious that Fe^{3+} ions can be used to regulate the size of ferrite MNPs and smaller-sized $MnFe_2O_4$ MNPs have potential supercapacitor applications.

11.3.1.4   Copper Ferrite ($CuFe_2O_4$)

Liang et al. reported copper-based magnetically ordered pseudocapacitor (MOPC) and copper ferrites ($CuFe_2O_4$), which display exciting magnetocapacitive effects [59]. The use of chelating molecules like ammonium salt of purpuric acid (ASPA) results in greater active mass loading, which in turn gives improved electrochemical performance. Electrodes were constructed using $CuFe_2O_4$ nanoparticles in combination with multi-walled carbon nanotubes (MWCNTs) conductive additives.

Electrodes were fabricated by varying the amount of MWCNTs, which resulted in specific capacitance values in the range of 0.04 to 2.76 F cm^{-2} at a scan rate of 2 mV s^{-1}. The electrode with the maximum capacitance of 2.76 F cm^{-2} exhibited 25.7% capacity retention at a scan rate of 100 mV s^{-1}. The improved pseudocapacity of $CuFe_2O_4$ nanoparticles in combination with magnetic properties established $CuFe_2O_4$ nanoparticles as a promising material for negative electrodes in supercapacitors. Piao et al. reported a copper ferrite-attached graphene nanosheet ($CuFe_2O_4$-GN) synthesized via a single-step solvothermal process in which the graphene oxide is reduced to graphene simultaneously with the production of $CuFe_2O_4$ nanoparticle [60]. The morphological analysis of the composite nanosheets confirmed that the $CuFe_2O_4$ nanoparticles of a diameter of ~100 nm were well grafted on GN. The reaction parameters that influence the shape and size of the nanoparticles are the concentration of the starting material, precipitating agent, stabilizing agent, and graphene oxide. The CV and GCD analysis for investigating the electrochemical properties of the $CuFe_2O_4$-GN composite displayed high specific capacitance of 576.6 F g^{-1} at a scan rate 1 A g^{-1}, good rate capability, and excellent cycling stability. The $CuFe_2O_4$–GN composite has promising applications in electrochemical capacitors due to the excellent synergism between graphene and the nanoparticles.

11.3.2   The Effect of External Magnetic Field

A stable magnetic field may alter the performance of a supercapacitor mainly constructed by magnetic transition metal oxide electrode by changing some parameters such as Lorentz force, Nernst layer, and magneto-dielectric constant. The Lorentz force applying on a moving charge particle can be expressed as follows:

$$\vec{F}_L = q\vec{E} + q\left(\vec{v} \times \vec{B}\right) \tag{11.1}$$

where q is the charge, $\vec{E}$ is the applied electric field, $\vec{v}$ is the velocity of the charge, and $\vec{B}$ is the applied magnetic field. In absence of magnetic field ($\vec{B}=0$), only $q\vec{E}$ contributes toward specific capacitance at applied potential. In presence of magnetic field, the charge particles experience an additional force, Lorentz force $\vec{F}_L$, which helps in easy intercalation of charge carriers into the electrode, resulting in the improvement of capacitance. With increasing the magnetic field, the magnetic domains present in electrode material start aligning, which reduces the magnetoresistance in ferromagnetic materials and improves the charge storage capacity [61]. The magnetic nano-leaflets consist of small domains revealing localized magnetic moments independently. The random domains are aligned under an external magnetic field and possess a net magnetization in the direction of the external field (Figure 11.5). This alignment of domains in one direction offers free movement of electrons from one domain to other and improves charge storage properties.

Furthermore, the magnetic attraction force increases the kinetic energy, resulting in the increment of the effective flux of the electrolyte ions under a magnetic field. The magnetohydrodynamic (MHD) effect enhances the limiting current by decreasing the Nernst layer. Therefore, MHD lowers the R_{ct} and modifies the diffusion current and the electrochemical performance as well [62].

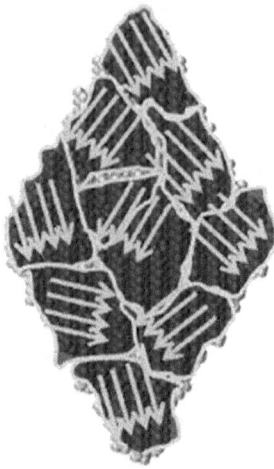

Domains in one nano-leaflate

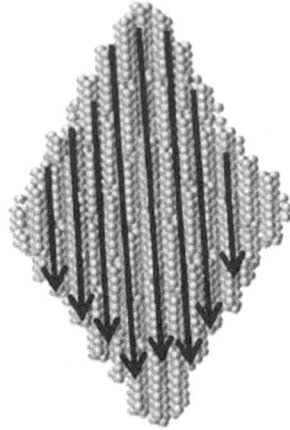

Net magnetic allignment

Randomized leaflets

Oriented spin alignment under magnetic field

FIGURE 11.5 Domain alignments under magnetic field. Adapted with permission from [61]. Copyright (2018) Wiley Publishing.

Various postulates have been considered for explaining the change in electrochemical performance of an electrode under magnetic field. The electrode having high specific surface area allows deeper penetration of electrolyte ions under a magnetic field resulting high capacitance. Reduction of the Nernst layer at the electrode–electrolyte interface under a stable magnetic environment results in enhancement of the capacitance. The Lorentz force, operated along the diagonal direction, enhances the electrolyte's convection, which decreases the electrolyte's resistance and therefore greater diffusion of the ions results in the enhanced pseudocapacitance. Furthermore, the gradient force accumulates electrolytic ions indicating improved double-layer capacitance.

So, not only the specific surface area and the pore structure of the electrode materials, but also electrical, internal magnetic properties and applied external magnetic field affect the electrochemical performance [63]. Along with this, the size, mobility, and diffusion of the electrolyte ions also play a crucial role in determining the overall performance of a supercapacitor device.

11.3.3 SPINEL OXIDE NANOPARTICLES

11.3.3.1 Iron Oxide (Fe_3O_4)

Magnetite (Fe_3O_4) has been extensively investigated due to its high natural abundance, unique electrical and magnetic properties, and eco-friendliness. The unique electrical and magnetic properties of the cubic inverse spinal magnetite are caused by the hopping of electrons between Fe^{3+} and Fe^{2+} at octahedral sites. Wang et al. reported AC/Fe_3O_4 NPs nanocomposites by simple hydrothermal method followed by ultrasonication [16]. Among the as-prepared magnetized and nonmagnetized electrodes, magnetized electrodes exhibited high electrochemical performance. The micro-magnetic field by each Fe_3O_4 NPs produced a stable magnetic field, which enhanced the energy state of electrons under the Lorentz force, and the transportation efficacy of the electrons was greatly improved as well. Furthermore, the stable magnetic field improves the electrical conductivity of the electrolytes by changing their microscopic structure. These two factors help to improve the overall performance of the supercapacitor by reducing the internal charge-transfer resistance. The specific capacitance improvement of magnetized active carbon/Fe_3O_4 NPs was enhanced by 33.1% at a specific current density of 1 A g^{-1}, and the specific energy was increased to 15.97 Wh kg^{-1}.

Pal et al. synthesized pure Fe_3O_4 nanoparticles and Fe_3O_4/reduced graphene oxide (rGO) hybrid by hydrothermal method and investigated its electrochemical performance in an external magnetic field (0.125 T) (Figure 11.6) [13]. The composite electrode exhibited specific capacitance of 451 and 868.89 F g^{-1} at a scan rate of 5 mV s^{-1} without and with the magnetic field, respectively. That means the hybrid electrode exhibited 1.93 times higher capacitance under magnetic field. The reason behind the improvement of electrochemical activity is described as that Fe_3O_4/rGO composite having high specific surface area allows the deeper penetration of electrolyte ions under the magnetic field. The composite achieved high specific energy of 120.68 Wh kg^{-1} and high specific power of 3.91 kW kg^{-1} in the presence of magnetic field.

Sinan et al. reported Fe_3O_4 nanospheres having a diameter of ~10 nm synthesized via coprecipitation method [64]. The as-prepared Fe_3O_4 nanospheres are bifunctional in nature and display a superparamagnetic behavior with saturation magnetization (M_s) value of 64 emu g^{-1} at 298 K and show soft ferromagnetic behavior with M_s value of 71 emu g^{-1} at 10 K. They integrated the pseudocapacitive Fe_3O_4 nanoparticles into hazelnut shell via hydrothermal carbonization process in presence of MgO ceramic templates resulting in a hierarchical porous structure of Fe_3O_4/C nanocomposite. The composite has a large specific surface area of 344 $m^2 g^{-1}$ is capable of combining the EDL characteristic properties of carbon with the faradic pseudocapacitance of Fe_3O_4. The material displayed a specific capacitance of 136 F g^{-1} at a scan rate of 1 A g^{-1} in 1 M Na_2SO_4 electrolyte. The resultant specific energy and specific power are

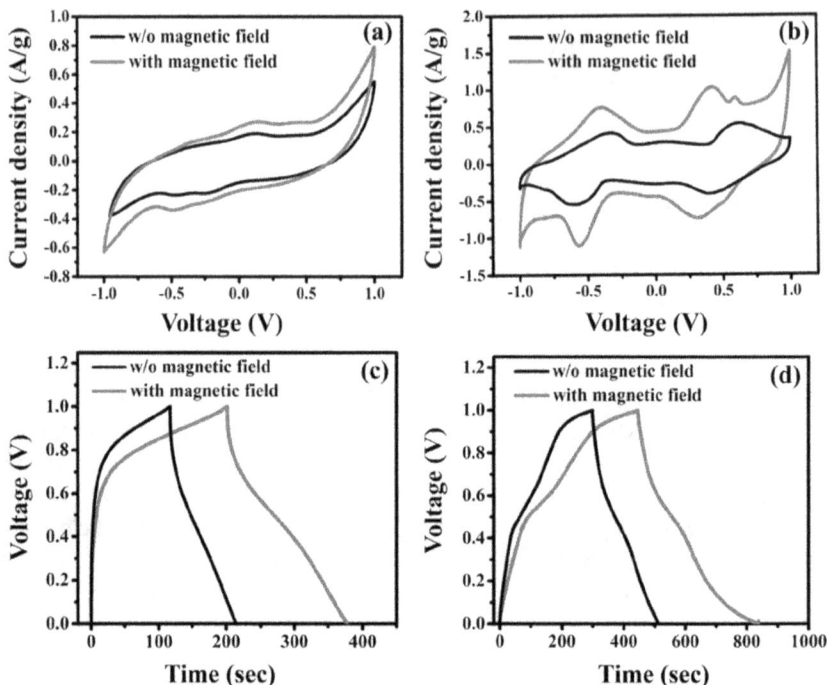

FIGURE 11.6 Various CV curves of (a) Fe_3O_4 NPs and (b) Fe_3O_4/rGO composite at scan rate of 5 mV s^{-1} with (0.125 T) or without magnetic field. GCD curves of (c) Fe_3O_4 NPs and (d) Fe_3O_4/rGO composite at specific current of 0.1 A g^{-1} with (0.125 T) or without magnetic field. Adapted with permission from [13]. Copyright (2018) IOP Publishing.

estimated to be 27.2 Wh kg^{-1} and 705.5 W kg^{-1}, respectively. Moreover, the composite exhibited improved cycling stability with no capacitance loss after 1,000 cycles at 2 A g^{-1}. The Fe_3O_4/C nanocomposite reported here produced higher specific energy in comparison to other reported composites of Fe_3O_4 with other carbonaceous materials like graphene, nanotubes, etc. The As Fe_3O_4/C nanocomposite is capable of working within a large negative potential of -1.2 to 0 V. Thus, it is possible to fabricate asymmetric supercapacitor devices by combining it with suitable positive electrodes.

The magnetic nanoparticles are mainly functionalized to improve their properties. Functionalized Fe_3O_4@D-NH$_2$ MNPs are reported combining Fe_3O_4 nanoparticles with10-arm–NH$_2$ terminated polyamidoamine (PAMAM) dendrimer having a diethylenetriamine core [65]. The facile electron transfer between the surface and the core of the dendrimer is useful for energy storage applications. When this dendrimer is combined with Fe_3O_4 nanoparticles having high porosity and larger surface area, it results in excellent electronic properties in Fe_3O_4@D-NH$_2$ nanocomposite. The Fe_3O_4@D-NH$_2$ electrode recorded a specific capacitance of 70–120 F g^{-1} depending on the material loading. The elevated specific capacitance and improved charge/discharge rate make the Fe_3O_4@D-NH$_2$ composite an excellent candidate for energy storage applications.

Octahedral magnetic Fe_3O_4 nanoparticles can also be synthesized by a simple chemical oxidation method [66]. Arun et al. have then modified the as-synthesized nanoparticles with carbon using sugar solution at different temperatures. Carbon-modified Fe_3O_4 nanoparticles prepared at 600°C, with an average grain size of 40 nm showed large magnetic saturation of 87 emu g^{-1}. The potential of the carbon-modified Fe_3O_4 MNPs for the negative electrode in supercapacitors was investigated. The uncovered Fe_3O_4 has a capacitance value of 148 F g^{-1} at a scan rate of 2 mV s^{-1}, whereas the carbon-modified hybrid materials exhibit a high specific capacitance of 274 F g^{-1}. The enhancement in capacitance is due to the synergistic effect of the EDLC property of carbon and faradic process at the Fe^{3+}/Fe^{2+} active sites in pseudo-capacitive iron oxide. The magnetite Fe_3O_4 particles display bifunctional material, and particle size plays a crucial role in regulating the superparamagnetic or ferromagnetic behavior [67].

11.3.3.2 Cobalt Oxide (Co_3O_4)

Cobalt oxide nanoparticles prepared by chemical reflux are reported to exhibit super-capacitive nature [68]. Packiaraj et al. fabricated Co_3O_4 nanoparticles with hexagonal flakes-like morphology. Magnetic characterization using a vibrating sample magnetometer (VSM) confirms the paramagnetic nature of the nanoparticles. The electro-chemical studies with Co_3O_4 nanoparticles-based electrodes record a high specific capacitance of 1,413 F g^{-1} at a specific current of 1 A g^{-1} and 98.4% retention after 1,000 cycles. The quantum confinement effect in cubic phase of Co_3O_4 nanoparticles is confirmed by a blue shift in the band gap energy calculated from the electronic absorption spectrum with a broad emission peak in UV/violet region [69].

Vijayakumar et al. reported a specific capacitance of 519 F g^{-1} for Co_3O_4 nanoparticles with 1.3% capacitance degradation after 1,000 cycles. UmaSudharshini et al. synthesized Co_3O_4 nanoparticles by a solvothermal technique by using a mixture of acetylacetonate salts of Co^{II} and Co^{III} ions [70]. The spinel Co_3O_4 shows nanosheet-like morphology for crystallites of size 40 nm. From UV–vis–DRS studies, two band gap energies are calculated at 1.33 and 2.25 eV, which are assigned to $O^{2-} \rightarrow Co^{2+}$ and $O^{2-} \rightarrow Co^{3+}$ possible charge-transfer processes. The occurrence of two band gaps confirms the nature of the as-prepared material to be a p-type semiconductor in nature. With increasing reaction temperature, the crystallite size increases, which in turn decreases the band gap energies. The material synthesized at low temperature records a specific capacitance of 778 F g^{-1}. Due to finite shape, size, and uncompensated spins on the surface, Co_3O_4 nanoparticles display weak ferromagnetic interaction, which is beneficial for electrical and electromechanical applications.

11.3.3.3 Manganese Oxide (Mn_3O_4)

Mn_3O_4 NPs have already been investigated as promising electrode materials for supercapacitors due to their low cost, easy synthesis, and high pseudocapacitance. Shah et al. reported a unique square-shaped Mn_3O_4 NPs produced by a simple hydrothermal process [71]. The porous architecture of Mn_3O_4 NPs helps to achieve high electrochemical performance with a high specific capacitance of 380 F g^{-1} at 1.0 mA cm^{-2} current. The Mn_3O_4 NPs also showed excellent electrochemical stability. The electrochemical performance can be further improved by combining it

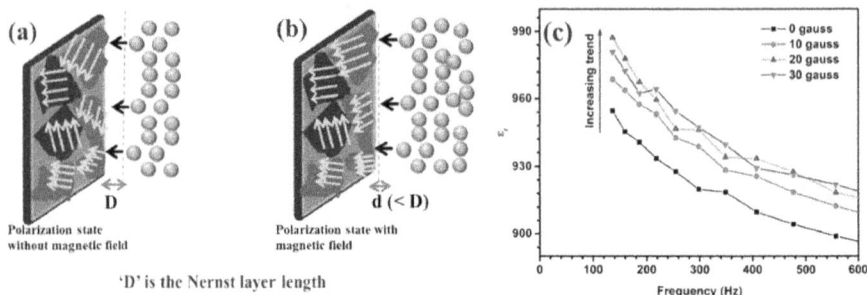

FIGURE 11.7 Schematic representation of electrolytes ion movement through the channel without or with magnetic field (a, b) and variation of dielectric constant under various magnetic fields. Adapted with permission from [11]. Copyright (2018) The Electrochemical Society.

with MWCNT (Mn_3O_4@MWCNT) [72]. The synergistic effect between the highly conductive EDLC-type material MWCNT and pseudocapacitive Mn_3O_4 results in improvement in overall performance.

Not always external magnetic field improves the specific capacitance; sometimes, it reduces the electrochemical performances. For example, Mn_3O_4 NPs show a decrease in the specific capacitance in presence of magnetic field [11]. It is discussed earlier that the dimension of Nernst's layer gets reduced in presence of external magnetic field (Figure 11.7a and b). Accordingly, the flow of electrolyte ions increased and the specific capacitance thereof. However, growth in the magneto-dielectric constant leads to a decrease in charge collection, thereby enhancing the insulating nature of the material (Figure 11.7c). The Mn_3O_4 nanoparticles reported by Haldar et al. show a specific capacitance of ~290 F g^{-1} at a scan rate of 10 mV s^{-1} and ~221 F g^{-1} at a specific current of 0.5 A g^{-1}.

11.3.3.4 Nickel Manganese Oxide (NiMn$_2$O$_4$)

The spinel $NiMn_2O_4$ possesses interesting magnetic properties along with electrical properties. The various kinds of magnetic interaction arise from the superexchange phenomena via the oxide ion present in Mn^{2+}-O^{2-}-Mn^{3+}, Mn^{3+}-O^{2-}-Mn^{3+}, and Mn^{3+}-O^{2-}-Mn^{4+} units. Moreover, the oxidation states of the manganese ion and thereby the oxygen deficiencies can be regulated by choosing appropriate synthetic techniques. Bhagwan et al. synthesized spinel ferromagnetic $NiMn_2O_4$ nanofibers via the electrospinning method [73]. The mesoporous nanofibers comprise interconnected $NiMn_2O_4$ nanoparticles, which act as intercalation sites for electrolytes ions and improve specific capacitance of 410 ($\pm$5) F g^{-1} at 1 A g^{-1}. Ray et al. reported the fabrication of spinel metal oxide $NiMn_2O_4$ nanoparticles and the investigation of the electrochemical properties [74]. They have used the sol-gel method for the synthesis of the $NiMn_2O_4$ nanospheres of ~8 nm diameter. The materials showed exceptional electrochemical properties associated with good rate capability. The electrode fabricated with the as-prepared nanomaterials showed a specific capacitance of 875 F g^{-1} at 2.0 mV s^{-1} scan rate. The asymmetric supercapacitor designed using the material is capable of working in a wide potential window of 1.8 V and shows a specific

TABLE 11.2
Various MNPs with Their Synthetic Routes, Morphology, Magnetic Properties, and Electrochemical Performance

Nanoparticles	Synthesis Method	Morphology and Particle Size (nm)	Magnetic Properties			Electrolyte	Capacitance	Ref.
			Saturation Magnetization (M_s, emu g⁻¹)	Remanent Magnetization (M_r, emu g⁻¹)	Coercive Field (H_c, Oe)			
$CoFe_2O_4$	Hydrothermal method	Square-shaped nanoplatelets (17 nm)	63.5	46.7	750	6 M KOH	429 F g⁻¹ at 0.5 A g⁻¹	52
$NiFe_2O_4$	Chemical oxidation method	Nanosphere (18.72 nm)	29	–	50	2 M KOH	277 F g⁻¹ at 1 A g⁻¹	55
$MnFe_2O_4$	Chemical oxidation method	Nanosphere (14 nm)	45	–	17	2 M KOH	415 F g⁻¹ at 1 A g⁻¹	58
Fe_3O_4/C	Coprecipitation + Hydrothermal method	Fe_3O_4 nanosphere (~10 nm) in C matrix	34	8	66	1 M Na_2SO_4	136.2 F g⁻¹ at 1 A g⁻¹	64
Fe_3O_4-C	Chemical oxidation method + Calcination	C decorated octahedral Fe_3O_4	87	–	124	6 M KOH	274 F g⁻¹ at 0.5 A g⁻¹	66
Fe_3O_4	Sol-gel method	Quasi-cube shaped nanocrystals (8±2 nm)	41.6	12.9	304	3 M KOH	185 F g⁻¹ at 1 mA	67
Co_3O_4	Chemical reflux method	Hexagonal-shaped nanoflakes	–	–	–	2 M KOH	1,413 F g⁻¹ at 1 A g⁻¹	68
$NiMn_2O_4$	Electrospinning	Interconnected nanoparticles (< 30 nm) in nanofibric network	34	13	96.7	1M KCl	410 (±5) F g⁻¹ at 1 A g⁻¹	73

energy of 75.01 W h kg^{-1} at specific power of 2,250.91 W kg^{-1}. The results conclude that the porous NiMn$_2$O$_4$ nanospheres are excellent candidates for energy storage applications. Dhas et al. reported a simple sol-gel method of preparation of NiMn$_2$O$_4$ and studied their electrochemical efficiency, which was found to be highly monitored by calcination temperature and electrolyte concentration [75]. All the electrochemical investigations were carried out in a three-electrode system using a KOH electrolyte. They prepared three different powder materials at three different annealing temperatures (500°C, 600°C, and 700°C) and named them as NMO1, NMO2, and NMO3, respectively. NMO1 having crystalline particles of 10 nm size shows a high specific capacitance of 571 F g^{-1} at a scan rate of 5 mV s^{-1}. The specific capacitance of the electrode further increases from 571 to 762 F g^{-1} when the concentration of the KOH electrolyte is changed from 1 to 6 M, respectively. This enhancement in capacitance value can be credited to excellent electrochemical utilization and effective charge storage phenomenon.

11.4   CORRELATION BETWEEN THE MORPHOLOGY, SIZE OF THE MNPs WITH THEIR MAGNETIC AND ELECTROCHEMICAL PROPERTIES

The various processing method produces MNPs with different size and morphology by simply altering the synthesis parameters. These MNPs show different physical properties including magnetic properties and electrochemical performances (Table 11.2). The specific capacitance of a material can be enhanced by creating unique morphology with large specific surface area (SSA) or by reducing the size of the material providing large SSA. Thirumurugan et al. showed that the reduction in size (50 nm to 14 nm) is beneficial for the improvement of capacitance (110 to 415 F g^{-1}) with the reduction of the value of M$_s$ and M$_r$ [58].

11.5   CONCLUSION AND FUTURE OUTLOOK

The use of MNPs as active electrode material in energy storage applications especially in supercapacitors was reviewed. The different synthetic routes for the preparation of various MNPs with different morphology and magnetic properties were discussed. The effect of the external magnetic field on the specific capacitance of various MNPs and their composite was reviewed. However, extensive research is needed to fully understand the mechanism behind the alteration of the electrochemical performance. More optimization is required in order to commercialize the MNPs as electrode materials for supercapacitor applications.

ACKNOWLEDGMENTS

MM thanks Sree Chaitanya College, Habra for giving permission to carry out the review work. KC acknowledges Ishaan Mandal for allowing enough time for literature survey and writing the draft.

REFERENCES

1. Liu, J., J.G. Zhang, Z. Yang, J.P. Lemmon, C. Imhoff, G.L. Graff, L. Li, J. Hu, C. Wang, J. Xiao, and G. Xia. 2013. Materials science and materials chemistry for large scale electrochemical energy storage: From transportation to electrical grid. *Advanced Functional Materials* 23:929–946.
2. Mandal, M., K. Chattopadhyay, S. Paria, A. Kundu, M. Chakraborty, S. Mal, W. Shin, S. Das, C. Nah, and S.K. Bhattacharya. 2022. Electrodeposited binder-free Mn-Co-S nanosheets toward high specific-energy aqueous asymmetric supercapacitors. *ACS Applied Electronic Materials* 49: 4357–4367.
3. Mandal, M., K. Chattopadhyay, M. Chakraborty, W. Shin, K.K. Bera, S. Chatterjee, A. Hossain, D. Majumdar, A. Gayen, C. Nah, and S.K. Bhattacharya. 2022. Room temperature synthesis of perovskite hydroxide, $MnSn(OH)_6$: A negative electrode for supercapacitor. *Electronic Materials Letters* 18: 559–567.
4. Mandal, M., R. Nagaraj, K. Chattopadhyay, M. Chakraborty, S. Chatterjee, D. Ghosh, and S.K. Bhattacharya. 2021. A high-performance pseudocapacitive electrode based on $CuO–MnO_2$ composite in redox-mediated electrolyte. *Journal of Materials Science* 56: 3325–3335.
5. Mandal, M., A. Maitra, T. Das, and C.K. Das. 2015. *Graphene and related two-dimensional materials* (3–23). John Wiley & Sons, Inc., Hoboken.
6. Ali, A., T. Shah, R. Ullah, P. Zhou, M. Guo, M. Ovais, Z. Tan, and Y. Rui. 2021. Review on recent progress in magnetic nanoparticles: Synthesis, characterization, and diverse applications. *Frontiers in Chemistry* 9: 629054.
7. Mohammed, L., H.G. Gomaa, D. Ragab, and J. Zhu. 2017. Magnetic nanoparticles for environmental and biomedical applications: A review. *Particuology* 30:1–14.
8. Abdel Maksoud, M.I.A., R.A. Fahim, A.E. Shalan, M. Abd Elkodous, S.O. Olojede, A.I. Osman, C. Farrell, A.A.H. Al-Muhtaseb, A.S. Awed, A.H. Ashour, and D.W. Rooney. 2021. Advanced materials and technologies for supercapacitors used in energy conversion and storage: A review. *Environmental Chemistry Letters* 19: 375–439.
9. Arun, T., S.S. Dhanabalan, R. Udayabhaskar, K. Ravichandran, A. Akbari-Fakhrabadi, and M.J. Morel. 2022. Magnetic nanomaterials for energy storage applications. In *Inorganic Materials for Energy, Medicine and Environmental Remediation* (131–150). Springer, Cham.
10. Kumar, N.S., R.P. Suvarna, K.C.B. Naidu, and B.V.S. Reddy. 2022. Magnetic nanoparticles for high energy storage applications. In *Fundamentals and Industrial Applications of Magnetic Nanoparticles* (601–618). Woodhead Publishing.
11. Haldar, P., S. Biswas, V. Sharma, and A. Chandra. 2018. Understanding the origin of magnetic field dependent specific capacitance in Mn_3O_4 nanoparticle based supercapacitors. *Journal of the Electrochemical Society* 165:A3230.
12. Wei, H., H. Gu, J. Guo, D. Cui, X. Yan, J. Liu, D. Cao, X. Wang, S. Wei, and Z. Guo. 2018. Significantly enhanced energy density of magnetite/polypyrrole nanocomposite capacitors at high rates by low magnetic fields. *Advanced Composites and Hybrid Materials* 1:127–134.
13. Pal, S., S. Majumder, S. Dutta, S. Banerjee, B. Satpati, and S. De. 2018. Magnetic field induced electrochemical performance enhancement in reduced graphene oxide anchored Fe_3O_4 nanoparticle hybrid based supercapacitor. *Journal of Physics D: Applied Physics* 51:375501.
14. Arun, T., K. Prakash, and R.J. Joseyphus. 2013. Synthesis and magnetic properties of prussian blue modified Fe nanoparticles. *Journal of Magnetism and Magnetic Materials* 345:100–105.

15. Arun, T., K. Prakash, R. Kuppusamy, and R.J. Joseyphus. 2013. Magnetic properties of Prussian blue modified Fe_3O_4 nanocubes. *Journal of Physics and Chemistry of Solids* 74:1761–1768.

16. Wang, G., H. Xu, L. Lu, and H. Zhao. 2014. Magnetization-induced double-layer capacitance enhancement in active carbon/Fe_3O_4 nanocomposites. *Journal of Energy Chemistry* 23:809–815.

17. Fecht, H.J., E. Hellstern, Z. Fu, and W.L. Johnson. 1990. Nanocrystalline metals prepared by high-energy ball milling. *Metallurgical Transactions A*, 21:2333–2337.

18. Benjamin, J.S.. 1970. Dispersion strengthened superalloys by mechanical alloying. *Metall Trans* 1:2943–2951.

19. Baláž, P., M. Achimovičová, M. Baláž, P. Billik, Z. Cherkezova-Zheleva, J.M. Criado, F. Delogu, E. Dutková, E. Gaffet, F.J. Gotor, and R. Kumar. 2013. Hallmarks of mechanochemistry: From nanoparticles to technology. *Chemical Society Reviews* 42: 7571–7637.

20. Österle, W., G. Orts-Gil, T. Gross, C. Deutsch, R. Hinrichs, M.A.Z. Vasconcellos, H. Zoz, D. Yigit, and X. Sun. 2013. Impact of high energy ball milling on the nanostructure of magnetite–graphite and magnetite–graphite–molybdenum disulphide blends. *Materials Characterization* 86:28–38.

21. Mohamed, A.E.M.A. and M.A. Mohamed. 2019. Nanoparticles: Magnetism and applications. In *Magnetic Nanostructures* (1–12). Springer, Cham.

22. Biehl, P., M. Von der Lühe, S. Dutz, and F.H. Schacher. 2018. Synthesis, characterization, and applications of magnetic nanoparticles featuring polyzwitterionic coatings. *Polymers* 10:91.

23. Shin, D.N., Y. Matsuda, and E.R. Bernstein. 2004. On the iron oxide neutral cluster distribution in the gas phase. II. Detection through 118 nm single photon ionization. *The Journal of Chemical Physics* 120:4157–4164.

24. Kurland, H.D., J. Grabow, G. Staupendahl, W. Andrä, S. Dutz, and M.E. Bellemann. 2007. Magnetic iron oxide nanopowders produced by CO_2 laser evaporation. *Journal of Magnetism and Magnetic Materials* 311:73–77.

25. Choi, D.S., A.W. Robertson, J.H. Warner, S.O. Kim, and H. Kim. 2016. Low-temperature chemical vapor deposition synthesis of Pt–Co alloyed nanoparticles with enhanced oxygen reduction reaction catalysis. *Advanced Materials* 28:7115–7122.

26. Yang, G.W. 2007. Laser ablation in liquids: Applications in the synthesis of nanocrystals. *Progress in Materials Science* 52:648–698.

27. Wang, Y., J.A. Pan, H. Wu, and D.V. Talapin. 2019. Direct wavelength-selective optical and electron-beam lithography of functional inorganic nanomaterials. *ACS Nano* 13:13917–13931.

28. Tyurikova, I.A., S.E. Alexandrov, K.S. Tyurikov, D.A. Kirilenko, A.B. Speshilova, and A.L. Shakhmin. 2020. Fast and controllable synthesis of core–shell Fe_3O_4–C nanoparticles by aerosol CVD. *ACS Omega*, 5: 8146–8150.

29. Reddy, L.H., J.L. Arias, J. Nicolas, and P. Couvreur. 2012. Magnetic nanoparticles: Design and characterization, toxicity and biocompatibility, pharmaceutical and biomedical applications. *Chemical Reviews* 112:5818–5878.

30. Kawamura, G., S. Alvarez, I.E. Stewart, M. Catenacci, Z. Chen, and Y.C. Ha. 2015. Production of oxidation-resistant Cu-based nanoparticles by wire explosion. *Scientific Reports* 5:1–8.

31. Kurlyandskaya, G.V., S.M. Bhagat, A.P. Safronov, I.V. Beketov, and A. Larrañaga. 2011. Spherical magnetic nanoparticles fabricated by electric explosion of wire. *AIP Advances* 1:042122.

32. Sandeep Kumar, V. and A. Venkatesh. 2013. Advances in graphene-based sensors and devices. *Journal of Nanomedicine & Nanotechnology* 4:e127.

33. Mosayebi, J., M. Kiyasatfar, and S. Laurent. 2017. Synthesis, functionalization, and design of magnetic nanoparticles for theranostic applications. *Advanced Healthcare Materials* 6:1700306.

34. Effenberger, F.B., R.A. Couto, P.K. Kiyohara, G. Machado, S.H. Masunaga, R.F. Jardim, and L.M. Rossi. 2017. Economically attractive route for the preparation of high quality magnetic nanoparticles by the thermal decomposition of iron(III) acetylacetonate. *Nanotechnology* 28:115603.

35. Lu, A.H., E.E. Salabas, and F. Schüth. 2007. Magnetic nanoparticles: Synthesis, protection, functionalization, and application. *Angewandte Chemie International Edition* 46:1222–1244.

36. Patsula, V., L. Kosinová, M. Lovrić, L. Ferhatovic Hamzić, M. Rabyk, R. Konefal, A. Paruzel, M. Šlouf, V. Herynek, S. Gajović, and D. Horák. 2016. Superparamagnetic Fe_3O_4 nanoparticles: Synthesis by thermal decomposition of iron(III) glucuronate and application in magnetic resonance imaging. *ACS Applied Materials & Interfaces* 8:7238–7247.

37. Kudr, J., Y. Haddad, L. Richtera, Z. Heger, M. Cernak, V. Adam, and O. Zitka. 2017. Magnetic nanoparticles: From design and synthesis to real world applications. *Nanomaterials* 7:243.

38. Faraji, M., Y. Yamini, and M. Rezaee. 2010. Magnetic nanoparticles: Synthesis, stabilization, functionalization, characterization, and applications. *Journal of the Iranian Chemical Society* 7:1–37.

39. Lu, T., J. Wang, J. Yin, A. Wang, X. Wang, and T. Zhang. 2013. Surfactant effects on the microstructures of Fe3O4 nanoparticles synthesized by microemulsion method. *Colloids and Surfaces A: Physicochemical and Engineering Aspects* 436:675–683.

40. Sobhani, A. and M. Salavati-Niasari. 2015. Synthesis and characterization of $FeSe_2$ nanoparticles and $FeSe_2/FeO(OH)$ nanocomposites by hydrothermal method. *Journal of Alloys and Compounds* 625:26–33.

41. Zahid, M., N. Nadeem, M.A. Hanif, I.A. Bhatti, H.N. Bhatti, and G. Mustafa. 2019. Metal ferrites and their graphene-based nanocomposites: Synthesis, characterization, and applications in wastewater treatment. In *Magnetic Nanostructures* (181–212). Springer, Cham.

42. Shukla, S., R. Khan, and A. Daverey. 2021. Synthesis and characterization of magnetic nanoparticles, and their applications in wastewater treatment: A review. *Environmental Technology & Innovation* 24:101924.

43. Karimzadeh, I., M. Aghazadeh, M.R. Ganjali, P. Norouzi, T. Doroudi, and P.H. Kolivand. 2017. Saccharide-coated superparamagnetic Fe_3O_4 nanoparticles (SPIONs) for biomedical applications: An efficient and scalable route for preparation and in situ surface coating through cathodic electrochemical deposition (CED). *Materials Letters* 189:290–294.

44. Karimzadeh, I., H.R. Dizaji, and M. Aghazadeh. 2016. Preparation, characterization and PEGylation of superparamagnetic Fe_3O_4 nanoparticles from ethanol medium via cathodic electrochemical deposition (CED) method. *Materials Research Express* 3:095022.

45. Komeili, A. 2012. Molecular mechanisms of compartmentalization and biomineralization in magnetotactic bacteria. *FEMS Microbiology Reviews* 36:232–255.

46. Xu, C., A.R. Puente-Santiago, D. Rodriguez-Padron, A. Caballero, A.M. Balu, A.A. Romero, M.J. Munoz-Batista, and R. Luque. 2019. Controllable design of polypyrrole-iron oxide nanocoral architectures for supercapacitors with ultrahigh cycling stability. *ACS Applied Energy Materials* 2:2161–2168.

47. Javed, M.S., A.J. Khan, M. Hanif, M.T. Nazir, S. Hussain, M. Saleem, R. Raza, S. Yun, and Z. Liu. 2021. Engineering the performance of negative electrode for supercapacitor

by polyaniline coated Fe_3O_4 nanoparticles enables high stability up to 25,000 cycles. *International Journal of Hydrogen Energy*, 46: 9976–9987.

48. Park, S., C.J. Raj, R. Manikandan, B.C. Kim, and K.H. Yu. 2020. Citric acid stabilized iron oxide nanoparticles for battery-type supercapacitor electrode. *Journal of Ceramic Processing Research* 21: 78–283.

49. Sathyan, M., P.J. Jandas, M. Venkatesan, S.C. Pillai, and H. John. 2022. Electrode material for high performance symmetric supercapacitors based on superparamagnetic Fe_3O_4 nanoparticles modified with cetyltrimetylammonium bromide. *Synthetic Metals* 287: 117080.

50. Ali, G.A., O.A. Fouad, S.A. Makhlouf, M.M. Yusoff, and K.F. Chong. 2014. Co_3O_4/SiO_2 nanocomposites for supercapacitor application. *Journal of Solid State Electrochemistry* 18:2505–2512.

51. Wang, L., F. Liu, A. Pal, Y. Ning, Z. Wang, B. Zhao, R. Bradley, and W. Wu. 2021. Ultra-small Fe_3O_4 nanoparticles encapsulated in hollow porous carbon nanocapsules for high performance supercapacitors. *Carbon* 179: 327–336.

52. Kennaz, H., A. Harat, O. Guellati, D.Y. Momodu, F. Barzegar, J.K. Dangbegnon, N. Manyala, and M. Guerioune. 2018. Synthesis and electrochemical investigation of spinel cobalt ferrite magnetic nanoparticles for supercapacitor application. *Journal of Solid State Electrochemistry* 22:835–847.

53. Elseman, A.M., M.G. Fayed, S.G. Mohamed, D.A. Rayan, N.K. Allam, M.M. Rashad, and Q.L. Song. 2020. $CoFe_2O_4$@carbon spheres electrode: A one-step solvothermal method for enhancing the electrochemical performance of hybrid supercapacitors. *ChemElectroChem* 7: 526–534.

54. Vijayalakshmi, S., E. Elaiyappillai, P.M. Johnson, and I.S. Lydia. 2020. Multifunctional magnetic $CoFe_2O_4$ nanoparticles for the photocatalytic discoloration of aqueous methyl violet dye and energy storage applications. *Journal of Materials Science: Materials in Electronics* 31: 10738–10749.

55. Arun, T., T. Kavinkumar, R. Udayabhaskar, R. Kiruthiga, M.J. Morel, R. Aepuru, N. Dineshbabu, K. Ravichandran, A. Akbari-Fakhrabadi, and R.V. Mangalaraja. 2021. $NiFe_2O_4$ nanospheres with size-tunable magnetic and electrochemical properties for superior supercapacitor electrode performance. *Electrochimica Acta* 399: 139346.

56. Bashir, A.K.H., N. Matinise, J. Sackey, K. Kaviyarasu, I.G. Madiba, L. Kodseti, F.I. Ezema, and M. Maaza. 2020. Investigation of electrochemical performance, optical and magnetic properties of $NiFe_2O_4$ nanoparticles prepared by a green chemistry method. *Physica E: Low-dimensional Systems and Nanostructures* 119: 114002.

57. Sivakumar, M., B. Muthukutty, G. Panomsuwan, V. Veeramani, Z. Jiang, and T. Maiyalagan. 2022. Facile synthesis of $NiFe_2O_4$ nanoparticle with carbon nanotube composite electrodes for high-performance asymmetric supercapacitor. *Colloids and Surfaces A: Physicochemical and Engineering Aspects* 648: 129188.

58. Arun, T., T.K. Kumar, R. Udayabhaskar, M.J. Morel, G. Rajesh, R.V. Mangalaraja, and A. Akbari-Fakhrabadi. 2020. Size dependent magnetic and capacitive performance of $MnFe_2O_4$ magnetic nanoparticles. *Materials Letters* 276: 128240.

59. Liang, W., W. Yang, S. Sakib, and I. Zhitomirsky. 2022. Magnetic $CuFe_2O_4$ Nanoparticles with Pseudocapacitive Properties for Electrical Energy Storage. *Molecules* 27: 5313.

60. Zhang, W., B. Quan, C. Lee, S.K. Park, X. Li, E. Choi, G. Diao, and Y. Piao. 2015. One-step facile solvothermal synthesis of copper ferrite–graphene composite as a high-performance supercapacitor material. *ACS Applied Materials & Interfaces* 7:2404–2414.

61. Sharma, V., S. Biswas, and A. Chandra. 2018. Need for revisiting the use of magnetic oxides as electrode materials in supercapacitors: Unequivocal evidence of significant variation in specific capacitance under variable magnetic field. *Advanced Energy Materials* 8:1800573.

62. Bund, A., S. Koehler, H.H. Kuehnlein, and W. Plieth. 2003. Magnetic field effects in electrochemical reactions. *Electrochimica Acta* 49:147–152.
63. Chowdhury, A., A. Dhar, S. Biswas, V. Sharma, P.S. Burada, and A. Chandra. 2020. Theoretical model for magnetic supercapacitors—From the electrode material to electrolyte ion dependence. *The Journal of Physical Chemistry C* 124: 26613–26624.
64. Sinan, N. and E. Unur. 2016. Fe_3O_4/carbon nanocomposite: Investigation of capacitive & magnetic properties for supercapacitor applications. *Materials Chemistry and Physics* 183:571–579.
65. Chandra, S., M.D. Patel, H. Lang, and D. Bahadur. 2015. Dendrimer-functionalized magnetic nanoparticles: A new electrode material for electrochemical energy storage devices. *Journal of Power Sources* 280:217–226.
66. Arun, T., K. Prabakaran, R. Udayabhaskar, R.V. Mangalaraja, and A. Akbari-Fakhrabadi. 2019. Carbon decorated octahedral shaped Fe_3O_4 and α-Fe_2O_3 magnetic hybrid nanomaterials for next generation supercapacitor applications. *Applied Surface Science* 485:147–157.
67. Mitchell, E., R.K. Gupta, K. Mensah-Darkwa, D. Kumar, K. Ramasamy, B.K. Gupta, and P. Kahol. 2014. Facile synthesis and morphogenesis of superparamagnetic iron oxide nanoparticles for high-performance supercapacitor applications. *New Journal of Chemistry* 38:4344–4350.
68. Packiaraj, R., P. Devendran, K.S. Venkatesh, A. Manikandan, and N. Nallamuthu. 2019. Electrochemical investigations of magnetic Co_3O_4 nanoparticles as an active electrode for supercapacitor applications. *Journal of Superconductivity and Novel Magnetism* 32:2427–2436.
69. Vijayakumar, S., A.K. Ponnalagi, S. Nagamuthu, and G. Muralidharan. 2013. Microwave assisted synthesis of Co_3O_4 nanoparticles for high-performance supercapacitors. *Electrochimica Acta* 106:500–505.
70. UmaSudharshini, A., M. Bououdina, M. Venkateshwarlu, C. Manoharan, and P. Dhamodharan, 2020. Low temperature solvothermal synthesis of pristine Co_3O_4 nanoparticles as potential supercapacitor. *Surfaces and Interfaces* 19: 100535.
71. Shah, H.U., F. Wang, A.M. Toufiq, S. Ali, Z.U.H. Khan, Y. Li, J. Hu, and K. He. 2018. Electrochemical properties of controlled size Mn_3O_4 nanoparticles for supercapacitor applications. *Journal of Nanoscience and Nanotechnology*, 18:719–724.
72. Mandal, M., D. Ghosh, K. Chattopadhyay, and C.K. Das. 2016. A novel asymmetric supercapacitor designed with Mn_3O_4@ multi-wall carbon nanotube nanocomposite and reduced graphene oxide electrodes. *Journal of Electronic Materials* 45:3491–3500.
73. Bhagwan, J., S. Rani, V. Sivasankaran, K.L. Yadav, and Y. Sharma. 2017. Improved energy storage, magnetic and electrical properties of aligned, mesoporous and high aspect ratio nanofibers of spinel-$NiMn_2O_4$. *Applied Surface Science* 426:913–923.
74. Ray, A., A. Roy, M. Ghosh, J.A. Ramos-Ramón, S. Saha, U. Pal, S.K. Bhattacharya, and S. Das, 2019. Study on charge storage mechanism in working electrodes fabricated by sol-gel derived spinel $NiMn_2O_4$ nanoparticles for supercapacitor application. *Applied Surface Science* 463:513–525.
75. Dhas, S.D., P.S. Maldar, M.D. Patil, M.R. Waikar, R.G. Sonkawade, S.K. Chakarvarti, S.K. Shinde, D.Y. Kim, and A.V. Moholkar. 2021. Probing the electrochemical properties of NiMn2O4 nanoparticles as prominent electrode materials for supercapacitor applications. *Materials Science and Engineering: B* 271: 115298.

12 Functionalized Magnetic Nanomaterials for Data Storage Applications

Sivanantham Nallusamy
K. Ramakrishnan College of Engineering

Vasanthi Venkidusamy
NIT-Tiruchirappalli

N. Chidhambaram
Rajah Serfoji Government College (Autonomous)

Durga Prasad Pabba
Universidad Tecnologica Metropolitana

Shajahan Shanavas
Khalifa University of Science and Technology

Arun Thirumurugan
Sede Vallenar, Universidad de Atacama

CONTENTS

DOI: 10.1201/9781003335580-12

12.1 INTRODUCTION

Current data creation and consumption trends indicate that the devices and storage media which we use would need further physical space. Every day, 2.5 quintillion bytes of human and machine-generated data were produced at our current rate, and the rate is increasing continuously. The amount of data generated in 2020 was expected to touch 35 trillion gigabytes (ZB), but nearly 33 ZB were touched in 2018 itself, prompting the International Data Corporation (IDC) to revise its estimate for 2025. Magnetic media such as magnetic drum memory, magnetic core memory, magnetic tape, hard disk drive (HDD), magnetic stripe, and floppy disk drives were used in the early days of data storage (FDD). Magnetic tapes and HDDs were widely used as magnetic storage devices with long term backup retention and large storage capacity after the evolution of magnetic data storage devices over the last few decades [1, "Advanced Control Systems for Data Storage on Magnetic Tape: A Long-Lasting Success Story [President's Message]," in IEEE Control Systems Magazine, 42 (2022) 8–11, doi: 10.1109/MCS.2022.3171472.]. Industrialists were constantly interested in the expansion of new materials and procedures for the fabrication of advanced magnetic tape and HDDs to increase storage capacity while reducing size, cost per bit of stored data, access time, and data rate. The breakthrough in magnetic materials and disk drive fabrication technology enabled large capacity at low cost, allowing room-size computers in "glass houses" to be replaced by portable devices such as iPods and laptops [2].

Magnetic memory devices in general used ferromagnetic materials that contained magnetic domains with unpaired electrons. By magnetizing the spins in the domains "ascending" and "descending," which correspond to the binary system's 1s and 0s, information in terms of bytes can be recorded on ferromagnetic media [3]. The beauty of magnetic materials is that they can store and retrieve data in the same way it was recorded. By shrinking to the nanoscale, magnetic material's remarkable properties and advantages over their bulk behavior can be realized. There are many functional magnetic materials with interesting properties such as magnetostriction, giant magneto resistance, magnetic refrigeration, and magnetic fluidity. The processing of bulk materials into various shapes with dimensions of a few nanometers resulted in materials with desired properties for various applications.

Advances in materials science have resulted in the development of new technologies in disk drive design, such as giant magnetoresistance (GMR), thin-film head, magnetoresistance head, magnetic recording media, and magnetic tunnel junction (MTJ) heads. As a result, a diverse range of magnetic materials, such as soft magnetic materials with high magnetization for writing heads, and antiferromagnetic alloys with high blocking temperature and low susceptibility for pinning films in GMR sensors are demonstrated for the advancement of magnetic recording. The development of nano-engineered organic, inorganic, and hybrid organic–inorganic composite magnetic materials based on spintronics/quantum developments has grown in popularity and has aided recent density advances [4].

12.1.1 MAGNETIC DATA STORAGE

A vast amount of data was stored on audio players, mobile phones, personal and commercial computers, Web 2.0 apps, corporate storage systems, and numerous similar applications using magnetic storage devices and related systems, which were virtually ubiquitous in our digital life. The rise of digital media storage and the need for low-cost storage have resulted in rapid increases in storage capacity as well as reductions in size and cost per GB of data [5]. Valdemar Poulsen invented the use of magnetic medium to record and reproduce information in 1898 for telegraphone.

The evolution of this fundamental device culminated in the HDD (IBM 350), which was commercialized in 1955 and reached its current stage after many years of technological advancements. Other emerging technologies, in addition to magnetic storage devices, include optical, magneto-optical, electronic, electro-mechanical, and molecular data storage. Among these, optical storage was widely used due to its low cost, portability, and market in audio and video prerecording. However, optical disk storage was inferior to magnetic hard disk storage in every way because of its low performance, high cost per read/write head, and limited capacity, making it unsuitable for the nonremovable online data storage slots occupied by magnetic hard disks [6].

Magnetic storage/recording technology showed numerous challenges, and the demand continues to rise because of the increase in digital data stored on the internet. As a result, storage management issues such as reliability, scalability, and performance have arisen in digital data storage and archival libraries that use hard disk drive arrays. In comparison to the original IBM 305 drive, modern hard disk drives have about 8 folds of magnitude greater bit areal densities and are 8 folds of magnitude lighter [7]. The progress in hard disk drive bit areal density over the last half-century has been faster than Moore's law and has continued to grow at a slow rate. After the beginning of thin-film-based writing heads, GMR sensors, and advanced digital recording tracks with thin sputtered media in the 1990s, bit areal density increased rapidly. Recent MRAM technologies based on magneto resistive (MR) effect provided nonvolatile magnetic recording media with the speed and density of semiconductor-based memory devices, static and dynamic RAM [8].

Further efforts to increase the bit areal density resulted in the concept of superparamagnetic grains, in which, each data set was made up of individual grains with diameters of 8 nm or less. With technologies such as exchange coupled composite (ECC) media patterned media, perpendicular recording media, HAMR, and MARM, the superparamagnetic limit, difficulties in scaling of bit size to grain size ratio, and less thermal stability, these limitations were overcome. Many of these hard disk drive technologies have got applications in magnetic tape machinery to increase storage performance, capacity, and utilization. The expansion of magnetic recording skill, which increases the total quantity of bits that can be recorded on a single drive, creates new difficulties for the efficient reading and writing of vast volumes of data that flow into and out of the read channel. Thus, it is worthwhile to discuss the major technological and materials advances that have already been made, as well as future technologies, to meet the challenges of producing magnetic storage devices with high capacity, low cost, and improved performance.

12.1.2 UTILIZATION OF MAGNETIC NANOMATERIALS

Magnetic nanoparticles (MNPs), which display a unique magnetic characteristics that distinguish them from their bulk counterparts, have received a significant consideration in the last 2 decades and were becoming increasingly significant in a variety of solicitations from storage and recording media to biomedical devices [9–14]. MNP's basic properties, such as saturation magnetization (Ms), susceptibility (χ), spin lifetime (τ), coercivity (H), and blocking temperature (T), were greatly influenced by nano-scaling laws [15]. The magnetization and control of magnetic characteristics from the ferromagnetic to the superparamagnetic regimes were determined by the size, shape, and volume of magnetic particles. Below a critical size, ranging from 2 to 20 nm in diameter, multi-domain MNPs were converted to single domains, resulting in unusual magnetic properties such as superparamagnetism and quantum tunneling of magnetization.

MNPs were developed by combining magnetic metal elements, composites, metal alloys, and/or their oxides [16–19]. Since 2000, the nanostructures of Fe, Co, and their alloys FePt and CoPt have been the most extensively studied magnetic materials for magnetic recording and storage applications. The conventional materials used in high-density magnetic recording devices were γ-Fe_2O_3, CrO_2, and Co-γ-Fe_2O_3. Another group of magnetic materials that drew a lot of attention was spinel ferrite MFe_2O_4 (M=Fe, Co, Ni, Mn, Zn, and so on) [20–23]. The MNPs can be developed from several metal precursors, composites, metal alloys and/or their oxides with magnetic characteristics. High-density recording applications for audio/video, high-density digital recording disks, and high-speed digital magnetic tape attracted a lot of interest in spinel ferrites. Furthermore, the nanoparticle's self-assembled two-dimensional structure has demonstrated their fundamental and technological interest in ultrahigh density recording media. The widely accepted goal of higher areal density necessitates the arrangement of extremely small magnetic bits and grains in two dimensions with out-of-plane magnetization, resulting in the development of a new concept of bit-patterned media in which function of individual nanoparticle as single bit [24].

12.1.3 MAGNETIC DATA STORAGE MECHANISM

The general mechanism of magnetic storage was magnetization of the recording medium with corresponding electric strength of the information using an electromagnet as a reading head, and the reverse in case of reading the stored information. Magnetic tape storage uses magnetic material in the form of tape that runs at a constant speed and was magnetized by a static writing head with a current equal to the amount of data to be stored. Tape storage has two data storage methods based on the direction of the head movement: linear serpentine and helical scanning. (Figure 12.1). The helical scanning process comprises laying down tracks side by side at a particular angle to the edge of a tape using head mounted on a rotating scanner, as opposed to the linear serpentine process, which involved laying down a significant number of tracks in a linear way alongside the length of the tape from one end to the other and then in the opposite direction after the head was readjusted laterally across the tape [25].

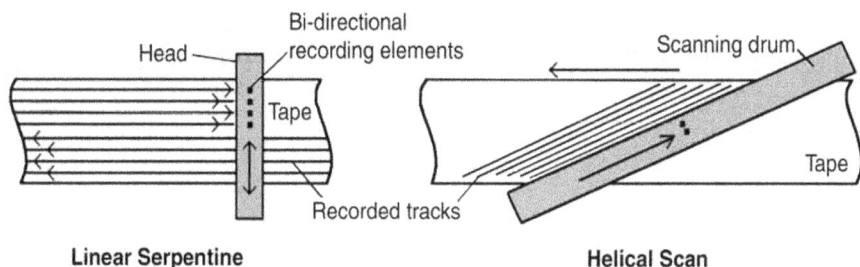

FIGURE 12.1 Magnetic tape recording using the linear serpentine and (right) helical-scan techniques. Reproduced with permission from [25] copyright (2006) Springer.

Magnetic hard drives were made up of two main components: magnetic medium and recording head. The recording head or slider was used to both record and read data from the disk. An electrical pulse equivalent to the digitalized data was applied over copper coil, which produces a magnetic flux that changes the state of the magnetization direction of a small area of magnetic medium, a magnetic layer on a rigid disk, below the write head as one bit, during the data recording process on hard disk. By using a massive magnetoresistance sensor to detect magnetism in the recording medium and generate an output voltage in the read head, the stored data was recovered. To store digital data, a magnetic recording system was used, and the current supplied to the write head was encoded as pulses to signify the digital data (1s or 0s) [26, 27]. A track was a series of bits recorded onto a rotating disk at a given radius. The linear density of the bits recorded down the track was decided by the intervals between the writer and reader. The width of the write head determined the track width, and the product of these two determined the areal storage density. Disk drives have distinct thin-film structures for the write and read heads that are placed on the backside of a mechanical slider that glides over the disk's surface utilizing a hydrodynamic air bearing. Without external power supply to the hard disk, the magnetization of the magnetic medium remained stable over years. MRAM's digital storage was analogous to magnetic hard disk drives and was considered a nonvolatile memory device. Two recording methods have been commercialized based on the direction of magnetization with respect to the surface of the magnetic medium. The conventional method of recording in a disk drive was longitudinal recording, in which the easy axis of magnetization of the recording medium was parallel to its surface, whereas the recently used perpendicular recording method (PMR) had the easy axis of magnetization perpendicular to the surface of the recording medium. PMR reduced the area of each bit, resulting in an increase in the recording medium's areal density [4].

Many technological advances in materials, mechanisms, and device structure have been observed to improve the capacity and performance of magnetic data storage devices. This chapter discusses the various fabrication routes of MNPs and thin films, as well as their functionalization with organic/inorganic molecules to tailor key magnetic parameters such as magnetic anisotropy, H_c, M_s, and susceptibility. Magnetic nanostructure functionalization has numerous proven advantages in biomedical applications. However, this is still a novel approach in the development of magnetic storage devices. As a result, this section provides insight into the use of functionalized magnetic materials in magnetic data storage devices.

12.2   SYNTHESIS/PREPARATION OF MAGNETIC NANOMATERIALS

Despite the rapid evolution of research in MNPs/thin films, more development in synthesis approaches is emerging to find the best synthesis methods that produce high-quality NPs. The last decade has seen extensive research into various approaches for the creation of MNPs. To produce MNPs, the common methods are ball milling, coprecipitation, hydrothermal, thermal decomposition, sol-gel method, microemulsion, and biological method. The schematic of the methods used for the formation of MNPs is shown in Figure 12.2.

12.2.1   THIN-FILM TECHNIQUES

Laser ablation deposition (LAD) in which a powerful laser beam impinges on a target, causing material to be ejected and gathered on an appropriate substrate, leading to the formation of a thin layer. When compared to sputtering, the laser ablation technique has some advantages, such as a faster deposition rate, a better reflection of the target composition, and the absence of gas stress during film formulation. The

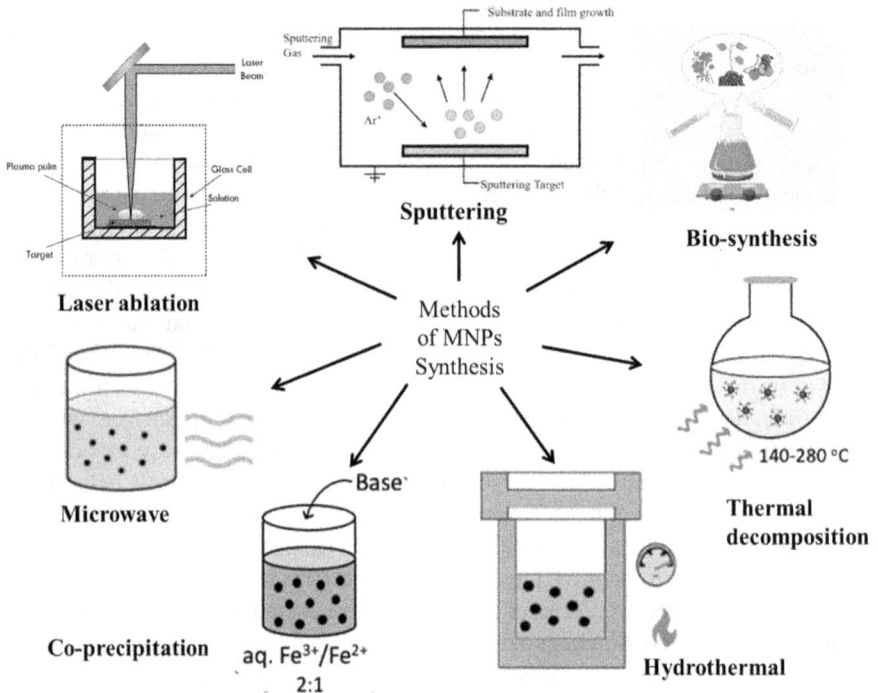

FIGURE 12.2   Methods for the preparation of magnetic nanoparticles

material, which is positioned at the bottom of a cell submerged in a liquid solution, absorbed the incident laser beam, causing a heat wave that resulted in the plume formation. This plume is ejected into the liquid, where it aggregates to form NPs. This method is less expensive than wet chemistry methods for synthesizing a range of particles with size control, and it does not require reducing agents or produce hazardous waste. This method is suitable for producing thin films/NPs with high magnetization of magnetic metal alloys, metal oxides, and ferrite.

Sputtering is a physical vapor deposition process that deposits materials onto a desired substrate by ejecting atoms from that material and condensing those ejected atoms under high vacuum conditions. This method yielded thin films ranging in thickness from a few nanometers to a few micrometers. The sputtering apparatus includes an annealing furnace for predeposition annealing of the NPs. The reduction in particle size required to achieve this goal has an impact on thermal instability and magnetic anisotropy. Sputtering is one of the most effective one-step methods for producing self-assembled metal and metal alloy NPs. Sputtering could be used to create high-density FePt nanodot arrays with porous aluminum templates at a density of 10^{12} dots/in^2. Porous anodic alumina (PAA) templates are made by dc anodizing aluminum foil in acidic solutions. Under the right conditions, hexagonally close packed pore arrays with adjustable size and periodicity could be produced and attached to glass or Si/SiO$_2$ substrates. Sputtering (Fe/Pt) n multilayers and removing the PAA templates produces FePt nanodot arrays and the thickness were varied between 500 and 2,000 nm by varying the number of recurrences [28]. Sputtering is used to prepare a patterned FePt nanodot array that stores each data bit in isolated dots at an ultrahigh density of beyond 1 Tbit/in^2. In this process, a 70 nm-thick CrV film is sputtered on glass as a seed layer, and a type of positive E-beam resist (ER) is encrusted on CrV film. E-beam exposure is used to define a regular array of circular holes with fixed sizes, and then a 7 nm FePt is sputtered-deposited at RT under high vacuum. After the acetone lift-off process, the regular array of FePt of uniform size is obtained. Postannealing at 400°C aids in the phase conversion of FePt from A1 to L10 [29]. Similarly, FePt/C and CoPt/C films are deposited onto Si (111) substrates by dc magnetron sputtering from solid CoPt (FePt) and C targets at a pressure of 5 m. Torr in a high purity Argon gas flow. FePt/C films with high magnetic ordering and distinctive characteristics were produced by direct deposition of FePt and C onto warmed substrates at temperatures exceeding 450°C, whereas ordered fct FePt particles are produced by sputtering at high substrate temperatures.

12.2.2 CHEMICAL SYNTHESIS METHODS

Coprecipitation is the most used solution-based method for preparing metal oxide, ferrites, and perovskites NPs. This method requires careful adjustment of the solution pH and reaction temperature. When large quantities of nanocrystals are required, the synthesis of MNPs via coprecipitation is accessible and simple. Metal nitrates, acetates, and chlorides are commonly used as precursors in this method, while ammonia and NaOH are frequently used as a precipitating agent and pH controller. For example, ferric, manganese (II) chloride as metal precursors and NaOH as precipitants could be used in the preparation of manganese ferrite (MnFe$_2$O$_4$) NPs. This method could be used to prepare NPs of $CoFe_2O_4$, $MnFe_2O_4$, and Fe_3O_4. The chemical composition, particle size, and shape of materials prepared using this method are defined by pH, the type and fraction of the salts, and the temperature.

Hydrothermal method is solution-based synthesis method for producing MNPs on a large scale at high pressure and temperature. Hydrolysis and oxidation reactions occur during the hydrothermal process to produce NPs. By choosing the suitable solvent combination and adjusting parameters such as pressure, temperature, and time, this method can produce fine MNPs with controllable size, size distribution, and shape. The hydrothermal method was used to create NPs of Fe_3O_4 NPs of 15 nm size and spherical shape, as well as Chitosan-coated Fe_4O_3 NPs of 25 nm. As shown in Figure 12.3, nanocrystalline and mesoporous $NiFe_2O_4$ particles of 5–8 nm were synthesized utilizing supramolecular assembly as a structure directing agent [31]. Under alkaline pH, an anionic surfactant assembly of lauric acid could work together with the positively charged metal centers of $NiFe_2O_4$ species through electrostatic exchanges, which is intended to stabilize the material's mesophase.

Other method used to prepare MNPs is sol-gel process and the whole chemistry of the sol-gel process involves the gel development at RT via hydrolysis and metal

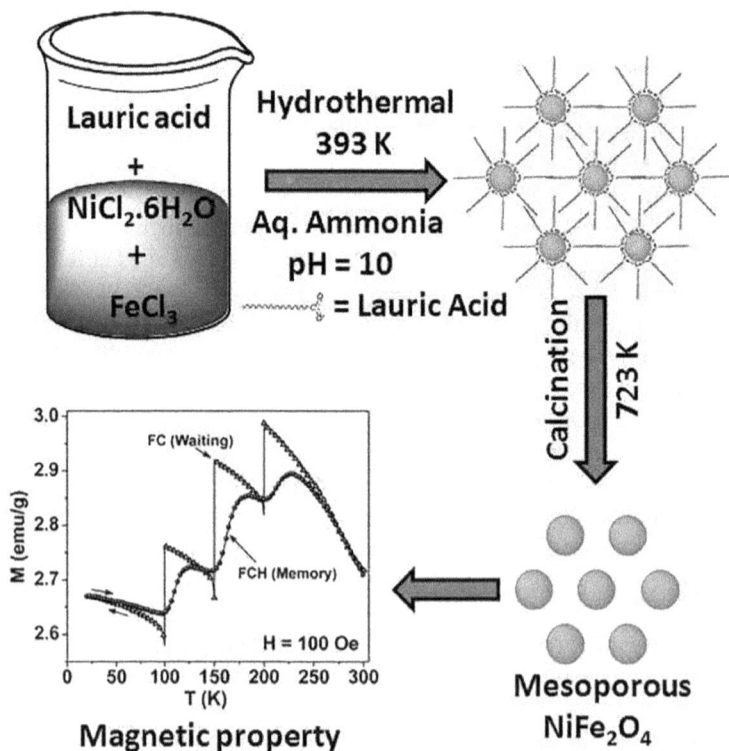

FIGURE 12.3   Steps involved in the formation $NiFe_2O_4$ NPs by using lauric acid as template and its interesting memory effect as a consequence of interparticle interaction of self-assembled NP. Reprinted with permission from [30] (2016) Royal Society of Chemistry.

alkoxide polycondensation reactions. To prepare a colloidal, metallic precursors are mixed in water or other solvents and homogeneously dispersed. Particles connect due to the van der Waals forces, and this contact gets stronger as the temperature increases. In order to create gel, the mixture must be heated until the solvent has been evaporated and the solution has dried. This method is useful for producing large quantities of MNPs with controlled size and well-defined shape. The benefits of this process, such as its simplicity, low cost, and low reaction temperature, made it more suitable for the formulation of NPs without the use of specialized equipment. However, occasionally, contamination from byproduct reactions occurs with this approach, necessitating retreatment to generate pure MNPs. The disadvantage of this method is that it requires a longer reaction time and uses toxic organic solvents.

Microwave-assisted synthesis is a relatively new technique for producing fine-grained NPs. Energy is delivered directly to materials in a microwave system via molecular interaction with electromagnetic radiation, which generates heat. The benefit of this method is that the temperature used is in the range of 100°C–200°C for a very short time (no more than 10 minutes), allowing for easy removal of vapor produced during microwave heating. This method can be used in industrial applications to mass produce NPs with uniform size distribution, high quality, and improved reproducibility. Microwave-assisted rapid synthesis was demonstrated to create fine NPs of spinel ferrites from 3 to 5 nm, stabilized by oleic acid, oleyl amine, and trioctylphosphine oxide [32]. This method could be used to create Fe_3O_4, γ-Fe_2O_3, and other oxide-based MNPs without the use of a stabilizing agent. Microwave heating provides a more homogeneous and shorter nucleation time than other conventional methods, making it a more innovative, efficient, cost-effective, and time-effective chemical method for the fabrication of MNPs.

The thermal decomposition technique is also could be used for producing MNPs. To produce NPs of preferred size and shape, organometallic precursors are decomposed in the existence of organic surfactants. Stabilizing agents such as fatty acids, hexadecyl amine, and oleic acid helps to control the growth of NPs during the deposition process and produce a particle with a desirable size. The thermal decomposition of $Fe(CO)_5$ results in the formation of iron NPs, which can then be oxidized to form high purity iron oxide NPs. The decomposition of the metal precursor $Fe(CO)_5$ with cationic metal centers can cause in the direct development of Fe_3O_4 NPs.

12.2.3 Other Methods

Biological synthesis also could be used for synthesizing MNPs using living organisms such as microorganisms (bacteria, viruses, fungi, and actinomycetes) and plants. The advantage of this method is its eco friendliness, and cleanliness. The synthesis of NPs from various plant parts has piqued the interest of researchers. Biological synthesis mechanism of NP formation using microorganisms and plants is even now underneath research. A few studies suggested that potential mechanisms for the mycosynthesis of MNPs include nitrate reductase activity, shuttle electrons quinones, and a miscellaneous mechanism. But, the process is not clear enough to acknowledge to prepare MNPs. As a catalyst in the Suzuki–Miyaura reaction and photocatalysis, a biologically synthesized Fe_3O_4 was used [33]. Some shortcomings of this method, such as yield and dispersion of MNPs, need further investigation.

12.3 FUNCTIONALIZATION OF MAGNETIC NANOMATERIALS

The proportion of magnetic grains needed to hold a bit of data has decreased from 1,000 to a few 100 in recent years due to improving data storage density, as well as a drop in grain size. Ferromagnetic NPs with uniform size and shape are important in the design and fabrication of self-assembled magnetic media including hard disks and magnetic tape. The particles' single-domain status is also critical for preventing undesirable magnetization reversal via domain wall motion [34]. However, strong magnetic anisotropy must be maintained to prevent thermal instability of magnetization. Thus, functionalization of MNPs with organic, polymer, or inorganic molecules reduces particle aggregation, increases particle stability and uniformity, and aids in the formation of single-domain particles and self-assembling of particles in magnetic media without interaction with neighboring particles. The list of magnetic nanoparticles functionalized with various molecules and the preparation methods are tabulated in Table 12.1.

High magnetic moment superparamagnetic NPs can be prepared in organic phase syntheses to achieve monodispersity in size, shape, and composition. They are typically stabilized by surfactants which are all hydrophobic. The water-soluble quantum dots were originally created using the surfactant addition technique, which was

TABLE 12.1
Magnetic Nanoparticles, Their Method of Preparation and Functionalized Molecules

Magnetic Nanoparticles	Functionalized Material	Method Used	Ref.
10 mol% Mn-doped ZnO	Amine	Magnetron sputtering	39
5 % Ni-doped ZnO	TOPO	Magnetron sputtering	40
5 mol% Ni-doped ZnO	Thiol	Magnetron sputtering	42
10 mol% Mn-doped ZnO	Thiol	Magnetron sputtering	41
FePt	Carbon	Sputtering	9
CoPt	Carbon	Sputtering	9
FePt	Carbon nanotubes	Polyol process	54
Fe_3O_4	APTS	Coprecipitation	43
Fe_3O_4	SiO_2	Modified reduction reaction	37
Fe_3O_4	SiO_2-NH_2	Coprecipitation	44
Fe_3O_4	Borophene	Hydrothermal	46
Fe_3O_4	Polyacrylic acid	Coprecipitation	36
Fe_3O_4	SiO_2	Sol-gel	51
Fe_3O_4	GO	Solvothermal method	53
FeCo	Graphene	Chemical vapor deposition	35
Cobalt	Polystyrene	Reflux method	55
FePt	Polyethylenimine	Solvothermal and spin coating	56

subsequently effectively proven for dispersing MNPs into aqueous solutions for biomedical purposes. However, reports on the use of organic-functionalized MNPs in magnetic data storage are uncommon. Nonetheless, some studies found a significant change in the magnetic behavior of MNPs that were functionalized with organic molecules. After coating of a PEG-modified phospholipid micelle, high magnetic moment superparamagnetic FeCo NPs were reported [35]. Iron oxide NPs coated with polyacrylic acid demonstrated superparamagnetic behavior, with significant changes in Hc and residual magnetization as polyacrylic acid concentration was increased [36]. There was no H_c or remanence witnessed for Fe_3O_4, Fe_3O_4/silica, and Fe_3O_4/silica-NH_2, indicating the superparamagnetic nature of materials with decreased M_s after silica modification on Fe_3O_4 [37]. Although amine functionalization on the surface of Fe_3O_4 reduced the magnetization, as a basic catalyst it controls particle size and inhibits hydrolysis to form particles with regular morphology [38].

Pure ZnO is paramagnetic by default; however, doping it with Ni and Mn causes it to become ferromagnetic. Surface functionalization with different organic molecules as thiol, amine, and trioctahedral phosphine oxide (TOPO) molecules is anticipated to further improve this ferromagnetic property. Both un-functionalized and functionalized Ni, Mn-ZnO exhibited a ferromagnetic characteristic (Figure 12.4). Bound magnetron polarons (BMP), which are created by the exchange interaction among oxygen vacancies and Mn ions, have been used to explain how oxygen vacancies' impact on ferromagnetism in Mn-ZnO films at different oxygen pressures. All ZnO: Mn film's Ms increased after being functionalized with amine, however the amine-functionalized 10 mol% ZnO: Mn film had the better Ms value of 2.1 emu/cm^3 (Figure 12.4a). The functional group features of N-added BMP model that resulted in the addition of N atom on the exterior of the Zn. The N atom stimulates the charge redistribution that leads to hybridization between 4s and 2p orbitals and resulted in the establishment of Zn–N and N–O bonds which enhanced the Ms [39]. The M–H loop of TOPO-functionalized Ni-ZnO films recorded at RT showed a ferromagnetic behavior for all the unfunctionalized and functionalized Ni-ZnO films (Figure 12.4b). The Ms of Ni-ZnO films are 5.7, 10.1, and 8.6 emu/cm^3 for 3, 5, and

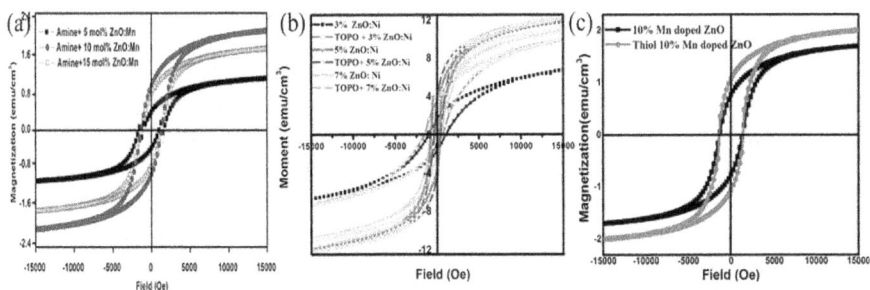

FIGURE 12.4 M–H curve of (a) amine-functionalized Mn-doped ZnO, (b) TOPO-functionalized Ni-doped ZnO, (c) thiol-functionalized Ni-doped ZnO, and (d) thiol-functionalized Mn-doped ZnO thin film. Reprinted with permission from [39] copyright (2021) Springer, [40] copyright (2019) Elsevier and [41] copyright (2018) through Copyright Clearance Center, Inc.

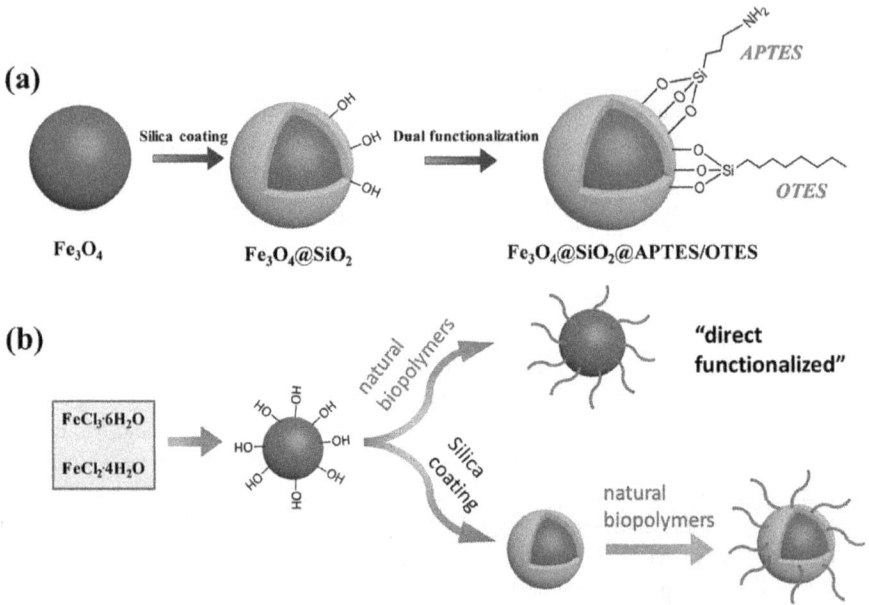

FIGURE 12.5 (a) Process of TOPO functionalization on ZnO thin film by spin-coating technique, Bound magnetic polaron model, (b) Ni-doped ZnO, and (c) TOPO-modified Ni-doped ZnO. Reproduced with permission from [40] copyright (2019) Elsevier.

7 mol%, respectively. Higher Ms was observed for 5 mol% of Ni-ZnO. The occurrence of extra grain boundaries, which order ferromagnetic like behavior, favors ferromagnetic behavior as crystallite size decreases. Due to the exchange interaction among oxygen vacancies and Ni ions in the Ni-ZnO, the development of BMP is the cause of the ferromagnetism in the Ni-doped films (Figure 12.5b). In host matrix, the oxygen vacancies are coupled with Ni ions and form the isolated magnetic polarons. The process of TOPO functionalization is schematically shown in Figure 12.5a. More polarons overlap as a result of more oxygen vacancies, which in turn mediates ferromagnetism.

Considering the TOPO-functionalized Ni-doped ZnO, the Ms was improved in all films but retains the same trend. The Ms values were measured as around 8, 12, and 10 emu/cm^3 for 3, 5, and 7 mol%, respectively. The enhanced Ms is due to the formation of Zn–P and P–O bonds upon surface functionalization. Due to TOPO functionalization, the P atom sits on the exterior of the Zn. Due to the P atom's lower electronegativity value (2.19), when compared to the O atom, the charge redistribution is caused (3.44). In this case, the ligands share the Zn site's charge, causing a charge imbalance between the Zn–O bonds and lowering the Zn–O bond as a result. As a result, the Zn–O bond close to the P site extended, while the Zn–O bonds next to it significantly shrank. The system becomes more ferromagnetic because of variations in bond length and charge redistribution that result in additional holes on the p orbital. Therefore, instead of oxygen vacancies as in Ni-ZnO film, the augmentation of ferromagnetism in TOPO-modified Ni-ZnO films could be caused by charge distribution (Figure 12.5c). Due to the hybridization of Zn-O-P orbitals after TOPO functionalization, the Hc was then reduced for all films, but it still follows the same pattern as unfunctionalized films. For spintronic memory devices, the low Hc and high Ms are beneficial. Given this, the TOPO-functionalized 5 mol% Ni-ZnO has the lowest Hc (290 Oe) and highest Ms (11.51 emu/cm^3) of all the films, making it suitable for usage in spintronic applications [40].

Functionalization of the same Ni-doped ZnO films with thiol molecules improved the Ms to 6.9, 13.2, and 11.2 emu/cm^3 for 3, 5, and 7 mol% of Ni concentration, respectively. This enhancement in magnetization can be attributed to thiol ("S" ligands) adsorption on the surface of Ni-ZnO films, which modifies the surface bonds of Zn-4s and O-2p states in conduction bands. Thiol molecules' alteration of the surface bond triggers the ferromagnetic properties. Thus, thiol functionalization in Ni-ZnO led to ferromagnetism as a result of modification in ZnO's conduction band, which in turn led to the formation of surface layers linking sulfur-related bonds on the exterior. Specifically, after adsorption on the ZnO surface, S-3p states are coupled with O-2p states and delocalized nearby the Fermi level. As a result, the delocalizations near the Fermi level and spin polarization of O-2p states are caused by the spin polarized S-3p levels. Hence, the magnetism exists in only on the exterior, and thus the thiol atoms disturb the surface atomic layers [42]. Similar impact on the magnetization was witnessed in the case of thiol functionalization Mn-ZnO, and for 10% Mn, a doped ZnO film, a maximum of 18% enhanced ferromagnetism was observed [41] (Figure 12.4c).

Surfactant and coupling agents were among the organic small molecules used to alter the exterior of Fe$_3$O$_4$ NPs, resulting in monodisperse well-defined particles with less interparticle interaction. The water-soluble quantum dot manufacturing technique was initially devised, and subsequently it was successfully tested for dispersing magnetic NPs into aqueous mediums. Silane coupling agent is commonly used surface alkylation modifying material such as 3-aminopropyltriethoxysilane (APTS), (3-aminopropyl) triethoxysilane (APTES), and octyl triethoxysilane (OTES). Due to the monodispersity in size and shape, superparamagnetic particles are made in organic phase synthesis. However, these SPM particles are normally stabilized by surfactants like oleic acid and/or oleyl amine and are hydrophobic molecules. The Fe$_3$O$_4$ NPs were created using the coprecipitation method, and then they were dispersed into a homogenous solution containing 3-aminopropyltriethoxysi-lane (APTS) to create magnetic Fe$_3$O$_4$ NPs (Fe$_3$O$_4$-APTS) modified by APTS [43]. Lee et al. coated SiO$_2$ on the outer layer of Fe$_3$O$_4$ NPs with TEOS as a silicon source first, followed by adding (3-amino-propyl) triethoxysilane (APTES) and octyltriethoxysilane (OTES) into the mixed solution, then the surface-modified magnetic nano-adsorbed material (Fe$_3$O$_4$@SiO$_2$@APTES/OTES) and the schematic preparation process is showed in Figure 12.6a. The possible two types for the functionalization of NMPs with natural biopolymer are shown in Figure 12.6b. Likewise, Zhang et al. formed a layer of SiO$_2$ on the outer layer of Fe$_3$O$_4$ NPs using sodium silicate as the silicon precursor in alkaline situations and then alkylated the exterior of Fe$_3$O$_4$@SiO$_2$ by adding APTS, yielding Fe$_3$O$_4$@SiO$_2$@APTS [44].

Wei et al. have functionalized the Fe$_3$O$_4$ NPs with borophene, which is a material with unique electronic structure, to develop high-performance rewritable memory

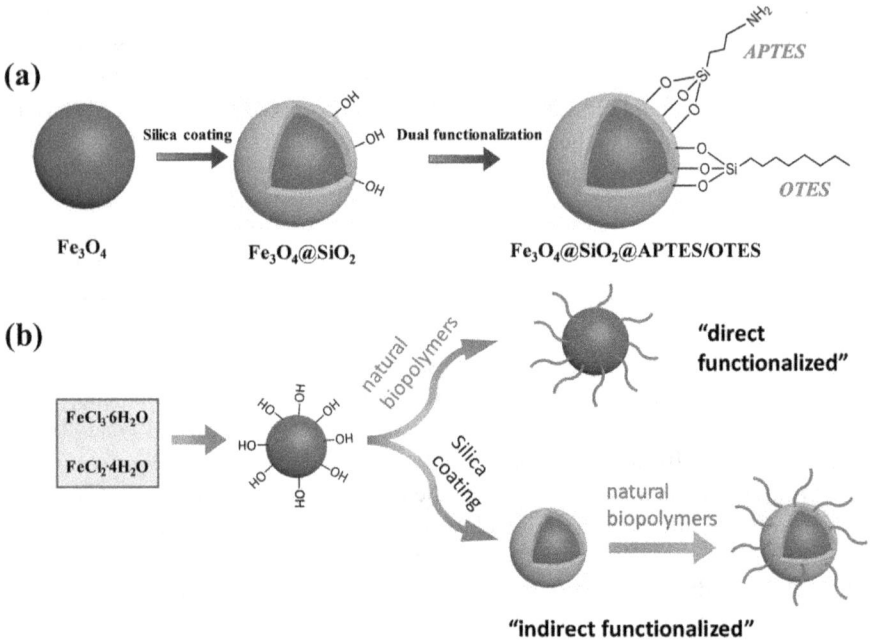

FIGURE 12.6 Synthesis of dual functional MNPs and (b) two different ways for the functionalization the MNPs with biopolymers. Reproduced with permission from [45] copyright (2015) American Chemical Society.

device. For this purpose, a metal–dielectric–insulator–metal (MDIM) arrangement was designated to form the memory device. The borophene-functionalized Fe_3O_4 core–shell NPs and poly (vinyl pyrrolidone) (PVP) were mixed by ethanol followed by ultra-sonication, where the core–shell NPs are the active layers and PVP are the insulators. Schematic illustration of fabrication of borophene-functionalized Fe_3O_4 NPs and the structure of borophene shell is shown in Figure 12.7a–c). The TEM micrograph of borophene-functionalized Fe_3O_4 NPs is shown in Figure 12.8(a). The schematic diagram of the developed memory device with borophene-functionalized Fe_3O_4 NPs is shown in Figure 12.8b.

A set voltage of 0.57 V and high on/off current ratio of 10^5 were measured from the I–V and retention time–current measurement, respectively. Its reset voltage of around 0.19 V is 20 folds lower than that of the MoS_2-PVP device and the on/off current ratio of 8.23×10^5 is over two orders of magnitude higher than the BQDs-PVP device (Figure 12.8c and d), implying that the borophene-Fe_3O_4 device has superior memory performance while consuming less power [46].

The foremost and easiest way for the modification of MNP is functionalization with inorganic small molecules such as SiO_2, Al_2O_3, graphene, and carbon, on the surface of Fe_2O_3 NP [47–50]. The chemical bonding interaction occurs on the interface between Fe_3O_4 NPs exterior and inorganic materials, causing in high bonding strength. Chen et al. demonstrated the activation of the exterior of Fe_3O_4 NPs in

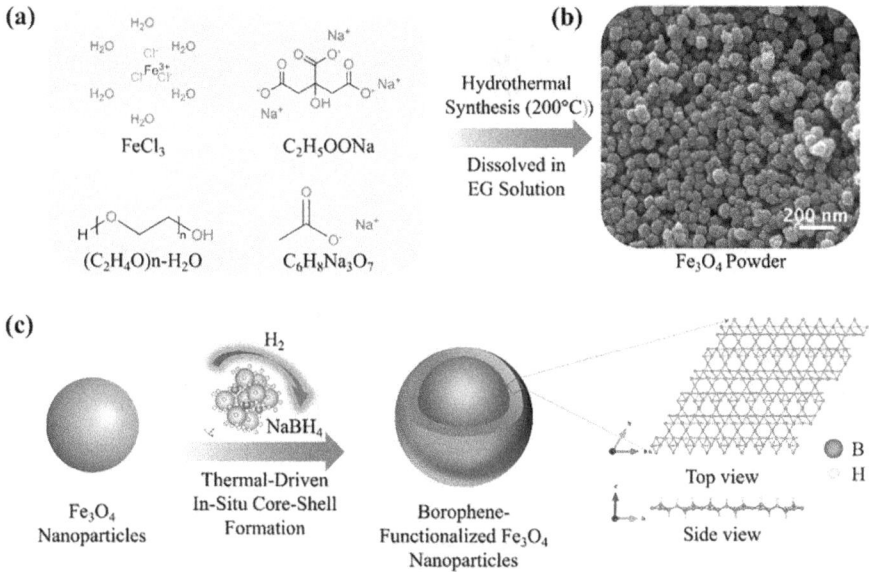

FIGURE 12.7 (a) Schematic illustration for the synthesis of Fe_3O_4 NPs. (b) SEM micrograph of Fe_3O_4 NPs (c) Schematic illustration for borophene-functionalized Fe_3O_4 NPs formation and borophene shell structure. Reproduced with permission from [46] copyright (2021) American Chemical Society.

HCl solution and then used TEOS as the silicon source, and MNPs ($Fe_3O_4@SiO_2$) coated with SiO_2 were obtained at RT [51]. The as-prepared $Fe_3O_4@SiO_2$ NPs were redispersed into a homogeneous solution comprising CTAB and TEOS, and finally, the MNPs coated with a layer of mesoporous SiO_2 ($Fe_3O_4@SiO_2@m\text{-}SiO_2$) were obtained [52]. Wei et al. [53] used a self-assembly process to prepare a core–shell magnetic Fe_3O_4/GO and demonstrated good water dispersibility, high M_s, and sensitive magnetic response.

The procedure for grafting inorganic small molecules on the exterior of MNPs is simple and straightforward. After surface functionalization, the structural stability and magnetic properties of magnetic nanomaterials are altered. The effect of substrate and oxidation during annealing was reduced in DC magnetron sputtered FePt and CoPt films encapsulated with 200 Å layers of C as under layer and over layer. Small FePt NPs form in carbon-rich samples and exhibit superparamagnetic behavior at RT. The Hc of FePt/C and CoPt/C could be organized by tuning the bilayer thickness and annealing environments. The Hc develops significantly more slowly in films with higher C content because the matrix material restricts the expansion of the particle size. CoPt/C developed with 700°C showed a Ms of around 750 emu/cc and an anisotropy constant (Ku) of around 9×10^6 ergs/cc at RT. At low temperature, the Hc increased from 5,500 to 9,200 Oe and Ms and K_u improved to 779 emu/cc and 1.2×10^7 ergs/cc respectively, which are nearer to the theoretical value of bulk CoPt [9]. Strong exchange contacts (M positive) between particles were seen in films with little carbon content, and this led to a more squared hysteresis loop. The exchange

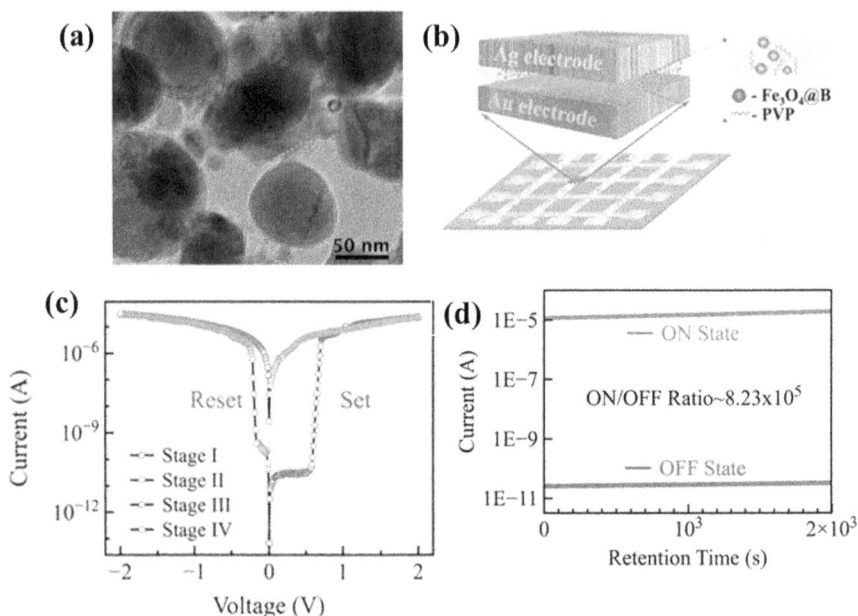

FIGURE 12.8 (a) TEM micrograph of borophene-functionalized Fe_3O_4, (b) schematic diagram of borophene-modified Fe_3O_4 memory device. (c–d) I–V curve and retention curve of the memory device of borophene-modified Fe_3O_4 NPs, respectively. Reproduced with permission from [46] copyright (2021) American Chemical Society.

contact between particles reduces as carbon content rises. The interaction between the particles is primarily dipolar (M negative) and a more slanted loop is seen when the carbon concentration is significantly higher than CoPt with a component ratio of roughly 6:9 showed M close to zero as a consequence of balancing among exchange and dipolar interactions. Given that the materials exhibited fairly squared loops while preserving particle isolation, it may be the best for magnetic recording applications. This results in a higher signal-to-noise ratio in the recording film. Strong exchange contacts ($\delta M_{positive}$) between particles were seen in films with little carbon content, and this led to a more squared hysteresis loop. The exchange contact between particles reduces as carbon content rises. The interaction between the nanoparticles is primarily dipolar ($\delta M_{negative}$) and a more slanted loop is seen when the carbon concentration is significantly higher than CoPt. Films with a component ratio of roughly 6:9 showed δM close to zero as a consequence of balancing among exchange and dipolar interactions. Given that the materials exhibited fairly squared loops while preserving particle isolation, it may be the best for magnetic recording applications. This results in a higher signal-to-noise ratio in the recording film.

$L1_0$ FePt NPs with face-centered tetragonal (fct) structure have high k_u of around 7×10^7 erg/cm^3 and high H_c makes the fct-structured NPs to be utilized for magnetic recording. During thermal annealing, the surfaces of the CNTs act as a substrate to

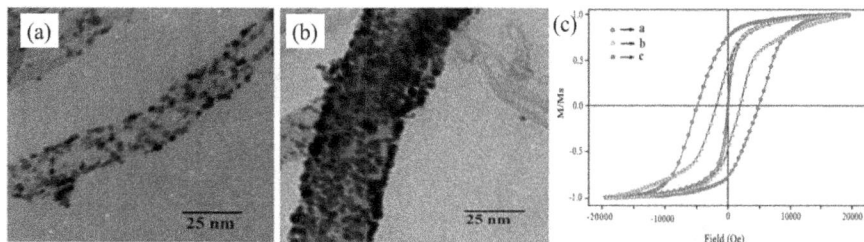

FIGURE 12.9 TEM micrograph of FePt NPs on CNTs which were prepared with (a) 2 h and (b) 4 h. (c) Hysteresis loops of FePt/CNT: a: as-synthesized, b: thermally treated at 600 C for 2 h, and c: at 700 C for 2h. Reproduced with permission from [54] copyright (2013) Springer.

stop the coalescence of the NPs. The thermal treatment of FePt NPs at 600°C for 2 h in reducing environment leads to phase transition from fcc to fct-L1$_0$ and most of the FePt NPs have converted to fct-L1$_0$ after 700 °C annealing. During functionalization processes, many active sites on CNTs are formed, which are useful for nucleation and FePt NPs growth (Figure 12.9a and b). The reaction between the surfaces of the NPs and the CNTs produces fixed size FePt NPs with a typical size of around 3.5 nm which are annealed at 600 °C and the higher H$_c$ of 5.1 kOe was observed for the samples annealed at 700 °C (Figure 12.9c) [54].

Magnetic storage media require ferro/ferri MNPs that act as individual magnets at room temperature. Magnetically induced particle aggregation and interparticle attractive force impede solution-based synthesis of stabilized self-assembled ferromagnetic NPs. Polymer functionalization of ferromagnetic NPs allows the particles to self-assemble and avoids the magnetically induced aggregation that is commonly associated with ferromagnetic NPs.

Pyun et al. synthesized the polystyrene-coated Co NPs with end groups that interact with the growing ferromagnetic Co NP exterior, acting as a steric buffer among the particles and reducing interparticle magnetic interactions [55]. Similarly, monodisperse 4 nm FePt NPs coated with a layer of polyethylenimine (PEI) and thermal treatment converts the fcc to fct structure, in which superparamagnetic thin FePt NPs assemblies change to ferromagnetic [56]. The TEM micrograph of polystyerene/cobalt and PEI-mediated FePt MNPs are showed in Figure 10a, b, d, and e. To form core–shell composites, the oleic acid surfactants were replaced with poly (acrylic acid)-b-poly(styrene) (PAA-b-PS) diblock copolymers to improve the solution stability of the CoFe$_2$O$_4$ NPs. The schematic process of self-assembled functionalized MNPs in thin film form and the physical re orientation to have a same alignment direction is shown in Figure 12.11a. When compared to oleic acid-stabilized CoFe$_2$O$_4$ NPs, the interparticle spacing for the core–shell complex was increased (Figure 12.11b and c).

The magnetic behavior of CoFe$_2$O$_4$ NPs with a size of 18 nm remains ferrimagnetic after polymer coating, indicating that changing the surface ligands has little effect on the magnetic properties. When polymer modified MNPs were placed onto

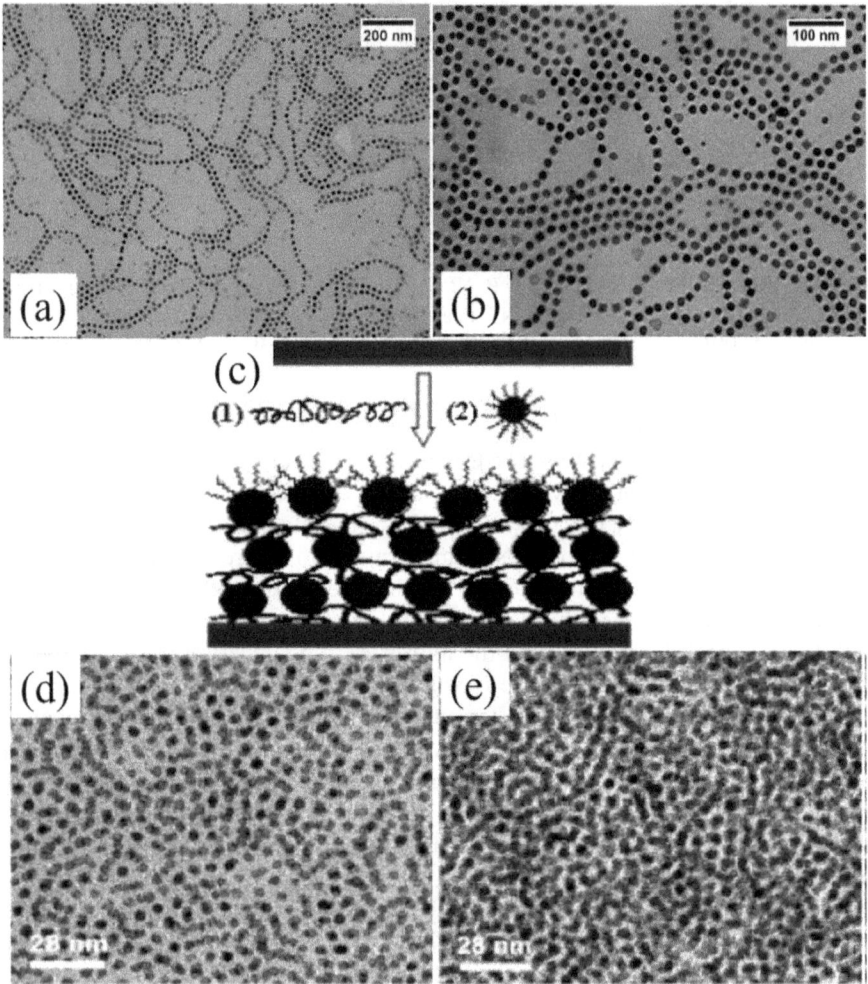

FIGURE 12.10 TEM images of (a–b) polystyrene-coated ferromagnetic Co NPs., Reproduced with permission from [55] copyright (2007), (c) schematic diagram, and (d–e) TEM micrograph of PEI-mediated assembly of 4 nm $Fe_{58}Pt_{42}$ NPs on silicon oxide coated Cu grids with single later and more number of layers. Reproduced with permission from [56] (2003) American Chemical Society.

a substrate in a magnetic field, packs of MNPs formed and the magnetic easy axes of the magnetic particles were physically reorientated to be aligned with magnetic field by heating the materials beyond the poly(styrene)'s glass transition temperature, which provided the particles with the necessary mobility (Figure 12.11d and e). The write/read experiments carried out for polymer-coated $CoFe_2O_4$ NPs revealed that self-assembled and magnetically orientated $CoFe_2O_4$ endure magnetization reversal shifts at moderate linear densities and can be read back nondestructively [57]. The

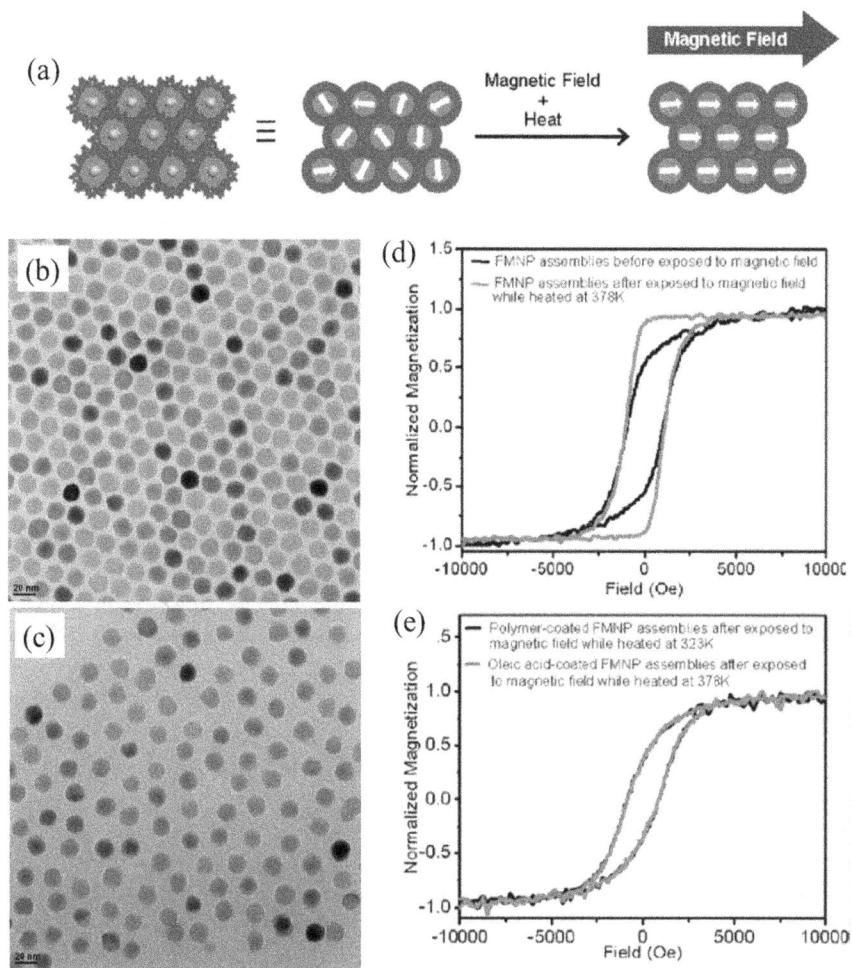

FIGURE 12.11   (a) A schematic illustration of magnetic alignment process of Co ferrite NPs modified with PAA_{33}-b-PS_{340}. (d) The hysteresis loops of polymer-coated $CoFe_2O_4$ NPs measured at RT (before) and at 378 K (after) magnetic alignment. (e) Hysteresis curves polymer-coated Co ferrite NPs heated to 323 K while exposed to a magnetic field and the oleic acid-Co ferrite NPs exposed to a magnetic field and heated at 378 K Reproduced with permission from [57] copyright (2010) American Chemical Society.

overall process involved in the fabrication of devices for magnetic data storage applications is shown in Figure 12.12.

12.4   SUMMARY AND SCOPE

The first commercial HDD used aluminum disks coated with γ-Fe_2O_3 magnetic particles and polymer binder to locate a single record in an average time of 600 ms. It

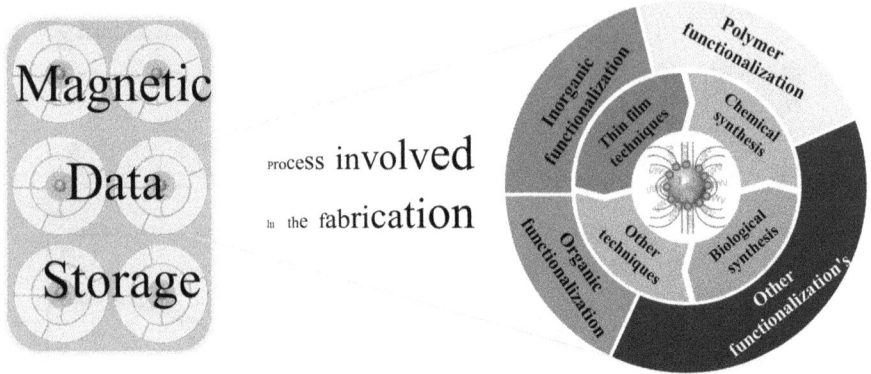

FIGURE 12.12   Process involved in the fabrication of magnetic data storage applications.

could store 5 MB of data at a density of 2 kbits/in^2 and required a 9×15 m room to be placed in. The disk size has continued to shrink, with a doubling of areal density every 18 months, a phenomenon known in the semiconductor industry as Moore's law. Every 18 months, the cost per MB of storage has decreased by a factor of two. In 2018, the disk diameters were 2.5 and 3.5 in., and the number of disks ranged from one to a few. The recording areal density was several TB/in^2 with a single 2.5 in. diameter disk in 1 TB portable disk storage. The advancement of magnetic reading/writing heads has contributed to the advancement of hard disk drive areal density. The progressive reduction in the length and thickness of the write/read head sensor resulted in higher areal density and, consequently, higher HDD capacity. This progress has continued for the past 3 decades and will continue in the future in the motivation of high-density data storage that will meet future needs in massive data storage.

To meet increased storage demand in the upcoming years, the industry would need to increase NAND flash memory production as well as hard disk drive capacity. Modern HDDs with perpendicular magnetic recording (PMR) and shingled magnetic recording (SMR) platters have an area density of about 0.95 Terabit per square inch (Tb/in^2) and can store up to 10 TB of data (on seven 1.43 TB platters). Two-dimensional magnetic recording (TDMR) technology has the potential to increase the areal density of HDD disks by 5% to 10%, which is a significant enhancement. Furthermore, Showa Denko K.K. (SDK), the world's largest independent hard drive platter manufacturer, has announced plans to begin mass production of ninth-generation PMR HDD media with areal density of up to 1.3Tb/in^2 in the upcoming years. Heat-assisted magnetic recording (HAMR) technology has the potential to significantly increase HDD capacity in the upcoming years. Unfortunately, mass production of actual hard drives featuring HAMR has been delayed for several years already and now it turns out that the first HAMR-based HDDs are in due. MAMR, or microwave-assisted magnetic recording, asserts several benefits, the majority of which are related to that all-important storage metric—reliability [58]. MAMR drives are expected to hit the market in early 2023.

MAMR could theoretically reduce the magnetic field required for magnetization reversal by one-third [59], whereas HAMR could theoretically reduce the magnetic field required for recording almost to zero by heating a recording medium to the Curie point. HAMR is a promising energy-assisted magnetic recording technology that has the potential to solve the trilemma associated with increasing HDD recording density. Toshiba is currently working on the commercialization of 3.5-inch nearline HDDs with MAMR read/write heads and media. To increase the recording density even further, the size and magnetic anisotropy Ku of magnetic grains on recording media must be reduced. Recently, the HDD industry has considered using an iron–platinum (FePt) ordered alloy that provides Ku several times to several tens of times higher than the currently used Co-Cr-Pt alloy [60]. According to Seagate and the ASTC, HAMR will result in HDD capacities approaching 50TB "early next decade" and combining HAMR with bit-patterned media to build heated-dot magnetic recording (HDMR) drives will allow us to reach the 100TB HDD milestone.

REFERENCES

[1] Historical evolution of magnetic data storage devices and related conferences | Microsystem Technologies, (n.d.). https://dl.acm.org/doi/abs/10.5555/3288646.3288703 (accessed October 15, 2022).

[2] H. Coufal, L. Dhar, C.D. Mee, Materials for magnetic data storage: The ongoing quest for superior magnetic materials, MRS Bull. 2006 315. 31 (2011) 374–378. https://doi.org/10.1557/MRS2006.96.

[3] A. Anžel, D. Heider, G. Hattab, The visual story of data storage: From storage properties to user interfaces, *Comput. Struct. Biotechnol. J.* 19 (2021) 4904–4918. https://doi.org/10.1016/J.CSBJ.2021.08.031.

[4] R.L. Comstock, Review modern magnetic materials in data storage, *J. Mater. Sci. Mater. Electron.* 139 (2002) 509–523. https://doi.org/10.1023/A:1019642215245.

[5] I.R. McFadyen, E.E. Fullerton, M.J. Carey, State-of-the-art magnetic hard disk drives, *MRS Bull.* 2006 315. 31 (2011) 379–383. https://doi.org/10.1557/MRS2006.97.

[6] D.A. Thompson, J.S. Best, Future of magnetic data storage technology, *IBM J. Res. Dev.* 44 (2000) 311–322. https://doi.org/10.1147/RD.443.0311.

[7] R. Wood, The feasibility of magnetic recording at 1 terabit per square inch, *IEEE Trans. Magn.* 36 (2000) 36–42. https://doi.org/10.1109/20.824422.

[8] Advances in magnetic data storage technologies, Proc. IEEE. 96 (2008) 1749–1753. https://doi.org/10.1109/JPROC.2008.2004308.

[9] X. Sun, Y. Huang, D.E. Nikles, FePt and CoPt magnetic nanoparticles film for future high density data storage media, *Int. J. Nanotechnol.* 1 (2004) 328–346. https://doi.org/10.1504/IJNT.2004.004914.

[10] T. Arun, R. Justin Joseyphus, Prussian blue modified Fe3O4 nanoparticles for Cs detoxification, *J. Mater. Sci.* 49 (2014) 7014–7022. https://doi.org/10.1007/S10853-014-8406-X/FIGURES/6.

[11] T. Arun, K. Prakash, R. Justin Joseyphus, Synthesis and magnetic properties of prussian blue modified Fe nanoparticles, *J. Magn. Magn. Mater.* 345 (2013) 100–105. https://doi.org/10.1016/J.JMMM.2013.05.058.

[12] T. Arun, K. Prabakaran, R. Udayabhaskar, R. V. Mangalaraja, A. Akbari-Fakhrabadi, Carbon decorated octahedral shaped Fe3O4 and α-Fe2O3 magnetic hybrid nanomaterials

for next generation supercapacitor applications, *Appl. Surf. Sci.* 485 (2019) 147–157. https://doi.org/10.1016/j.apsusc.2019.04.177.

[13] T. Arun, S.K. Verma, P.K. Panda, R.J. Joseyphus, E. Jha, A. Akbari-Fakhrabadi, P. Sengupta, D.K. Ray, V.S. Benitha, K. Jeyasubramanyan, P. V. Satyam, Facile synthesized novel hybrid graphene oxide/cobalt ferrite magnetic nanoparticles based surface coating material inhibit bacterial secretion pathway for antibacterial effect, *Mater. Sci. Eng. C.* 104 (2019) 109932. https://doi.org/10.1016/J.MSEC.2019.109932.

[14] S.K. Verma, A. Thirumurugan, P.K. Panda, P. Patel, A. Nandi, E. Jha, K. Prabakaran, R. Udayabhaskar, R. V. Mangalaraja, Y.K. Mishra, A. Akbari-Fakhrabadi, M.J. Morel, M. Suar, R. Ahuja, Altered electrochemical properties of iron oxide nanoparticles by carbon enhance molecular biocompatibility through discrepant atomic interaction, *Mater. Today Bio.* 12 (2021) 100131. https://doi.org/10.1016/J.MTBIO.2021.100131.

[15] Y.W. Jun, J.W. Seo, J. Cheon, Nanoscaling laws of magnetic nanoparticles and their applicabilities in biomedical sciences, *Acc. Chem. Res.* 41 (2008) 179–189. https://doi.org/10.1021/AR700121F/ASSET/IMAGES/MEDIUM/AR-2007-00121F_0014.GIF.

[16] A.T. Ravichandran, K. Catherine Siriya Pushpa, K. Ravichandran, T. Arun, C. Ravidhas, B. Muralidharan, Effect of size reduction on the magnetic and antibacterial properties of ZnO:Zr:Mn nanoparticles synthesized by a cost-effective chemical method, *J. Mater. Sci. Mater. Electron.* 27 (2016) 5825–5832. https://doi.org/10.1007/s10854-016-4498-1.

[17] K. Ravichandran, K. Nithiyadevi, B. Sakthivel, T. Arun, E. Sindhuja, G. Muruganandam, Synthesis of ZnO:Co/rGO nanocomposites for enhanced photocatalytic and antibacterial activities, *Ceram. Int.* 42 (2016) 17539–17550. https://doi.org/10.1016/j.ceramint.2016.08.067.

[18] S. Chandra, K. Ravichandran, G. George, T. Arun, P. V. Rajkumar, Influence of Fe and Fe+F doping on the properties of sprayed SnO2 thin films, *J. Mater. Sci. Mater. Electron.* 27 (2016) 9558–9564. https://doi.org/10.1007/s10854-016-5008-1.

[19] J.S. Anandhi, T. Arun, R.J. Joseyphus, Role of magnetic anisotropy on the heating mechanism of Co-doped Fe3O4 nanoparticles, *Phys. B Condens. Matter.* 598 (2020) 412429. https://doi.org/10.1016/j.physb.2020.412429.

[20] P. Thandapani, M. Ramalinga Viswanathan, M. Vinícius-Araújo, A.F. Bakuzis, F. Béron, A. Thirumurugan, J.C. Denardin, J.A. Jiménez, A. Akbari-Fakhrabadi, Single-phase and binary phase nanogranular ferrites for magnetic hyperthermia application, *J. Am. Ceram. Soc.* 103 (2020) 5086–5097. https://doi.org/10.1111/jace.17175.

[21] T. Arun, T. Kavinkumar, R. Udayabhaskar, R. Kiruthiga, M.J. Morel, R. Aepuru, N. Dineshbabu, K. Ravichandran, A. Akbari-Fakhrabadi, R. V. Mangalaraja, NiFe2O4 nanospheres with size-tunable magnetic and electrochemical properties for superior supercapacitor electrode performance, *Electrochim. Acta.* 399 (2021) 139346. https://doi.org/10.1016/J.ELECTACTA.2021.139346.

[22] T. Arun, T. Kavin Kumar, R. Udayabhaskar, M.J. Morel, G. Rajesh, R. V. Mangalaraja, A. Akbari-Fakhrabadi, Size dependent magnetic and capacitive performance of MnFe2O4 magnetic nanoparticles, *Mater. Lett.* 276 (2020) 128240. https://doi.org/10.1016/j.matlet.2020.128240.

[23] T. Arun, K. Prakash, R. Kuppusamy, R.J. Joseyphus, Magnetic properties of prussian blue modified Fe$_3$O$_4$ nanocubes, *J. Phys. Chem. Solids.* 74 (2013). https://doi.org/10.1016/j.jpcs.2013.07.005.

[24] R.A. Griffiths, A. Williams, C. Oakland, J. Roberts, A. Vijayaraghavan, T. Thomson, Directed self-assembly of block copolymers for use in bit patterned media fabrication, *J. Phys. D. Appl. Phys.* 46 (2013) 503001. https://doi.org/10.1088/0022-3727/46/50/503001.

[25] R.H. Dee, Magnetic tape: The challenge of reaching hard-disk-drive data densities on flexible media, *MRS Bull.* 31 (2006) 404–408. https://doi.org/10.1557/MRS2006.102.

[26] R.L. Comstock, *Introduction to Magnetism and Magnetic Recording*, (1999) 487. https://www.wiley.com/en-us/Introduction+to+Magnetism+and+Magnetic+Recording-p-9780471317142 (accessed October 15, 2022).

[27] S.X. Wang, A.M. Aleksandr M. Taratorin, Magnetic information storage technology, (1999) 536.

[28] C. Kim, T. Loedding, S. Jang, H. Zeng, Z. Li, Y. Sui, D.J. Sellmyer, FePt nanodot arrays with perpendicular easy axis, large coercivity, and extremely high density, *Appl. Phys. Lett.* 91 (2007) 172508. https://doi.org/10.1063/1.2802038.

[29] J.S. Noh, H. Kim, D.W. Chun, W.Y. Jeong, W. Lee, Hyperfine FePt patterned media for terabit data storage, *Curr. Appl. Phys.* 11 (2011) S33–S35. https://doi.org/10.1016/J.CAP.2011.07.006.

[30] V. Kumari, K. Dey, S. Giri, A. Bhaumik, Magnetic memory effect in self-assembled nickel ferrite nanoparticles having mesoscopic void spaces, *RSC Adv.* 6 (2016) 45701–45707. https://doi.org/10.1039/C6RA05483H.

[31] M. Zahid, N. Nadeem, M.A. Hanif, I.A. Bhatti, H.N. Bhatti, G. Mustafa, Metal Ferrites and Their Graphene-Based Nanocomposites: Synthesis, Characterization, and Applications in Wastewater Treatment, *Nanotechnol. Life Sci.* (2019) 181–212. https://doi.org/10.1007/978-3-030-16439-3_10/COVER.

[32] M. Harada, M. Kuwa, R. Sato, T. Teranishi, M. Takahashi, S. Maenosono, Cation Distribution in onodispersed MFe2O4(M = Mn, Fe, Co, Ni, and Zn) nanoparticles investigated by X-ray absorption fine structure spectroscopy: Implications for magnetic data storage, catalysts, sensors, and ferrofluids, *ACS Appl. Nano Mater.* 3 (2020) 8389–8402. https://doi.org/10.1021/ACSANM.0C01810/SUPPL_FILE/AN0C01810_SI_001.PDF.

[33] Q. Zhang, X. Yang, J. Guan, Applications of magnetic nanomaterials in heterogeneous catalysis, *ACS Appl. Nano Mater.* 2 (2019) 4681–4697. https://doi.org/10.1021/ACSANM.9B00976/ASSET/IMAGES/MEDIUM/AN9B00976_0016.GIF.

[34] A.R. Mohtasebzadeh, L. Ye, T.M. Crawford, Magnetic nanoparticle arrays self-assembled on perpendicular magnetic recording media, *Int. J. Mol. Sci.* 16 (2015) 19769. https://doi.org/10.3390/IJMS160819769.

[35] W.S. Seo, J.H. Lee, X. Sun, Y. Suzuki, D. Mann, Z. Liu, M. Terashima, P.C. Yang, M. V. McConnell, D.G. Nishimura, H. Dai, FeCo/graphitic-shell nanocrystals as advanced magnetic-resonance-imaging and near-infrared agents, *Nat. Mater.* 2006 512 971–976. https://doi.org/10.1038/nmat1775.

[36] L.M. Sanchez, D.A. Martin, V.A. Alvarez, J.S. Gonzalez, Polyacrylic acid-coated iron oxide magnetic nanoparticles: The polymer molecular weight influence, *Colloids Surfaces A Physicochem. Eng. Asp.* 543 (2018) 28–37. https://doi.org/10.1016/J.COLSURFA.2018.01.050.

[37] B. Jia, L. Gao, Silica shell cemented anisotropic architecture of Fe3O4 beads via magnetic-field-induced self-assembly, *Scr. Mater.* 56 (2007) 677–680. https://doi.org/10.1016/J.SCRIPTAMAT.2006.12.045.

[38] N. Zhu, H. Ji, P. Yu, J. Niu, M.U. Farooq, M.W. Akram, I.O. Udego, H. Li, X. Niu, Surface modification of magnetic iron oxide nanoparticles, *Nanomater.* 2018, Vol. 8, Page 810. 8 (2018) 810. https://doi.org/10.3390/NANO8100810.

[39] S. Nallusamy, G. Nammalvar, Triggering of ferromagnetism by amine functionalization on ZnO:Mn films grown by RF magnetron sputtering, *J. Mater. Sci. Mater. Electron.* 32 (2021) 1623–1630. https://doi.org/10.1007/S10854-020-04931-1/FIGURES/7.

[40] S. Nallusamy, G. Nammalvar, Enhancing the saturation magnetisation in Ni doped ZnO thin films by TOPO functionalization, *J. Magn. Magn. Mater.* 485 (2019) 297–303. https://doi.org/10.1016/J.JMMM.2019.04.089.

[41] S. Nallusamy, G. Nammalvar, Enhancement of ferromagnetism in Thiol function-alized Mn doped ZnO thin films, *Mater. Res. Express.* 5 (2018) 026418. https://doi.org/10.1088/2053-1591/AAAF56.

[42] S. Nallusamy, G. Nammalvar, Fabrication of Thiol-functionalized Ni-doped ZnO thin films for room-temperature ferromagnetism, *IEEE Magn. Lett.* 8 (2017). https://doi.org/10.1109/LMAG.2017.2753177.

[43] M. Ma, Y. Zhang, W. Yu, H.Y. Shen, H.Q. Zhang, N. Gu, Preparation and characterization of magnetite nanoparticles coated by amino silane, *Colloids Surfaces A Physicochem. Eng. Asp.* 212 (2003) 219–226. https://doi.org/10.1016/S0927-7757(02)00305-9.

[44] B. Zhang, J. Xing, Y. Lang, H. Liu, Synthesis of amino-silane modified magnetic silica adsorbents and application for adsorption of flavonoids from *Glycyrrhiza uralensis* Fisch, *Sci. China Ser. B Chem.* 51 (2008) 145–151. https://doi.org/10.1007/S11426-007-0104-Y.

[45] F. Gao, An overview of surface-functionalized magnetic nanoparticles: Preparation and application for wastewater treatment, *ChemistrySelect.* 4 (2019) 6805–6811. https://doi.org/10.1002/SLCT.201900701.

[46] W. Shao, G. Tai, C. Hou, Z. Wu, Z. Wu, X. Liang, Borophene-functionalized magnetic nanoparticles: Synthesis and memory device application, *ACS Appl. Electron. Mater.* 3 (2021) 1133–1141. https://doi.org/10.1021/ACSAELM.0C01004/SUPPL_FILE/EL0C01004_SI_001.PDF.

[47] T. Arun, S.S. Dhanabalan, R. Udayabhaskar, K. Ravichandran, A. Akbari-Fakhrabadi, M.J. Morel, Magnetic nanomaterials for energy storage applications, *Inorganic Materials for Energy, Medicine and Environmental Remediation* (2022) 131–150. https://doi.org/10.1007/978-3-030-79899-4_6.

[48] A. Thirumurugan, R. Udayabhaskar, T. Prabhakaran, M.J. Morel, A. Akbari-Fakhrabadi, K. Ravichandran, K. Prabakaran, R. V. Mangalaraja, Magnetic and electrochemical characteristics of carbon-modified magnetic nanoparticles, *Fundam. Prop. Multifunct. Nanomater.* (2021) 235–252. https://doi.org/10.1016/B978-0-12-822352-9.00010-9.

[49] A. Thirumurugan, A. Akbari-Fakhrabadi, R.J. Joseyphus, Surface modification of highly magnetic nanoparticles for water treatment to remove radioactive toxins, *Green Methods for Wastewater Treatment.* (2020) 31–54. https://doi.org/10.1007/978-3-030-16427-0_2.

[50] T. Arun, A. Mohanty, A. Rosenkranz, B. Wang, J. Yu, M.J. Morel, R. Udayabhaskar, S.A. Hevia, A. Akbari-Fakhrabadi, R. V. Mangalaraja, A. Ramadoss, Role of electrolytes on the electrochemical characteristics of Fe3O4/MXene/RGO composites for supercapacitor applications, *Electrochim. Acta.* 367 (2021) 137473. https://doi.org/10.1016/j.electacta.2020.137473.

[51] X. Chen, Core/shell structured silica spheres with controllable thickness of mesoporous shell and its adsorption, drug storage and release properties, *Colloids Surfaces A Physicochem. Eng. Asp.* 428 (2013) 79–85. https://doi.org/10.1016/J.COLSURFA.2013.03.038.

[52] X. Wang, Y. Dai, J.L. Zou, L.Y. Meng, S. Ishikawa, S. Li, M. Abuobeidah, H.G. Fu, Characteristics and antibacterial activity of Ag-embedded Fe3O4@SiO2 magnetic composite as a reusable water disinfectant, *RSC Adv.* 3 (2013) 11751–11758. https://doi.org/10.1039/C3RA23203D.

[53] H. Wei, W. Yang, Q. Xi, X. Chen, Preparation of Fe3O4@graphene oxide core–shell magnetic particles for use in protein adsorption, *Mater. Lett.* 82 (2012) 224–226. https://doi.org/10.1016/J.MATLET.2012.05.086.

[54] R. Moradi, S.A. Sebt, H. Arabi, M.M. Larijani, Decoration of carbon nanotube with size-controlled L10-FePt nanoparticles for storage media, *Appl. Phys. A Mater. Sci. Process.* 113 (2013) 61–66. https://doi.org/10.1007/S00339-013-7856-3/FIGURES/6.

[55] P.Y. Keng, I. Shim, B.D. Korth, J.F. Douglas, J. Pyun, Synthesis and self-assembly of polymer-coated ferromagnetic nanoparticles, *ACS Nano.* 1 (2007) 279–292. https://doi.org/10.1021/NN7001213/ASSET/IMAGES/MEDIUM/NN-2007-001213_0014.GIF.

[56] S. Sun, S. Anders, T. Thomson, J.E.E. Baglin, M.F. Toney, H.F. Hamann, C.B. Murray, B.D. Terris, Controlled Synthesis and Assembly of FePt Nanoparticles, J. *Phys. Chem. B.* 107 (2003) 5419–5425. https://doi.org/10.1021/JP027314O.

[57] Q. Dai, D. Berman, K. Virwani, J. Frommer, P.O. Jubert, M. Lam, T. Topuria, W. Imaino, A. Nelson, Self-assembled ferrimagnet-polymer composites for magnetic recording media, Nano Lett. 10 (2010) 3216–3221. https://doi.org/10.1021/NL1022749/SUPPL_FILE/NL1022749_SI_001.PDF.

[58] *Seagate's HDD roadmap teases 100TB drives by 2025- Storage - News - HEXUS.net*, (n.d.). https://hexus.net/tech/news/storage/123953-seagates-hdd-roadmap-teases-100tb-drives-2025/?print=1 (accessed October 16, 2022).

[59] J.G. Zhu, X. Zhu, Y. Tang, Microwave assisted magnetic recording, *IEEE Trans. Magn.* 44 (2008) 125–131. https://doi.org/10.1109/TMAG.2007.911031.

[60] D. Weller, A. Moser, L. Folks, M.E. Best, W. Lee, M.F. Toney, M. Schwickert, High ku materials approach to 100 gbits/in2, *IEEE Trans. Magn.* 36 (2000) 10–15. https://doi.org/10.1109/20.824418.

Index